HUMAN FACTORS
IN SYSTEMS ENGINEERING

WILEY SERIES IN SYSTEMS ENGINEERING

Andrew P. Sage

ANDREW P. SAGE and JAMES D. PALMER
Software Systems Engineering

WILLIAM B. ROUSE
Design for Success: A Human-Centered Approach to Designing Successful Products and Systems

LEONARD ADELMAN
Evaluating Decision Support and Expert System Technology

ANDREW P. SAGE
Decision Support Systems Engineering

YEFIM FASSER and DONALD BRETTNER
Process Improvement in the Electronics Industry

WILLIAM B. ROUSE
Strategies for Innovation

ANDREW P. SAGE
Systems Engineering

HORST TEMPELMEIER and HEINRICH KUHN
Flexible Manufacturing Systems: Decision Support for Design and Operation

WILLIAM B. ROUSE
Catalysts for Change: Concepts and Principles for Enabling Innovation

LIPING FANG, KEITH W. HIPEL, and D. MARC KILGOUR
Interactive Decision Making: The Graph Model for Conflict Resolution

DAVID A. SCHUM
Evidential Foundations of Probabilistic Reasoning

JENS RASMUSSEN, ANNELISE MARK PEJTERSEN
and LEONARD P. GOODSTEIN
Cognitive Systems Engineering

ANDREW P. SAGE
Systems Management for Information Technology and Software Engineering

ALPHONSE CHAPANIS
Human Factors in Systems Engineering

HUMAN FACTORS IN SYSTEMS ENGINEERING

ALPHONSE CHAPANIS

A Wiley-Interscience Publication
JOHN WILEY & SONS, INC.
New York / Chichester / Brisbane / Toronto / Singapore

Figure 5.14 from Chapanis, Ethnic Variables in Human Factors Engineering, 1975. Reprinted by permission of the Johns Hopkins University Press.

This text is printed on acid-free paper.

Copyright © 1996 by John Wiley & Sons, Inc.

All rights reserved. Published simultaneously in Canada.

Reproduction or translation of any part of this work beyond that permitted by Section 107 or 108 of the 1976 United States Copyright Act without the permission of the copyright owner is unlawful. Requests for permission or further information should be addressed to the Permission Department, John Wiley & Sons, Inc., 605 Third Avenue, New York, NY 10158-0012

Library of Congress Cataloging-in-Publication Data

Chapanis, Alphonse.
    Human factors in systems engineering / Alphonse Chapanis.
      p. cm. — (Wiley series in systems engineering)
    "A Wiley-Interscience publication."
    Includes index.
    ISBN 0-471-13782-0 (cloth : alk. paper)
    1. Systems engineering. 2. Human engineering. I. Title.
II. Series.
TA168.C42   1996
620.8'2—dc20   95-46163
                                                 CIP

Printed in the United States of America
10 9 8 7 6 5 4 3 2 1

Contents

Preface		ix
1	**Introduction**	**1**
	About This Book	1
	A Historical Perspective	3
	The Government's Role	10
	Human Factors	11
	Human Involvement with Machines and Systems	16
	Some Human-Factors Postulates	18
	Summary	19
	References	20
2	**Systems and Systems Engineering**	**21**
	Systems	21
	Systems Engineering	26
	System Life Cycle	28
	The Systems-Engineering Process	38
	Concurrent Engineering	52
	Speaking Practically	54
	Summary	55
	References	55

vi CONTENTS

3 Standards, Codes, Specifications, and Other Work Products — 58
- Standards and Codes — 58
- Guidelines — 63
- Human Engineering Program Plan (HEPP) — 64
- Systems-Engineering Work Products — 69
- Summary — 76
- References — 77

4 Human-Factors Methods — 79
- Basic versus Applied Research — 80
- Some Additional Preliminary Remarks — 81
- The Methods — 84
- Summary — 138
- References — 139

5 Human Physical Characteristics — 142
- The Starting Point — 142
- Anthropometry: The Science of Human Dimensions — 158
- The Skeletal System — 170
- Skeletal Muscles — 172
- Other Machinery of the Body — 179
- Interactions with the Environment — 180
- Work, Rest, and Work–Rest Cycles — 198
- Summary — 202
- References — 202

6 Human Mental Characteristics — 206
- The Human Subsystem — 206
- Attending — 209
- Sensing — 211
- Remembering — 235
- Decision-Making — 238
- Learning — 240
- Responding — 247
- Summary — 255
- References — 256

7	**Personnel Selection and Training**	**260**
	Relationships Between System Design, Selection, and Training	260
	Mandatory Personnel Requirements	262
	Reducing System Complexity	263
	Human-Factors Contributions to Training Programs	264
	Training Equipment	265
	Summary	268
	References	268
8	**System Requirements**	**270**
	Specifying Human–System Requirements	271
	Tradeoffs	281
	Summary	286
	References	286
9	**Postscript**	**288**
	Documentation	288
	What Makes a Successful Human-Factors Program?	291
	The Final Word	294
	References	295
Appendix A	**Acronyms and Abbreviations**	**297**
Appendix B	**Some ANSI and International Standards**	**307**
	ANSI Standards	307
	International Standards	309
Index		**311**

Preface

June 30, 1982 is a critical and traumatic date in my life. After 36 years at The Johns Hopkins University, where I had taught human factors and conducted and directed research in the field, I was being retired. It was heart-wrenching to have to leave my familiar office, the well-equipped and furnished Communications Research Laboratory I had established, and the graduate students, on whom I had come to rely for their sharp criticism of my writings, their assistance in the research I was conducting, their curiosity and imagination, and, perhaps most important, their spirit and humor.

My anguish at being separated from that familiar and comfortable existence was, fortunately, mitigated by my being retained as a consultant on a fairly regular basis by what was then IBM's Federal Systems Division (FSD). Although FSD was, by IBM's standards, a rather small division with total revenues of only a few billion dollars a year, it was always a profitable division and was one of IBM's most exciting ones. It was there that large military and commercial systems—LAMPS, TRIDENT, space and banking systems, the automated postal system and, most recently, the FAA's air-traffic-control system—were developed. It also had the highest proportion of engineers and programmers to total employees of any IBM division. And it was there that I came into contact with the real world of system development and the way human factors really fit into that development process.

My work at FSD was almost exclusively with engineers—sharp and highly skilled professionals—who instilled in me new respect for that profession. Of the many with whom I worked, two were especially influential in shaping my new perspectives. William J. Budurka, a superb systems engineer with an unusually sympathetic view of human factors, taught me more about the real business of engineering than I ever could have learned from reading any book or taking any formal course. John B. Shafer, an engineer later trained as a psychologist, was a long-time employee of FSD in Owego, New York. He had worked on most of the major systems developed there, and was a colleague in many joint ventures with me. A valued friend, he gave me valuable new insights into the way human factors was really practiced, and provided me with several of the practical examples you will find in this book. My debt to both men is more than I can express in words. Although both have read and critiqued parts of this manuscript, neither can be held responsible if I have distorted or misrepresented what they have taught me.

At the time I joined FSD, the engineers at headquarters were engaged in defining the field of system engineering and in writing a week-long course titled "Systems Engineering Principles and Practices." John Shafer and I were invited to write a segment on human factors for that course. Our short introduction to human factors was so well received by engineers everywhere we taught it, that I suggested to Budurka that John and I could write a separate, two-day course on "Human Factors in Systems Engineering." With Budurka's encouragement and almost continuous advice and criticisms, John and I completed the first version of our course in June 1985.* It has subsequently gone through no less than eight revisions as we continued to learn from the people we taught. The course has been taught to several thousand engineers, programmers, and human-factors professionals in Bethesda, Gaithersburg, and Rockville, Maryland; Manassas, Virginia; Houston, Texas; Owego, New York; and Boulder and Colorado Springs, Colorado. The critiques were almost uniformly laudatory. Obviously we were doing something right.

With encouragement from Budurka, I decided to write a book, modeled after but modified substantially from the IBM course. This is the result. It's quite a different kind of human factors book than any other I know. For one thing, during all those years when I taught and wrote about human factors I had never appreciated, seldom heard about, and almost never read about the enormous amount of paperwork—work products, as it is referred to in engineering circles—involved in systems engineering. FSD taught me that engineers are driven by requirements, standards, and specifications, and they, in turn, generate a large number of documents—requirements, specifications, design documents, and rationale reports—most of which have substantial human-factors components. Yet these topics are conspicuously absent from virtually all textbooks on human factors.

In addition, most textbooks of human factors look at the problem of design from the standpoint of the human user or operator. They concentrate on the way the body functions from psychological, physiological, and biomechanical points of view, and summarize research on the ways in which people interact with various kinds of tools, machines, and systems, while performing tasks in diverse environments. They also provide numerous general guidelines and recommendations about user requirements based on those research findings. The guidelines and recommendations are written with the implicit, sometimes explicit, assumption that designers

---

*Chapanis, A., & Shafer, J. B. (1986). Factoring humans into FSD systems. *IBM Technical Directions, 12(1)*, 15–22.

will read them and figure out for themselves how to design things that match human capabilities and limitations.

The thing wrong with that approach is that, by and large, it doesn't work. Engineers, designers, and programmers don't read our textbooks, don't understand our guidelines and recommendations if they should happen to read them, and don't know how to design to satisfy our guidelines if they should happen to read and try to follow them. There is no reason why they should. We should not expect designers to do jobs for which we have been trained, and they have not.

In this book I have taken a completely different orientation. I start with systems, and the way in which systems are designed, and show how human factors can contribute at each stage of the system-development process. Those human-factors contributions are specifications, statements of precise, specific requirements that a particular system—not systems in general—must meet so that it will be usable by human operators and maintainers. In other words, I show that the task of the human-factors professional is to translate general guidelines and users' needs into project-specific specifications of systems requirements that engineers, designers, and programmers can meet without any further human-factors expertise. That is a responsibility few human-factors professionals have been willing or able to accept.

For years we have been telling, asking, even imploring, designers and engineers to understand human factors and to read the articles and books we write. That's wrong. Designers have to worry about too many other things that are important to their work. It's time for us to learn a little bit about engineering and the way designers and engineers go about their business. By doing so, we can learn exactly what human-factors inputs they need, when those inputs are needed, and in what form those inputs will be most useful. This doesn't mean that we have to become engineers. What it does mean is that we have to have some appreciation of what system design really entails.

Some have been saying for a long time that we need to learn to speak to engineers in their own language. In this book I have tried to tell you about the various documents that drive the system-development process, and that engineers are required to write. I have also tried to provide at least some basic vocabulary that will make it easier for you to talk to engineers and to be accepted as a genuine partner on a system-design team.

Although this is a book on systems engineering, it applies equally to the engineering of all kinds of products, even, as I have tried to show, to the design of something as simple as a can opener. The differences are primarily matters of degree, not of kind.

This is an introduction to the field. It makes no pretense of being a complete and thorough coverage of the topic. I make no assumptions about the reader's background, other than a genuine interest in learning how to design things better, and an open, inquiring mind. Nor is this book a substitute for any of several very good conventional textbooks on human factors/ergonomics. My approach is so different from the conventional that I think of this book as supplemental to, not a replacement for, other books.

This book was a long time in the making. It was interrupted by my mother's death and five serious surgical operations—one of them life-threatening. During long, seemingly endless periods of convalescence, I often despaired of ever being able to finish. Throughout these vicissitudes and grievous assaults on my psyche and soma, one person sustained, supported, and ever kept faith in me. And so, I dedicate this book to Viv, with love.

ALPHONSE CHAPANIS

Towson, Maryland

HUMAN FACTORS
IN SYSTEMS ENGINEERING

Chapter 1

Introduction

> The disregard for human factors in the control rooms was appalling. In some cases, the distribution of displays and controls seemed almost haphazard. It was as if someone had taken a box of dials and switches, turned his back, thrown the whole thing at the board, and attached things wherever they landed. For instance, sometimes 10 to 15 feet separated controls from the displays that had to be monitored while the controls were being operated. Also, sometimes no displays were provided to present critical information to the operators. There were many instances where information was displayed in a manner that was not usable by the operators, or else was misleading to them. A textbook example of what can go wrong in a man-machine system when people have not been taken into account.*

That opening quotation was taken from remarks made by Charles O. Hopkins, Technical Director of the Human Factors Society Study Group that investigated the Three Mile Island nuclear power plant after the accidental near-catastrophe of March 28, 1979. How could something like that happen? Even more important: How can we keep things like that from ever happening again?

ABOUT THIS BOOK

This book tries to answer that last question because it is about *systems engineering* —the development of tools, machines, and systems so that they will be safe, comfortable, and easy to use. Although it is about engineering and system design, it is not an engineering book in the conventional sense. It approaches design from the standpoint of those human considerations—the human factors—that have to be

*Machine Design, 55(2) 4. Reprinted by permission of the author and publisher.

taken into account to make machines and systems match us, the people who have to use and service them. Since that approach is not the usual way most systems are designed, this book is more about how systems *should* be designed rather than a description of current practices.

At the same time, this book is not like most conventional human factors books, which, although they contain a great deal of useful recommendations and guidelines, never come to grips with the practical realities of how one gets those recommendations implemented. To contribute effectively to design, human-factors professionals need to know something about systems and how they are developed. That knowledge helps us understand how system developers go about their work, and what human factors inputs are needed at each stage of development. Thus, Chapter 2 defines systems, points out some of their main characteristics, describes system life cycles, and explains the process by which systems are designed and built.

System design is largely managed with words—in requests for proposals (RFPs), operational-need documents, operational-concept documents, requirements, specifications, standards, and reviews. These are the documents that drive the engineer and they are the places where human-factors professionals make their real impact. These documents form two broad classes—those that are imposed on the engineer (and that includes the human-factors engineer) and those that the engineer must prepare. Chapter 3 is devoted to these work products, to the human-factors requirements contained in the former, and to the human-factors inputs required in the latter.

The next chapter, Chapter 4, is about the human-factors methods that are used to develop the human-factors inputs needed for specifications, reviews, and other work products. The emphasis here is not on the discovery of basic information that merely adds to our store of knowledge or settles some theoretical argument, but rather on the production of specific information that can be used in design.

The next two chapters are concerned with human characteristics, the raw material with which human-factors professionals work. The human-factors methods discussed in Chapter 4 are applied to these human characteristics to produce the inputs that are required for the specifications, reviews, and other work products described in Chapter 3. The chapters on human characteristics are highly selective, concentrating on those characteristics that have direct relevance to design. Chapter 5 first describes some very basic, but frequently overlooked characteristics that distinguish people from the hardware and software with which engineers work and then concentrates on the human operator as a biological and structural entity. Chapter 6 is concerned with the human operator as an information processor.

No system is complete until people have been selected and trained to operate it once the system has been constructed. Chapter 7 shows how system design impacts personnel requirements, both in terms of the numbers of personnel and the skills that will be required to operate the system. The chapter also shows the contributions that human-factors professionals can make to selection and training.

Chapter 8 is the most important one in this book, and the one to which all the preceding chapters have been leading. It comes to grips with the question: How do you design systems so that they will be compatible with human needs and capabilities? I point out that the proper role of the human-factors professional is to prepare precise application-specific requirements that can be met by designers and engineers without any human-factors expertise. These requirements are the inputs to the specifications, reviews, and other documents described in Chapter 3. Designers continually have to make choices among alternative concepts or conflicting requirements, and many of these choices involve human factors. Trade-offs, or trade studies, are methods for systematically choosing among conflicting design alternatives. They are continually required in systems development and I illustrate the procedure with an example.

The last chapter, Chapter 9, ties up some loose ends. First, I emphasize the importance of documentation to the system-development process and describe some of the clients served by documentation. Next, I describe the ingredients that are necessary for a successful human-factors program, and make some sobering observations about the reasons systems sometimes fail despite our best efforts. I conclude on a more optimistic note: In spite of all the hurdles and uncertainties, usable systems do get built, and there are enough success stories in the literature to allow one to conclude that human factors does make a difference.

The Scope of This Book

Although this book is titled *Human Factors in Systems Engineering,* and talks mostly about systems, the principles discussed apply not only to hardware, but also to software systems and to the design of simpler devices such as tools and machines. The same basic thought processes and procedures, differing only in amount of detail, formality, and rigor, apply to the design of all mechanical and electro-mechanical products.

Why Such a Book as This?

People have been designing systems for centuries, and many of them have been successful. What, then, is the rationale for writing a book on such a specialized topic as this one? To answer that question, it is helpful to make a brief excursion back in time.

A HISTORICAL PERSPECTIVE

Everything we invent or make is ultimately designed for human use and benefit. Even when our Stone Age ancestors first began to fashion their crude implements, they shaped axes to fit their hands and selected sizes and weights of stones they could wield easily and effectively. That same basic approach, elaborated with ever-increasing sophistication, has been used throughout all of human history.

For thousands of years, matching utensils, implements, and machines to human capabilities was done largely by common sense. Devices were successively modified and improved through trial and error, and lessons learned were passed on from master to apprentice. In many cases, few or no details about design methods and procedures used by our predecessors were preserved in archival form, leaving modern historians free to speculate, for example, about how, exactly, the great amphitheaters, baths, markets, and temples of imperial Rome were constructed (Mark, 1987). Nonetheless, the many imaginative, useful, and impressive products created throughout the ages provide ample testimony that this unstructured process served mankind well. That way of designing things, however, changed about a century ago, when the pace of technological innovation accelerated.

The Twentieth Century

It is safe to say that technology has advanced more since 1900 than it has in all of the preceeding history of mankind. Indeed, most technological marvels that we take for granted have been invented within the lifetimes of some persons alive today.

The first automobile had been invented by Karl Benz in 1885, but, by 1900, there were only about 14,000 cars in America and barely 144 miles of paved roads. Although telephones had been invented in 1876 and had been introduced into commercial use the following year, it was impossible for a subscriber in New York to call someone in San Francisco, because transcontinental telephone service would not be available until 1915. The first commercial electric power plant had been invented by Thomas Edison in 1882, but in 1900 our grandparents did without the electric washing machine, toaster, can opener, dishwasher, refrigerator, vacuum cleaner, razor, and microwave oven. The Wright brothers ushered in the "air age" with the first sustained manned flight in 1903. The world's first radio station began broadcasting in 1920. The first regularly scheduled television program was broadcast in 1928. The first commercial nuclear power plant came on line in 1963 (Trager, 1979).

Early Beginnings

Throughout most of mankind's involvement with machines, operators contributed power—raw muscle power—to their machine counterparts. In the early part of this century there were, to be sure, many powered machines like lathes, milling machines, cranes, conveyor systems, and hoists that extended human capabilities and made work easier. But it was largely human muscle power that still brought raw materials to machines, fed machines, operated controls, joined and fastened parts together, carted finished products away, packaged them, and loaded them onto vehicles. To integrate workers more effectively into such work environments, time-and-motion engineers of that period formulated and applied a number of principles of motion economy. Since they dealt almost exclusively with human movements, the principles were aptly named. Nonetheless, they can be said to have marked the beginnings of human factors, because they were the first attempts to systematize information about ways of designing jobs so that workers could do them more effectively, and with less strain.

World War II

Principles of motion economy are still valid today, because many jobs still require the exertion of human muscle power. World War II, however, ushered in new kinds of machines—radar and sonar, for example. The job of a radar or sonar operator does not require the exertion of much muscular effort. Instead, the operator has to perceive a changing pattern of information on a screen, and make decisions about what appears there. These are mental operations and the principles of motion economy were no longer appropriate to deal with the new problems created by such machines.

In addition, the machines of the World War II era exposed people to stresses that never before had been experienced. Before 1939 we had no really high-flying aircraft. But the war needed and produced aircraft that could routinely fly to higher altitudes, where the air was too thin for a pilot to survive for more than a few minutes without supplemental oxygen. Although the aircraft of those days seem slow by our modern standards, they could still expose a pilot to g-forces that would drain the blood from his head and leave him blind. These were powerful environmental forces such as had never been experienced before, requiring new data and new principles from the research of human scientists—physiologists, physicians, and psychologists. The findings of these scientists, as they applied to aircraft design, were published in 1946 by McFarland, whose book was the first to use the words "Human Factors" in its title, and in the same sense as I now do in this book.

The war also saw another development. The harnessing of nuclear energy, first for weapons, and shortly thereafter for power plants, was revolutionary.

Post–World War II and the Space Age

The period after World War II saw another momentous development—the space age. Space flights have become so commonplace that it is hard to remember the thousands of analyses, studies, and experiments that were required to clear the way. There were unique problems of vehicle design, involving exotic displays and controls. There were problems of vibration, of g-forces, and of weightlessness, that had to be explored and solved. For extravehicular activity, an entire, self-contained environment had to be designed for astronauts (Fig. 1.1). Torqueless tools had to be designed for use by men who were floating freely, encumbered by space suits with limited mobility. There were problems of nutrition, waste disposal, and work–rest cycles. There were the problems of selection, training, and simulator design. A catalog of these problems encompasses the entire field of human factors.

The Five Elements of System Design Perhaps more than anything else, the development of space systems brought into sharp focus the way five components must be successfully managed and balanced-off in system design. They are: *personnel selection, personnel training, machine design, job design,* and *environmental design*. These five components all interact with each other in complex ways. Systems that are designed for use by highly selected or skilled persons—such as astronauts—can be complex. Training may be much more sophisticated and jobs may be more complicated. On the other hand, systems such as telephones, VCRs, and personal computers, designed for use by people at large, have to be much simpler. A designer cannot assume that users of many devices will be selected in any special way, or that they will undergo any training at all.

6   1: INTRODUCTION

Figure 1.1 Hundreds of biomedical and human-factors studies made it possible for astronauts like Edwin A. Aldrin, Jr. (shown here descending from the Lunar Module) to walk on the surface of the moon. (*Photo courtesy of NASA*)

The environment figures into all this, too. Systems that have to be used in special or exotic environments—such as underwater, underground, in arctic regions, or in space—must accommodate operators who may be encumbered with special protective suits, and who may, therefore, be less mobile or flexible.

Trade-offs are almost always involved in dealing with these five components. A simple example concerns the use of colored lights to regulate traffic. About eight percent of otherwise perfectly normal males have reduced sensitivity to colors. Although these persons are popularly called "color blind," that term is really incorrect, because these persons do see some colors. One way of designing for that reality is by selection: administer tests of color vision to all potential drivers, and deny driver licenses to those persons who do not pass the test. Alternatively, you can design the system so that *even* persons with color-deficient vision can drive.

One way of doing that is to use position coding, that is, standardize the positions of the red, yellow, and green lights in traffic signals. Another alternative, often overlooked, is to use only those colors that virtually all color-deficient persons can discriminate. The same kinds of trade-offs apply, or should apply, for all systems that use color, for example, computer systems, display panels for control rooms, or color codes on pipes and wires.

These considerations are all involved in current efforts aimed at designing for the handicapped. Until recently, most handicapped persons were excluded from many activities because they could not cope with the world as it was designed. Today, however, many state and federal laws and regulations, for example, the Federal Rehabilitation Act of 1973 and the Americans with Disabilities Act of 1990, mandate that the environment and the devices we all have to use should be designed so that even handicapped persons can use them. This has resulted in the redesign of a host of devices (see, for example, McQuistion, 1993) and the design of barrier-free environments. This has not only greatly enhanced the quality of life for many handicapped persons, but also has added to our productive workforce.

Similar changes have been brought about by the Federal Age Discrimination in Employment Act of 1967. Elderly workers can no longer be barred from many jobs just because they happen to have lived longer. Instead, jobs must often be redesigned, to take into account subtle changes that occur as people age, for example, presbyopia. Fortunately, meeting these requirements does not only benefit the elderly. Systems that are well designed for the handicapped and elderly are often easier for normal and younger people to operate as well.

The Computer Revolution

Space exploration would not have been possible if it had not been accompanied by another development—the computer revolution. Computers probably have had a more profound effect on society, on our ways of living, and on our ways of doing business, than any other technological creation in this century. Computers help manage our checking accounts and our charge accounts. They help manufacture our goods, raise our crops, manage our farms, sort and distribute our mail, schedule our rail and air travel, book our theater tickets, check-out our groceries, diagnose our illnesses, teach our children, and amuse us with sophisticated games. Computers make it possible to erase time and distance in our telecommunications, thereby giving us the freedom to choose the times and places at which we work. They help guide our planes, direct our missiles, guard our shores, and plan battle strategies. In doing all these things, computers have created new industries and, at the same time, have spawned new forms of crime. To sum up, computers have become so intricately woven into the fabric of our daily lives that, if all the computers on earth were to vanish suddenly, our civilization would be thrown into immediate chaos.

Almost without realizing, we have changed from an agricultural and manufacturing society to an information society (Naisbitt, 1982). This is shown dramatically by the changes in the composition of the U.S. labor force. Until the early 1900s, the largest single class of American workers was engaged in agriculture. Ninety years later, farmers and farm managers had dropped to two percent of the work force. Then, from the early 1900s until about the end of World War II (roughly 1944), industry was the largest employer. Today, however, more working Americans spend their time creating, processing, or distributing information than in any other kind of productive activity. In 1960, for example, there were only a handful of computer programmers and operators in the United States, mostly working in scientific settings in academia and in industry. By 1991, almost 1.3 million people worked as programmers and computer operators, and another 900,000 were described as "mathematical and computer scientists" in a Bureau of the Census report on employment. The technology responsible for these dramatic changes has been developed within the past few decades.

The Seamy Side of Modern Technology

Although the breathtaking pace of technological change in the last hundred years has yielded great benefits for mankind, it has also produced machines and systems of enormous complexity, with frightening potential for destruction. Perhaps the most insidious feature of some of these systems is that they contain hidden hazards. There is no way, for example, that a person can see, hear, smell, taste or feel harmful radiation, yet the release of radiation from a nuclear power plant accident, such as occurred at Three Mile Island in Pennsylvania, or Chernobyl in the Ukraine, may affect multitudes of people and contaminate the environment for hundreds, perhaps thousands, of years. From July 24, 1959 to May 1, 1986, there were 18 major nuclear accidents, of which the accidents at Three Mile Island and Chernobyl are the most infamous. Atmospheric radiation from Chernobyl spread over several countries in Europe, resulting in the issuance of health precautions throughout Scandinavia, Poland, and Austria. The long-term effects of Chernobyl may not be known for decades.

Much more disturbing is that, in the ten-year period after 1979, the date of the accident at Three Mile Island, there have been nearly 30,000 safety "mishaps" at nuclear power plants in America and, of these 30,000, at least 1,000 have been labeled as *serious* by the American Nuclear Regulatory Commission. So-called "personnel errors" have been identified as the causes of nearly 75 percent of reported safety mishaps at nuclear power plants in 1987 (Young, 1989). The real extent of these dangers worldwide cannot even be estimated. Whatever they are, they must be enormous, and so represent quite a challenge for the human-factors discipline.

Not only nuclear power plants have the potential for catastrophe. Casey (1993) gave a gripping and alarming account of 17 disasters caused by design-induced human errors: among them, a radiotherapy machine that ran amok and killed a patient, a supertanker that ran aground because of navigation errors, a computer error that almost caused the stock market to crash, a 25-hour power failure in Manhattan because of an operator's inability to take action at the right time, and a young girl's death, resulting from improperly connected electrodes during a routine electrocardiogram. Stories such as these illustrate dramatically how new technologies, though they can bring us great benefits, can also result in deaths and misfortunes if they are not designed to be used correctly by people with human abilities and limitations.

The Hazards of Everyday Things

While nuclear power plant accidents, ship collisions, and aircraft accidents are dramatic and appalling because of the devastating toll they take, thousands of accidents occur daily with much smaller machines and systems that we use every day. In industry, and in our daily living, these are power presses, band saws, lathes, stud guns, automobiles, lawnmowers, snowmobiles, ladders, and stoves. These accidents usually involve only a single person, and occur so commonly that they do not command the attention given accidents involving larger systems. Yet the cumulative toll in human lives from our mishaps with smaller systems is far greater than that from the more dramatic, larger ones. In the decade from 1982 to 1993, for example, 8,200 persons lost their lives in all commercial and civilian air and space transport accidents, but 460,038 died as a result of automobile accidents, 106,900 were killed in industrial accidents, and 214,300 deaths were attributable to accidents in the home (National Safety Council, 1993).

Our interactions with modern technology do not always result in accidents, of course. Many, like trying to find the controls in a rented or unfamiliar automobile, trying to find out how to replace a burned-out light bulb in a kitchen appliance, or trying to program a VCR or use many modern computers, merely add to the inefficiencies, frustrations, and annoyances that plague all of us. Norman (1988) catalogs some of the everyday things that are confusing or difficult to use because of the way they are designed. His examples range from simple things like doors and faucets to more complex devices such as modern telephone systems and VCRs. Though less serious than accidents that result in injuries and death, our interactions with many such devices and machines result in somatic and psychosomatic insults that are crippling to both body and mind.

For years, when people made mistakes or misused the devices they worked with, it was customary to label the reason as *human error*. Even today, when people have difficulties programming a VCR, using a computer, or trying to use some of our new telephone systems, they are apt to blame themselves. "I guess I'm just dumb." Or, "That was stupid of me." And we still see newspaper accounts attributing many automobile, aircraft, and nuclear power plant accidents to human error.

Everyone makes mistakes and, to say that accidents are caused by human error, gives us no useful information. What human-factors research has shown is that people make more mistakes with some kinds of equipment than with others, and that many devices to which I have referred as "error provocative" (Chapanis, 1980) almost literally invite people to make mistakes because of the way they are designed. Although the realization has been slow to come, people are gradually beginning to understand that many of the difficulties we have with machines are directly related to the way they are designed. Newspaper and magazine stories (see, for example, Nussbaum & Neff, 1991), articles in professional journals, and our own experiences provide us with numerous examples of the ways in which machines often fail to match our needs, abilities, and limitations. Look around and you are sure to find numerous examples—labels you cannot read or understand, computer commands that are almost impossible to learn and remember, controls that move in unexpected directions, seats that give us backaches, and noise and glare that make it difficult for us to do our jobs.

The Need for Human Factors Perhaps the most important point to make about all these mismatches between people and the products they use is that the products were almost certainly designed with good intentions. Designers do not, I am sure, deliberately try to make things confusing, hard to use, unsafe, or uncomfortable. After all, they try to design things that will sell. But their own common sense, intuitions, experiences, training, or using themselves as design models, are clearly not enough to enable them to design products that avoid the problems they often cause.

Good design means not only designing for normal human use, but also designing against misuse, unintended uses, and abuses. It also means designing for all the sizes, shapes, attitudes, and personalities people have. People are so complex, so difficult to study, and sometimes so perverse in their behavior that designers often do not know how to factor human characteristics into the design process. As a result, they give up trying to do so. Those difficulties have been responsible for the emergence of the profession of human factors, a profession that uses information about how we see, how we hear, how we think, and how we act, to help design things that will serve us better.

THE GOVERNMENT'S ROLE

Although laws concerning disabled and elderly workers have focused attention on human issues, the U.S. government has had an even greater role in the development and definition of systems engineering and the promotion of human factors—and for good reason. The government is the largest purchaser and user of large systems. Everyone is familiar with the many military systems—aircraft, ships, tanks, missile systems—that have made this country into a major military power. Developing those systems required a structured approach—a set of formal procedures to instruct and guide engineers and designers and to provide the means by which developments could be monitored. At the same time, the realization that machines do not fight alone resulted in a number of stringent requirements that military systems be designed to make them usable by people with "only-human" abilities and capacities. This total approach taken by the government has been codified in numerous standards, specifications, and other documents that are discussed in Chapter 3.

Not so well publicized, however, is the government's role in purchasing and using a number of very large systems for nonmilitary purposes. NASA's space exploration program is perhaps the best known of these. Another equally large, some would say larger, system is the Federal Aviation Administration (FAA), whose air-traffic-control system monitors and directs all air traffic, and which has helped to make air travel one of the safest methods of long-distance travel. The Social Security Administration has an immense computerized system that keeps records on the systematic contributions made by all wage earners in this country and payments made to all retirees and beneficiaries of Medicare and Medicaid. The Internal Revenue Service is another large system that not only keeps financial records on individuals but on corporations and businesses as well. The United States Postal Service, with its huge system for the collection, sorting, and distribution of mail, and the Nuclear Regulatory Commission, with its role in the development of nuclear power for commercial purposes, are further examples.

The lessons learned in the development of large military systems have been carried over to commercial ones. Many military standards and specifications in their original, or slightly modified form, have been used to guide, direct, and monitor many non-military projects.

The success of the government's efforts has been slowly filtering down to industry. AT&T, Boeing, IBM, Lockheed-Martin, LORAL, NCR Corporation, and Westinghouse, for example, have developed their own systems-engineering procedures adapted from those of the government. We can see the benefits of applying those procedures in our telephone and telecommunications systems and in many localized ones, like the computerized Seagirt Marine Terminal in Baltimore, Maryland, that handles and directs ships, immense cranes, and trucks in the efficient reception, processing, and distribution of containerized products.

One must not get the idea, though, that the methods and principles of systems engineering apply only to very large systems. Later I shall show that the basic ideas are equally valid, although on a greatly reduced scale, to the development of automobiles, small computer systems, farm machinery, and hundreds of other products that we use in our daily lives. Keep this broader perspective in mind. First, a closer look at human factors.

HUMAN FACTORS

Human factors has been defined in many ways. One simple definition is *designing for human use*. Another simple definition is *humanizing technology*. However appealing these brief definitions may be, they do not convey the full flavor of the field. My definition is:

> Human factors is a body of information about human abilities, human limitations, and other human characteristics that are relevant to design.

Related to that is my definition of human-factors engineering:

> Human-factors engineering is the application of human factors information to the design of tools, machines, systems, tasks, jobs, and environments for safe, comfortable and effective human use.

The significant word in those definitions is *design,* because it is this that separates human factors from such purely academic disciplines as psychology, physiology, and anthropology. Our aim is to apply *what we know* to the design of practical things—of things we have to do, or have to use, because of our occupations, or things we want to do, or want to use, because of our inclinations.

Some Other Names

The literature also contains a number of other names that are often a source of confusion. Some of these are briefly defined in Table 1.1. One term I want to address in somewhat greater detail is *ergonomics.* This word has been appearing more and more often these days, particularly in advertisements about ergonomic chairs, ergonomic work stations, and ergonomic computer terminals. The implication of many of these advertisements is that ergonomics is concerned only with the

TABLE 1.1 Definitions of Some Terms Related to Human Factors

Applied experimental psychology—the body of scientific facts from experimental psychology that contributes to and is part of human factors.

Bioastronautics—a term used primarily by the Air Force and National Aeronautics and Space Administration (NASA). For all practical purposes, it is equivalent to human factors with emphasis on aerospace applications.

Biomechanics—a subspecialty of human factors, primarily concerned with human movements, muscle strength, and muscle power.

Engineering psychology—a subspecialty of human factors covering psychological facts and principles useful in design. It is synonymous with applied experimental psychology.

Ergonomics—a term used mainly in Europe and Asia. It is also used by the Federal Aviation Administration (FAA) and the popular press to refer to the more physiological, biomechanical, and anthropometric aspects of human factors, but professionals regard ergonomics as synonymous with human factors.

Human engineering—the application of human-factors facts and principles to design. It is generally used by all the military services and the FAA. It is equivalent to human-factors engineering.

Life sciences engineering—a term sometimes used by NASA. For all practical purposes, it is equivalent to human-factors engineering.

Life support—that part of human factors primarily concerned with the maintenance of health and the biomedical aspects of safety, protection, sustenance, escape, survival, and recovery of personnel.

Man–machine system design—for all practical purposes, this term is equivalent to human factors engineering.

MANPRINT—manpower and personnel integration, an Army plan to impose human factors, manpower, personnel, and training considerations throughout the entire materiel acquisition process.

Personnel subsystem design—that part of human-factors engineering that applies to personnel as contrasted with hardware and software.

Software psychology—a term that encompasses software development, query facility usage, and computer interface design to achieve ease of use.

physiological, anthropometric, or biomechanical aspects of machine design. That is not true.

Ergonomics is a word compounded of two Greek words: *ergon*—meaning "work" and *nomos*—meaning "laws of." The term originated in England and is widely used in Europe and Asia, whereas human factors originated at about the same time in the United States, and has been used mostly on the North American continent. Despite the difference in the way these words look and sound, those of us in the profession regard human factors and ergonomics as equivalent, an equivalence that was acknowledged in 1993 when the Human Factors Society officially changed its name to the Human Factors and Ergonomics Society. Our ergonomic colleagues in Europe and elsewhere in the world are just as much concerned as we are with the design of displays, signs, kitchen appliances, sports equipment, medical instrumentation, computer languages, manuals, and all the other things people use at play or work. Although I shall use the term *human factors* almost exclusively in this book, I could equally well have used ergonomics.

The Bottom Line In any case, you should not worry about the subtle differences among the names and definitions in Table 1.1. They are *all* concerned with designing to accommodate the human user and operator.

The Multi-Disciplinary Nature of Human Factors

Human factors, or ergonomics, is a multi-disciplinary field. Figure 1.2 shows some of the disciplines contributing to it. Most human-factors professionals in the United States come into the field by way of industrial engineering or psychology. Of the 58 human-factors graduate training programs offered by 51 academic institutions in the United States (Sanders & Smith 1988) 25, or about 43 percent, are in industrial engineering or in a department of industrial engineering combined with something else, for example, systems engineering. Twenty-three, or 39 percent of programs, are listed as being solely in departments of psychology, and the remaining 11, or 19 percent, are in other departments, such as mechanical engineering, operations research, or management science.

Psychology To elaborate on the disciplines that contribute to human factors, there are psychologists who study rats, psychologists who study children, and psychologists who study mentally disturbed persons. The kinds of psychologists who are in human factors are primarily experimental psychologists who study people at work. They provide data on such things as:

- Human sensory capacities;
- Psychomotor performance;
- Human decision making;
- Human error rates;
- Selection tests and procedures;
- Learning and training;

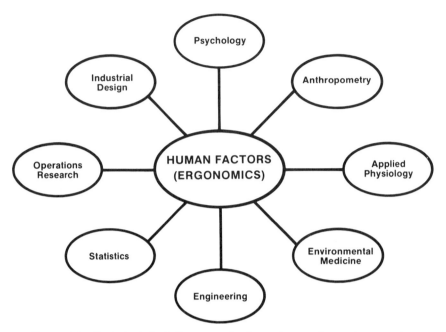

Figure 1.2 Technical disciplines contributing to human factors/ergonomics.

- Individual differences and their measurement;
- Team performance; and
- Methods for studying human performance.

Anthropometry Anthropometry is an applied branch of anthropology, concerned with the measurement of the physical features of people. Anthropometrists measure how tall we are, how far we can reach, how wide our hips are, how our joints flex, and how our bodies move. Measurements have been made on literally hundreds of body dimensions and body movements and for diverse populations. These are available in handbooks (for example, NASA Reference Publication 1024, 1978), and are used in the design of seats, chairs, tables, computer consoles, automobile interiors, airplane cockpits, and many other work stations.

Applied Physiology Applied physiology concerns the vital processes, such as cardiac function, respiration, oxygen consumption, and electromyographic activity, and the responses of these vital processes to work, stress, and environmental influences.

Environmental Medicine Environmental medicine is concerned with such environmental factors as noise, illumination, temperature, humidity, g-forces, and noxious gases and fumes, and their effects on health and human performance. Physicians also contribute valuable information on nutrition, which is, for example, essential to long-duration space flights.

Engineering Engineering provides information on electrical, mechanical, and chemical characteristics of elements and systems and principles of design, construction, and operation of structures, equipment, and systems.

Statistics Statistics are used in two important ways. First, they are used for summarizing large amounts of data on human measurements and human performance. Second, statistics are used to design sampling schemes and experiments for the conduct of human studies and performance measurements.

Operations Research Operations research is concerned with quantitative methods for the analysis of the performance of manpower, machinery, equipment, and policies in government, military, or commercial spheres. It is also responsible for the development of models, such as queueing and allocation models, for describing operations.

Industrial Design Finally, industrial design is concerned with the design, color, arrangement, and packaging of equipment to combine functionality and an aesthetically satisfying appearance.

Differences among Human Factors Professionals The Human Factors and Ergonomics Society, the principal organization representing this profession in the United States, has among its members representatives of all the disciplines mentioned in Figure 1.1. Not one of them, however, is trained in *all* these disciplines. The literature in each of the disciplines is so voluminous, with such a large amount of data and principles, that no one person could be conversant in all of it. To a considerable extent, the kind of expertise one has depends on where one studied and the kinds of work experience one has had. The important point to remember is that we are not all equivalent. There are differences among us, just as there are differences among engineers, designers, and managers.

Because of differences in training and experience, human-factors professionals often approach projects in somewhat different ways. Someone who is a specialist in anthropometrics is likely to concentrate on the dimensions of work spaces. A person trained in applied physiology is likely to be most concerned about the physiological effects produced by work or the work environment. A psychologically trained human-factors professional may downplay anthropometric, biomechanical, or physiological considerations. For these reasons, responsible professionals concede honestly their own limitations of expertise. When necessary, they call on assistance from human-factors professionals having other kinds of competence, to supplement what they do in design.

Human Factors Objectives

Human factors has a number of objectives, and Table 1.2 lists them in four general categories. Parenthetically, they are systems-engineering objectives as well. This is a formidable list of objectives. Two things help us achieve them.

16   1: INTRODUCTION

TABLE 1.2   Objectives of Human Factors

Basic operational objectives
   Reduce errors
   Increase safety
   Improve system performance
Objectives bearing on reliability, maintainability, and availability (RMA), and integrated logistic support (ILS)
   Increase reliability
   Improve maintainability
   Reduce personnel requirements
   Reduce training requirements
Objectives affecting users and operators
   Improve the working environment
   Reduce fatigue and physical stress
   Increase human comfort
   Reduce boredom and monotony
   Increase ease of use
   Increase user acceptance
   Increase aesthetic appearance
Other objectives
   Reduce losses of time and equipment
   Increase economy of production

Specialization According to Areas of Work   First, objectives differ according to where human-factors professionals work. Those who work with consumer products —such as kitchen appliances or home computers—focus largely on increasing safety, ease of use, and increasing user acceptance. Human-factors professionals who work with the design of office environments are most likely concerned with improving the working environment, reducing fatigue and physical stress, reducing boredom and monotony, and the reliability and maintainability of products. Those who work with military systems are generally most concerned with errors, safety, system performance, reliability, maintainability, and personnel and training requirements.

Correlation Among Objectives   The second thing that helps us meet multiple objectives is that they are correlated. A well human-engineered machine or system will meet several objectives at the same time. For example, well-designed computer programs are easier to learn, easier to use, and result in fewer mistakes and in better performance (Chapanis 1991b, pp. 364–367).

HUMAN INVOLVEMENT WITH MACHINES AND SYSTEMS

People are involved with machines and systems in three important ways—as users and operators, as maintainers, and as designers. All three classes of people determine the effectiveness with which machines can be used, and the extent to which accidents and other difficulties can be mitigated during machine use.

Users and Operators

In the final analysis, machines and machine systems are designed and built for only one purpose, namely, to make it possible for people—*users and operators* (and this includes supervisors and managers, as well)—to do something easier, simpler, faster, or better. The distinction between users and operators is not a sharp one, but, in general, we reserve the word *users* for people who interact with machines or small systems—such as stoves, power lawnmowers, and personal computers—and *operators* for people who interact with larger systems—such as automobiles, airplanes, ships, power plants, and large computers. For our purposes, however, the distinction is relatively unimportant. The important point is that there is a large class of people who have to use machines or systems that designers give them as part of their normal activities.

Maintainers

A machine is of no use unless it is in good working order. Although it is easy to recognize the importance of designing to meet user or operator needs, designers sometimes forget about *maintainers*—the people who have to service, maintain, and repair machines (Simmons, 1991). Many maintainer needs are identical with user needs, because maintainers are, after all, another class of users. However, some maintainer needs—for example, the need for simple diagnostic procedures, easy accessibility to areas where maintenance must be performed, special tools for maintenance—are unique to this class of users and should not be overlooked during the design process. The following case study (Howell, 1990, p. 6) illustrates this point clearly:

> The C-17, the Air Force's new heavy-duty airlifter originally had a three-person ground refueling crew. Three people were required because the fuel boost pump switches were in the aircraft cockpit. One person had to sit in the cockpit during refueling just to operate the switches. In an emergency, the ground crew would be dependent on a speedy reaction from the member in the cockpit to turn the switches off.
>
> When the switches were relocated to the wheelwell, refueling could be carried out more safely by only two persons. If this problem had been identified during design the change could have been made at far less cost.

A somewhat different kind of maintenance problem is illustrated by the following case study (also from Howell):

> The B-1B Bomber's fuel tanks are sealed with a gel-type sealant which has to be replaced periodically to avoid fuel leaks. Due to the size of the tank this internal tank maintenance can only be performed by fuel specialists who are 5'4" or smaller. Since

18   *1: INTRODUCTION*

the Air Force doesn't have many fuel specialists who are that small, this maintenance job had to be subcontracted out to civilian contractors. The problem was identified only after the initial prototype had been built and redesign at that stage was deemed too costly.

Twelve percent of large commercial airplane accidents are attributable to deficiencies in maintenance and inspection (Sheppard, 1994). Maintenance errors are also responsible for additional costs to airlines and for aggravating delays to passengers. In the military arena, the costs of maintaining some systems may be over ten times the initial cost of the system. As much as 80 percent of a maintainer's time may be spent in diagnosing a difficulty, and only 20 percent of the time in correcting a difficulty once it has been diagnosed. Clearly, very substantial savings may potentially be gained through proper design for maintenance.

Installation   Since some maintenance personnel install and set-up equipment as well as maintain it, installation may be regarded as a sub-category of maintenance. This is another activity that is sometimes overlooked in design. The difficulties of unpacking and assembling even common objects, such as toys, have been the source of many satirical articles and cartoons. (For an impressive example of substantial cost savings that resulted from a human-factors redesign of procedures for installing large computer systems, see Chapanis, 1991a, pp. 64–67.)

Designers

The third, and perhaps most important, group of people involved with machines are *designers*. Designers includes systems engineers, industrial engineers, programmers, architects, and human-factors professionals—everyone, in short, who has a direct role in the design and construction of machines. Human-factors professionals are included with designers because human-factors data and principles are useful only to the extent that they can be translated into design. Whenever glaring examples are discovered in which machines or systems are poorly designed for human use, there is a tendency to blame the engineer who was responsible for its design. Yet sometimes the fault can be traced to human-factors professionals who could not provide engineers with the kind of information they needed for design. When that happens, the underlying reason often is that human-factors professionals did not understand the design process and how human factors can contribute to that process. The aim of this book is to provide that kind of understanding.

SOME HUMAN-FACTORS POSTULATES

Underlying all the activities in which human-factors professionals engage during the design of machines and systems are some basic postulates that provide a rationale for what they do. These postulates also provide a view of the goals toward which human-factors efforts, and the contents of this book, are directed.

- The performance of a machine or system depends on the operator, not just the hardware or software.
- The maintainability of a machine or system, and hence its availability, depends on the maintainer, not just the machine.
- People are adaptable and they can learn to operate a poorly designed machine or system, but this usually results in
  - increased training time;
  - increased stress for operators;
  - increased human errors under stress; and
  - unused machine capabilities.
- You cannot assume that users, operators, or maintainers can do the jobs or functions you assign to them simply because you can do them.
  - You have to validate those assignments by appropriate human-factors techniques.
  - That validation should be in quantitative terms.
- Validation and verification of human functions is done throughout the entire development cycle.
- Some human-factors validation can be done with paper-and-pencil methods, some with simulations, some with experimentation, and some with other methods. A common feature of them all is that they take time. A human-engineering plan must be prepared early in the development cycle so that:
  - the engineer in charge can plan adequate funding for the effort; and
  - human-factors professionals can do their work in time to incorporate their findings into design before designs are frozen.
- A major objective of human factors is to define user requirements and specify human–system interfaces that will lead to effective system performance.
  - The purpose of final test and evaluation is to verify that designs have met requirements.
  - If major human factors problems are discovered during final test and evaluation, the engineers, the human-factors professionals, or both, did not do their jobs well.

SUMMARY

This chapter began with a preview of the contents of this book and then led into a brief history of some of the technological changes that have led to the emergence of the field of human factors. The terms *human factors* and *human factors engineering* were then defined. The technical specialties that contribute to the discipline were identified and described. The several objectives of human factors were listed and discussed, and the chapter concluded with some basic human-factors postulates that provide a rationale for the work that human-factors professionals do.

The next chapter defines and elaborates on systems and the systems-engineering process.

REFERENCES

Anonymous (1983). TMI: A human-factors mess. *Machine Design, 55(2),* 4.

Casey, S. (1993). *Set Phasers on Stun: And Other True Tales of Design Technology and Human Error.* Santa Barbara, CA: Aegean.

Chapanis, A. (1980). The error-provocative situation: A central measurement problem in human factors engineering. In W. E. Tarrants (Ed.), *The Measurement of Safety Performance* (pp. 99–128). New York: Garland/STPM Press.

Chapanis, A. (1991a). The business case for human factors in informatics. In B. Shackel & S. J. Richardson (Eds.), *Human Factors for Informatics Usability* (pp. 39–71). Cambridge, England: Cambridge University Press.

Chapanis, A. (1991b). Evaluating usability. In B. Shackel & S. J. Richardson (Eds.), *Human Factors for Informatics Usability* (pp. 359–395). Cambridge, England: Cambridge University Press.

Howell, E. (1990). IMPACTS: People make the difference. *CSERIAC Gateway, 1(4),* 6–7.

Mark, R. (1987). Reinterpreting ancient Roman structure. *American Scientist, 75(2),* 142–150.

McQuistion, L. (1993). Ergonomics for one. *Ergonomics in Design, January,* 9–10.

Naisbitt, J. (1982). *Megatrends: Ten New Directions Transforming Our Lives.* New York: Warner.

National Safety Council (1993). *Accident Facts, 1993 Edition.* Itasca, IL: Author.

Norman, D. A. (1988). *The Psychology of Everyday Things.* New York: NY: Basic.

Nussbaum, B., & Neff, R. (1991). 'I can't work this thing!' Frustrated by high tech? Designers are getting the message. *Business Week, 3211,* 58–62, 66.

Sanders, M. S., & Smith, L. (Eds.) (1988). *Directory of Human Factors Programs in the United States and Canada.* Santa Monica, CA: Human Factors Society.

Sheppard, W. T. (1994). Human factors and ergonomics in maintenance and inspection symposium abstract. In *Proceedings of the Human Factors and Ergonomics Society 38th Annual Meeting* (p. 100). Santa Monica, CA: Human Factors and Ergonomics Society.

Simmons, C. (1991). Maintainers—the forgotten users. *Ergonomist, 252,* 1.

Trager, J. (Ed.) (1979). *The People's Chronology: A Year-by-Year Record of Human Events from Prehistory to the Present.* New York: Holt, Rinehart, and Winston.

Young, L. (1989). Nuclear power reforms: How safe is safe enough? *The Sun* (Baltimore, MD), *304(112),* 1A and 9A (March 27).

Chapter 2

Systems and Systems Engineering

To contribute effectively to system design, human-factors professionals need to know something about systems and how they are developed. That knowledge helps you understand how systems developers go about their work and what human-factors inputs are needed at each stage of system development. This chapter defines systems, points out some of their main characteristics, describes system life cycles, and explains the process by which systems are designed and built. This chapter also defines some of the engineering terminology one should know in working with system developers.

Before proceeding, however, it is important to emphasize that systems engineering is not yet well established as an engineering discipline, even though many people are doing it. Because engineers have not yet been able to arrive at a consensus definition of systems engineering, textbooks on the subject treat it in different ways (see, for example, Wymore (1976), and Blanchard & Fabrycky (1990)). What follows in this chapter, and throughout the rest of this book, is only one of several current conceptions of systems, and the way in which systems engineering is, or should be, practiced.

SYSTEMS

The word *system* is used in many different ways, and it means different things to different people. We speak of the *solar system,* the *capitalistic system,* a *system for betting* on horses, or the *nervous system* of the human body—just to take a few examples. This book, however, is concerned only with *equipment systems,* which are defined in this way:

21

A system is an interacting combination, at any level of complexity, of people, materials, tools, machines, software, facilities, and procedures designed to work together for some common purpose.

Although my definition is a paraphrase of definitions that appear in several standards (for example, MIL-STD-721C, MIL-STD-882B), not everyone will agree with it. For one thing, there is a considerable amount of debate about whether people are inside or outside systems. For example, one could conceive of a project to design and construct the hardware and software for a surveillance and security system, or an orbiting satellite communication system, without any reference to the people who will use it. In fact, many so-called system projects today are contracted that way. However that may be, all systems ultimately function for some human purpose, and so have human–system interfaces, whether internal or external. These interfaces have to be designed and constructed at some point. In this book, however, I shall be concerned with systems in which people are clearly internal to them. In other words, these systems all include people in some way, and the design of their human–system interfaces is an integral part of system development.

Figure 2.1 is one way of representing the kinds of systems that are the primary concern of this book. A system (the large box on the right of Figure 2.1) receives inputs (information, raw materials) from external systems or entities, and delivers outputs (information, finished products, movement, services) to external entities. A major component of a system is the machine. For our purposes, machines include

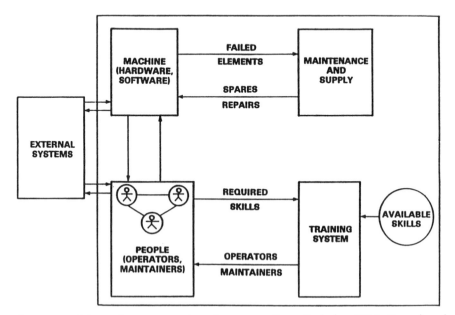

Figure 2.1 Schematic representation of a system. (From Budurka, 1984. Reproduced by permission of the author and publisher)

both hardware and software. Machines fail from time to time and defective components are delivered to maintenance and supply to be repaired or replaced by serviceable components. This maintenance-and-supply function constitutes what was referred to as integrated logistic support in Table 1.2 (see Chapter 1).

Another major component of any system is people—users, operators, and maintainers. Even the most highly automated system requires people if, for no other reason, to start, stop, and monitor the system. Sometimes human roles are combined. For example, users or operators may also service and maintain the machines with which they work. To do their jobs effectively, people must have certain skills. If selection alone cannot provide users, operators, and maintainers with requisite skills, these individuals must be trained, and the training system becomes part of the overall system.

Examples of systems that conform to the definition above are:

1. *Production systems*—systems for building and assembling products, such as computers and automobiles.
2. *Transportation systems*—such as automobiles, aircraft, ships, and railroads.
3. *Information-processing systems*—such as telephone systems, air-traffic-control systems or, the system used repeatedly in this book, automated bank tellers.
4. *Weapon systems*—such as missiles, bombers, and tanks.
5. *Service systems*—such as police and fire systems, gas stations, and warehouses.

These categories are not mutually exclusive. For example, police and fire systems contain or make use of information-processing systems, and production systems produce weapon systems as well as commercial products. Think of these categories merely as convenient groupings.

Levels of Organization

The amount of human involvement in systems varies widely and we can distinguish at least three very broad levels of organization:

1. *User-tool combinations*—for example, a carpenter using a saw, a farmer using a rake, or a mechanic using a wrench.
2. *Operator-machine combinations*—for example, a teenager using a power lawnmower, an operator using a computer terminal, or a driver operating an automobile.
3. *Multi-person systems*—for example, banking systems, air-traffic-control systems, or postal systems.

The distinctions between these levels are, by no means, sharp. For example, some so-called "tools" in industry are very large and complex machines (Figure 2.2).

Figure 2.2 A Wolverine VS Series Vertical CNC Copy and Profile Milling Machine—technically a machine tool. (*Photo courtesy* of *Cincinnati Milacron Marketing Company*)

Likewise, it is not always easy to decide whether to classify a machine and its operator as a system. Do a computer and its operator constitute a system, or not? Fortunately for our purposes, this taxonomic problem is not really of very great practical importance, because the basic principles of system development apply equally to machines and systems.

Systems Within Systems

Very large systems are usually combinations of smaller ones, or *subsystems,* and those subsystems may be made up of still smaller units that are still large enough to be classified as systems. Figure 2.3 is a diagram of a generic, four-level system hierarchy. Depending on their level of complexity, however, systems may have fewer or more than four levels.

To illustrate, the interconnected banking system of the world may be considered a system. Subsystems are the banking systems of individual countries or geographic areas, for example, the United States, Germany, or Switzerland. Each of these subsystems consists of still smaller systems, sub-subsystems, individual banks within a single country or geographic area, say, the United States. Examples of such sub-subsystems in the United States are the First National Bank of Maryland, the Irving Trust Company of New York, and the First Chicago Trust Company. Each of these banks has, in turn, a number of automated banking terminals which, in this

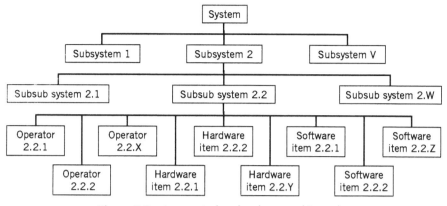

Figure 2.3 A generic four-level system hierarchy.

book, I shall also call systems—small ones, to be sure, but systems nonetheless. People, users (clients), maintainers, managers, all figure in every level of these banking systems and subsystems. Examples of other very large systems with several layers of subsystems are air-traffic-control systems, postal systems, and telephone systems.

Although the basic idea is simple enough, different organizations and authors have different names for systems and their parts. The terminology I shall use in this book is generally in accord with that of Blanchard and Fabrycky (1990). Table 2.1 compares it with the terminology used in the Federal System Company (FSC) and by government and military agencies. You can find still other names used by other authors or organizations. This diversity of names is not as troublesome as it might first appear. The practicing human-factors professional quickly adapts to the terminology of the organization for which he or she is working. The size of the project, whether for a total system or some part of it, is spelled out explicitly in Requests for

TABLE 2.1 Some Nomenclature Used in Referring to Systems

Terms Used Here	FSC Nomenclature	Military Terms
System	System	Prime item
subsystem	segment	critical item
sub-subsystem	element	—
—		
component:	hardware	hardware
human	configuration	configuration
hardware	item	item
software	software	computer
	configuration	software
	item	configuration
		item

Proposals (RFPs), Statements of Work (SOWs), or other planning documents that define the project.

The one really important difference between the terminology Blanchard and Fabrycky and I use, and that used by many other organizations, is that we explicitly identify the human (whether user, operator, or maintainer) as a system component. The failure of many engineers to consider the human as a system component and to design for the human user is one of the reasons why so many machines and systems are unsafe or difficult or inconvenient to use. An aim of this book is to help rectify that.

SYSTEMS ENGINEERING

Reduced to its essentials, systems engineering is the process by which systems are analyzed, designed, and constructed. More formal definitions are given in various books. The following is taken from Blanchard and Fabrycky (1990):

> Systems engineering is a process that . . . Transform(s) an operational need into a description of system performance parameters and a preferred system configuration through the use of an iterative process of functional analysis, synthesis, optimization, definition, design, test, and evaluation; Incorporate(s) related technical parameters and assure(s) compatibility of all physical, functional, and program interfaces in a manner that optimizes the total system definition and design; and Integrate(s) performance, producibility, reliability, maintainability, manability, supportability, and other specialties into the overall engineering effort (pp. 20–21).

Although Blanchard and Fabrycky do not specifically define "manability," they appear to equate it with operability or usability, and they use the term to encompass some typical human-factors considerations. What I have done, in essence, is to expand their brief treatment of manability into a book.

Although I agree with Blanchard and Fabrycky's definition, I prefer the following, simpler one from Budurka (1984):

> System engineering is the iterative but controlled process in which user needs are understood and evolved, through incremental development of requirements specifications and system design, to an operational system. Systems engineering includes the control and integration of all disciplines throughout the system life cycle in a manner so as to assure that all user requirements are satisfied (p. 41).

Systems Are Designed to Meet User Needs

Budurka's definition first calls attention to something that is sometimes overlooked, namely, that systems are designed to serve some human purpose or to fulfill some human need. We build systems to manufacture products, to transport us from place to place, to enable us to communicate quickly with one another, to help us to compute more effectively, and to serve many other purposes. Understanding and defining the reasons to develop a new system are a necessary first step. As the definition implies, it is also the last step, because how well it meets the initially established user requirements is an important criterion used in the final evaluation of a system.

System Development is Iterative

Next, Budurka's definition specifically emphasizes the iterative, nonlinear nature of system development, which is unlike highly linear manufacturing processes. Typically, ideas or designs are tried and modified. When found to be unsatisfactory, they are abandoned in favor of other alternatives. In fact, most system developments might be more properly thought of as following an upward spiral, rather than a straight line [see, for example, Boehm (1988)].

System Development Follows Specifications

Next, the definition calls attention specifically to specifications. As I shall point out in Chapter 3, system design is driven by specifications. Not only are specifications imposed on the engineer, but the engineer must also *prepare* specifications. Many of these specifications, whether imposed on or written by the engineer, require human-factors inputs. One of the substantial contributions human-factors professionals can make to the development process is to take responsibility for the specification of those parts of systems that affect or are affected by humans. In the past, these have often been specified without the benefit of human-factors expertise.

System Engineering Is Multi-Disciplinary

Finally, Budurka's definition implies that systems development is a multi-disciplinary activity. Figure 2.4 shows some of the technical disciplines that may form part of a systems-engineering team on a large project. These diverse groups bring different, sometimes conflicting points of view to a project. For example, human-factors requirements may sometimes have to be modified because they are too costly to implement, because they are too difficult to manufacture, because they are too unreliable, or for any of a number of other considerations. Conflicting requirements are usually resolved in *trade-off*, or, more simply, trade studies (see Chapter 8). Such studies identify candidate concepts, weight each of them according to quantifiable criteria, and select the concept that provides the best compromise among the various requirements. The human-factors professional can provide important contributions to trade studies, such as the establishment of trade study criteria, provision of quantitative data on human performance in each of the candidate concepts, evaluation of results, and resolution of conflicting requirements between human factors and other disciplines.

28   2: SYSTEMS AND SYSTEMS ENGINEERING

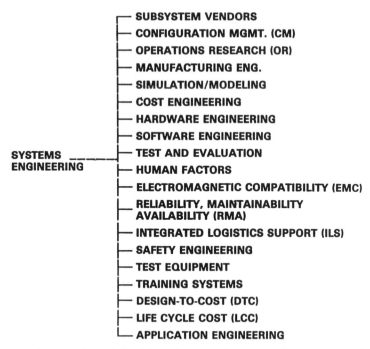

Figure 2.4 Some technical discipines that may form part of a systems-engineering team.

SYSTEM LIFE CYCLE

The life cycle of a system is a sequence of stages or phases in the life of a system. It is the time between the recognition of a need for a new system, and the retirement, scrapping, or removal of the system from service. In other words, it is the life history of a system from desire to burial.

Opinions differ on how many stages there are in a system life cycle and how the various stages should be identified and labeled. (Table 2.2). The general format I find congenial, and the one I shall follow (Figure 2.5) most closely matches Blanchard and Fabrycky's, and is a slightly modified version of the one given in DOD-HDBK-763. It is also used in some major industries in the United States and adapts itself well to the development of any kind of machine or system.

Although the diverse names and stages in Table 2.2 cannot be matched with one another exactly, do not be dismayed by the apparent disparities. Underlying them all are two fundamental characteristics of system life cycles, on which all would probably agree: (1) they proceed from the general to the specific; and (2) they start with needs and plans, progress through design at successively increasing levels of detail, go into production, are delivered to their customers for use, and are finally disposed of. As a practical matter you will find you can quickly pick up and adapt to the particular terminology used in the organization with which you are associated.

Operational-Need Determination

System development starts with the recognition of a need. This results in a memorandum or document that says why someone needs or wants a new system (or machine), or an improved version of an old one. Examples of customers who may want a new or improved system are government agencies that want an improved air-traffic-control tower or new combat aircraft, industries that want a new production line or accounting system, or banks that want automated bank tellers. The statement of operational need prepared by these organizations may then be included in a Request for Proposal (RFP), a formal invitation to prospective firms to bid on the development of the system, and the Statement of Work (SOW) that forms part of the contract made with successful bidders.

Sometimes operational-need statements are prepared by the same organization that is going to build the machine or system. Automobile manufacturers, for example, prepare their own statements of need in planning for a new line of automobiles. Similarly, computer manufacturers usually prepare their own statements of need before undertaking the development of new computers and computer systems. These in-house statements often use information supplied by marketing departments, which may have studied the field with focus groups and consumer and dealer surveys.

Human-factors professionals may be involved in the preparation of an

TABLE 2.2 Some Descriptions of System/Product Life Cycle Phases*

Kirk (1973)	Clark, Cramer et al. (1986)	Blanchard & Fabrycky (1990)	Cushman & Rosenberg (1991)	Kirwan & Ainsworth (1992)
Formation of objectives	Planning	Identification of need	Product planning	Concept
Definition	Definition	Conceptual design		Flow sheeting
Preliminary design	Design	Preliminary design	Design	Preliminary design
Detail design		Detail design and development		Detailed design
	Integration			
Testing	Acceptance		Testing and verification	
Conversion		Production and/or construction	Production	Construction
	Delivery		Marketing and evaluation	Commissioning
Operation	Product/system sales	Utilization and support		
Evaluation				Operating and maintaining
Maintenance and support				
		Phaseout and disposal		

*Approximately equivalent stages are aligned horizontally.

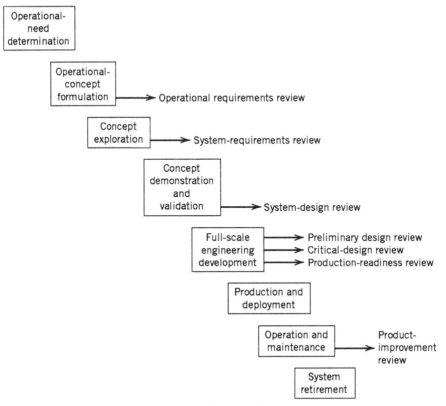

Figure 2.5 Major system life cycle phases with some representative reviews.

operational-need statement. They may, for example, analyze existing systems through any of a variety of techniques [among them, critical incident studies, accident studies, questionnaires, interviews, fault-tree analyses (see Chapter 4)] to identify difficulties with an existing system that need to be corrected in a new one.

At the time I was writing this book, for example, I analyzed problems in a production line that was manufacturing sections of automobile bodies. Using a combination of interviews, task analyses, critical incident accounts from workers, analyses of accident records, and inspections of the production line, I was able to prepare a catalog of human-factors design deficiencies in the current line, many of them almost classic textbook examples. My report constituted a major part of the statement of operational need for the design of a new production line, aimed at eliminating those deficiencies and making the line safer and more efficient.

In a market-driven economy such as we have, statements of operational need often concentrate on costs and marketability. Human-factors professionals can ensure that the needs of users, operators, and maintainers are not ignored.

Examples Operational needs and their corresponding statements vary greatly, depending on the product or system, intended use of the product, environment in which the product will be used, economic conditions, and a number of other factors. Automobile manufacturers need to change models yearly to be competitive. New household appliances may be developed for essentially the same reasons. Of course, developments of new machines or systems are sometimes undertaken primarily to incorporate new technology.

In the case of very simple devices, operational needs are sometimes not written down, but it is usually much better to have a written statement of the need. This avoids later misunderstandings about why a new system was being developed. Operational-need documents may be no more than a single page in length for small items such as household appliances, or as large as whole books for large systems, such as nuclear power plants or telephone exchanges. Whatever their size, they should identify existing and planned operational capabilities, exploitable technology, and constraints and deficiencies that need to be corrected, such as:

- excessive manpower (operators, maintainers);
- excessive operating costs;
- inadequate operational performance;
- inadequate system reliability or operational availability; and
- excessive operational complexity or susceptibility to error.

Table 2.3 is an example of an operational-need document for an automated bank teller.

As an exercise prepare an operational-need statement for an improved version of some familiar machine, say a power lawnmower, or a home copier.

Preparation of an Operational Concept

Often referred to as an operational scenario, an operational concept is a high-level description of the intended use of the system or equipment by its operator(s). It identifies major functions and activities and provides a common basis of understanding among the various members of the system-design team about what kind of solution they are looking for. Normally written in narrative form, it may include an operational profile, a graphical, or a pictorial representation of the operations the system is expected to perform. Figure 2.6 is an example of such an operational profile for a search-and-rescue helicopter. Note that this example depicts succinctly the operational need, and illustrates a probable operation, as well as some of the conditions under which the helicopter may have to perform.

The operational concept should address such questions as:

- What will the system look like?
- How will the system be used?
- In what ways can the system be misused or abused?
- What hardware and software will it require?
- What will be the user–system interfaces?

**TABLE 2.3   Operational-Need Document:
The Fourth National Bank and Trust Co., Anycity**

1. **Existing and Planned Operational Capabilities**

    A survey conducted by the Fourth National Bank and Trust Company has determined that a growing number of its current and potential customers would like to conduct selected banking transactions:
    a. at times when the bank's branch offices are normally closed, such as evenings, weekends, and holidays;
    b. at more convenient locations in the suburbs, such as shopping malls.

    From the bank's standpoint, however, it is too expensive to keep branch offices open for longer periods. Also, although new branch offices are planned in some suburban areas, the number will be limited by the large capital outlay required.

2. **Additional Operational Capability Required**

    The bank would like to provide additional banking hours for its customers without increasing the amount of time branch offices are open, at many more locations than those of existing and future branch offices.

3. **Exploitable Technology**

    Digital computer technology could be employed to conduct banking transactions automatically with sufficient levels of reliability and security. This would enable bank customers to conduct transactions on a 24-hour-per-day basis, every day, at almost any location that could be linked by communication lines to the bank's processing facilities.

4. **Constraints**
    a. Operational capability should be introduced on a trial basis in 24–36 months. Development costs could be offset by increased bank earnings in five years.
    b. Development approach should involve minimal technical risk.
    c. New facilities should interface with those now in use by bank tellers.
    d. Facilities should provide privacy, safety, security, and protection from the elements for bank customers, in conducting transactions.
    e. Due to the large amount of cash involved, security (both physical and data-access) is critical.
    f. Customer-interface equipment will be minimally protected from the environment and must, therefore, be "ruggedized."
    g. All-weather protection should be provided for the users.
    h. Facilities should be designed to be easy to use for a large proportion of potential customers.
    i. Facilities should be designed to withstand malicious use.

- For what kinds of users is the system designed?
- How many users will be required to operate the system?
- What skill levels must the users have?
- What training must users have?
- In what kinds of environments will the system be used?
- How will the system be serviced and maintained?
- What other constraints or special considerations apply to the system?

Notice that, at this stage, the system has not yet been constructed, much less designed. For that reason, the answers to all these questions can only be tentative

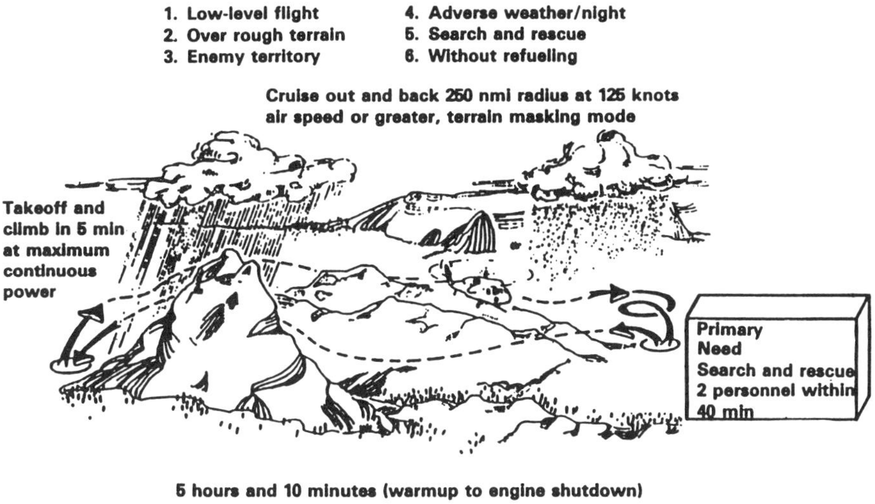

1. Low-level flight
2. Over rough terrain
3. Enemy territory
4. Adverse weather/night
5. Search and rescue
6. Without refueling

Cruise out and back 250 nmi radius at 125 knots air speed or greater, terrain masking mode

Takeoff and climb in 5 min at maximum continuous power

Primary Need
Search and rescue 2 personnel within 40 min

5 hours and 10 minutes (warmup to engine shutdown)

Figure 2.6 An operational profile for a search-and-rescue helicopter. (Reproduced by permission from Figure HFM-6, IBM 1993)

predictions, likely to be modified several times during system development. In addition, the results of any analyses that may be performed can only be preliminary. Despite the provisional nature of this document, it serves as a source of design requirements that must be met during later stages of development. The human-factors professional will also have used the document to derive training requirements and perhaps training materials, too.

As was the case with operational-need documents, operational concepts may vary in length from a single page to a small book, depending on the size and complexity of the system to which they apply. Table 2.4 is an example of an operational concept that would apply to an automated bank teller. Notice that it describes major functions that will be performed by bank customers, for example:

- Activate the system.
- Supply a personal identification code and account number(s).
- Identify the desired banking transaction,

and so on, but it does not say exactly how the customer will perform each of these functions, or what kinds of devices will be used in carrying out these functions.

Similarly, the operational concept describes functions that will be performed by the system:

- Request the customer to provide a personal identification code and account number(s).
- Validate the customer's inputs.
- Request the customer to identify the desired banking transaction,

34   2: SYSTEMS AND SYSTEMS ENGINEERING

TABLE 2.4   Operational Concept That Satisfies the Operational Need for Automated Bank Tellers: Fourth National Bank and Trust Co., Anycity

1. **Bank Customer Transaction Concept.** Bank customers will approach the facility either on foot or in their cars. At some locations, both kinds of facilities will be provided; at others, only one of the two. The customer will activate the system and, if available for use, the system will request the customer to provide a personal identification code and account number(s). The customer will supply the code and number(s) to the system and the system will determine their validity. If the customer's inputs are valid, the system will request the customer to identify the desired banking transaction. If either the code or account number(s) is invalid, the system will so inform the customer and turn itself off. After the customer identifies the desired banking transaction, the system verifies that the requested transaction is valid.

    If the requested transaction is valid, the system provides the customer, in English or Spanish, with step-by-step instructions for completing the transaction. In either language, the vocabulary will be at the eighth-grade reading level. Instructions and procedures will be so simple that 90 percent of first-time users of the system will be able to complete any transaction within M minutes.

    If a requested transaction is invalid, the system informs the customer why the transaction is invalid and turns itself off. If the customer makes errors in carrying out the step-by-step instructions, the system will give customer information about the error and ways of rectifying it. When the customer follows, without error, the step-by-step instructions provided by the system, the system completes the requested transaction. After the transaction is completed, the system:
    a. immediately updates all appropriate bank records in accordance with the completed transaction;
    b. immediately provides the customer with a hard-copy record of the completed transaction;
    c. includes the transaction in the customer's monthly bank statement(s); and
    d. immediately indicates that funds deposited are "not available" pending verification.
2. **Allowable Bank Customer Transactions.** The system will provide bank customers with the opportunity to conduct the following transactions:
    a. Withdraw funds from either the customer's checking or savings account.
    b. Deposit funds into either the customer's checking or savings account.
    c. Transfer funds from the customer's checking account to savings account, or vice versa.
    d. Determine the balance in the customer's savings or checking account.
3. **Allowable Number of Transaction and Withdrawal Amounts.** Bank customers will be able to conduct an unlimited number of transactions per day, except that the maximum amount of cash they may withdraw will be limited to $XXX per day. Withdrawals must be in multiples of $10.
4. **Logistics and Maintenance Concept.** The money supply and supply of receipts at each facility will be replenished daily. The maximum money supply will vary from location to location between the limits of $YY,YYY to $ZZZ,ZZZ, depending on predicted and measured demand. Deposits will be retrieved, verified, and credited to the customer's account(s) at least twice daily. Routine maintenance will be performed during the same time period, when the money supply is replenished. When the money supply is being replenished the system will be made unavailable for use and bank customers will be so

(continued)

TABLE 2.4 (*Continued*)

notified. Money replenishment and routine maintenance will be performed by a single bank employee.

5. **Management Concept.** Each day the system will provide the manager at each location with a hard-copy printout, summarizing the number of times the facility was used, the times at which the facility was used, and the average and total amounts of money involved in each of the allowable kinds of transaction.
6. **Training Concept.** Customers will receive no formal training. Brochures describing the location of facilities and services provided will be mailed with customers' bank statements. Service and bank maintainer personnel will receive a booklet describing service and routine maintenance procedures in a one-day tutorial. Major repairs and heavy maintenance will be performed by the manufacturer. Maintenance personnel will receive a booklet describing repair and maintenance operations in a four-day, hands-on training course. Managers will receive a booklet describing the operation and management of the system in a one-day briefing.

---

and so on, but once again does not specify exactly how each of these functions will be executed. More exact details about how functions will be performed are arrived at in the next stage of the system life cycle—the concept-exploration phase.

One thing more: Notice that the operational-concept document identifies users other than customers, namely, maintainers and managers. Most large systems have several classes of users, and it is important not to overlook them. In the case of our bank teller machine, we can identify four important user classes:

1. *Bank customers*—people who have a checking account, savings account, or both, and want to use the facilities to carry out one or more allowable transactions;
2. *Service personnel*—people who load the facilities periodically with money and transaction-confirmation slips;
3. *Maintenance personnel*—people who service, adjust, maintain, and repair equipment; and
4. *Supervisors* or *managers*—people who oversee the operation of the facilities, validate funds supplied to facilities, receive and evaluate tallies of transactions and amounts of money dispensed.

To sum up, human-factors professionals can make important contributions to the preparation of an operational-concept document. These ensure that the system scenarios are user-oriented, and that all classes of users have been identified.

As an exercise, prepare an operational-concept statement for the same machine for which you prepared an operational-need statement.

Concept Exploration

This phase sometimes requires a considerable amount of time and creative effort from the members of the design team. I think of it as the "brainstorming" phase, during which alternative solutions are proposed and tested. Some are modeled, tested, and discarded. Others are modeled, tested, modified, and tested again. Eventually, one solution is found that appears to satisfy the need and design requirements established earlier. For most large systems, some of the major human-factors activities during this phase are:

- Assisting in the allocation of functions to people, hardware, software, or some combination of them (see Chapter 4);
- Conducting trade studies to evaluate the costs and benefits of various alternatives involving human users or operators (see Chapter 8);
- Developing user–system interface requirements; and
- Prototyping and evaluating user interfaces.

The end of this phase is normally marked by the completion of a draft of system requirements, which is discussed in Chapter 3.

Concept Demonstration and Validation

The design concept selected during the previous phase is now breadboarded, mocked up, or simulated, and tested to verify that it does indeed satisfy the requirements established in the first two phases. Ensuring that user requirements are met is a major human-factors activity at this time. Does the design concept meet requirements related to staffing, operating, maintaining, and supporting the system? Does it meet dimensional requirements for workspaces, ingresses, egresses, and accesses for maintenance? Does it meet safety, personnel, and training requirements?

Full-Scale Engineering Development

During this phase the development leader or team specifies, designs, and builds (or procures) all the component items that are needed for the system, assembles them, and verifies that the system specifications and operational need have been satisfied. The component items may be hardware and software, or subsystems in the case of very large systems.

This is the first point at which hands-on, or real hardware and software, are normally available. A major human-factors activity is assessing operator performance with a complete product and its documentation. This involves operational tests, with representative users performing typical tasks, to determine whether the system meets performance requirements established earlier. These tests may reveal last-minute needed changes to make the system work as desired.

This phase is terminated when the development team secures approval for production of the system.

Production and Deployment

At this point the system goes into production. Multiple copies, when required, are produced, often with an initial, limited distribution to the customer, or for sale, to establish its acceptability in the marketplace. In this phase, human-factors activities include training users and operators in the operation of the system, evaluating the usability or operability of the system, and identifying changes that should be incorporated in the larger production run, or for follow-on versions.

Operation and Maintenance

With systems delivered and in use, human-factors professionals may now conduct follow-on tests of operator satisfaction with the use and maintenance of the system. The results of such tests may serve as guides to engineering changes that may be made in subsequent versions of the system.

System Retirement

In this phase the system is retired from use, scrapped, or replaced with a new system if needed. This activity is no longer as simple as it once was. Increasing consciousness on the part of both governments and individuals has led to concerns about environmental conservation and recycling of materials. As a result, ease and expense of disposal have joined the list of major system requirements, and a number of companies now consider disassembly, as well as assembly, during product and system design. Brennan (1994) discusses these issues in greater detail, and provides some human-factors guidelines for ease of disassembly.

Life Cycle Control Points

For virtually all development projects there are what engineers refer to as *control points* at various times in the system life cycle. These are milestones, times at which evaluations are made, to determine whether performed work is satisfactory, whether any changes need to be made in the development, and whether budgets and timetables of completion are realistic.

The evaluations consist of *reviews* (see Figure 2.5) and *audits,* and are required by customers for whom the system is being developed. In essence, the system engineer has to meet with the customer's representatives, describe what has been done, and defend those activities. For those systems being developed in-house, the systems engineer has to go through essentially the same kinds of reviews and audits for management. Human-factors professionals may also have to participate in these reviews and audits to defend decisions they have made about designs that affect usability, safety, or other human-factors concerns. These reviews can be penetrating and devastating if the human-factors professional has not done a good job up to that point, and carefully documented the reasons for the various decisions that were made.

Variations

Although the development of all products, tools, machines, and systems follows the same general pattern, there is enormous disparity in the lengths of their life cycles and in the relative amounts of time spent in each phase. For example, about 12 years elapsed from the time development of the USAF B-2 aircraft was authorized until the first aircraft was delivered. By contrast, for very simple products, some life cycle phases may be very brief, lasting no more than a few hours or minutes. For example,

- You see a potential market for an improved can opener (a need).
- You visualize how your can opener will work (operational concept).
- You try sketching various kinds of can openers, try various kinds of materials, and try putting them together (concept exploration).
- You see more clearly how your improved can opener is going to be made. You make one, and have users try it out on several cans (concept validation).
- You turn your design over to a machine shop to be made in several thousand copies (production).
- You market your can openers (distribution).
- People use them and clean them from time to time (use and maintenance).
- When it wears out, people toss your can opener into a trash can or buy a newer or different model (product retirement).

Although you can identify all the phases in the development of a product as simple as this one, the first three phases may overlap and almost seem to be one process. Keep in mind, too, that the design of even very simple products, such as tools, may be iterative. Scherzinger and colleagues (1994), for example, describe several cycles of prototype design–test–redesign they went through for something that, at first glance, seems extremely simple—an improved tool for raising manhole covers.

THE SYSTEMS-ENGINEERING PROCESS

Roughly paralleling the system life cycle phases is a series of activities that is sometimes referred to as the *systems-engineering process*. Engineers who may read my delineation of this process will find it strongly biased. As I pointed out in Figure 2.4, systems engineers have to accommodate a large number of technical considerations in their work and, to them, human factors often seems like a relatively minor one. The major thesis of this book, however, is that, for a system to be successful, three lines of development—the *user, hardware,* and *software*—have to be managed and woven into an integrated product throughout this process (Figure 2.7).

The so-called waterfall model (Royce, 1970) shown in Figure 2.7 characterizes

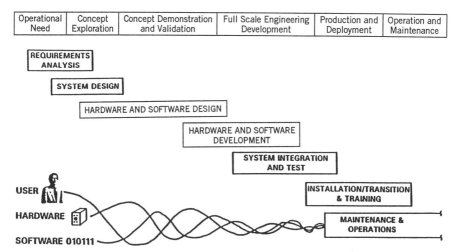

Figure 2.7 The systems-engineering process weaves three lines of development together as design progresses. Activities in the system-engineering process (*center*) are shown in their approximate relationships to life cycle phases (*top*), but the sizes of the rectangles are not intended to represent the amount of time taken up by each life cycle phase or activity. (Adapted with permission from Figure SS&R-6, IBM 1993).

the systems-engineering process as a series of steps organized serially, one leading to the next, but, as Royce points out, feedback loops and iterations almost invariably characterize development efforts. This has led Boehm (1988) and others to describe the process as an upward spiral that becomes increasingly tighter as development progresses (similar to what I have tried to show in Figure 2.7). I shall follow the waterfall model because I find it useful. Remember that there are usually iterations among the several steps.

Figure 2.8 elaborates the process with more details about the human-factors activities that should be performed at each step. The figure also shows some of the reviews, audits, and tests that are typically required at various times.

Requirements Analysis and Elaboration

The first step in the process is an enumeration of requirements the system must meet if it is to satisfy its operational need. The operational need and operational concepts documents, supplemented with additional information derived from analyses that follow from them, are the primary source of those requirements. These requirements provide the information necessary for the engineer to write specifications, a topic discussed in detail in Chapter 3. They also serve as a set of criteria that will be used in final test and evaluation, to determine whether the system that has been produced does, in fact, solve the problem that was initially identified in the operational need.

40   2: SYSTEMS AND SYSTEMS ENGINEERING

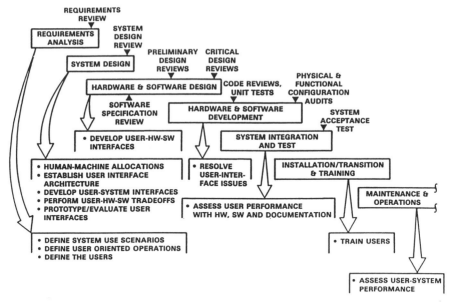

Figure 2.8 Some human-factors contributions to the systems-engineering process. (Reproduced by permission from Figure SS&R-7, IBM 1993)

Three Views of a System   A requirements analysis typically produces information that describes the system from three distinct points of view: the physical, functional, and operational (Figure 2.9).

The physical view, sometimes called the architectural view, focuses on what the system contains, that is, how it is constructed and how many people it will contain.

The functional view focuses on what the system does, that is, what the system must do to produce the required operational behavior. It shows inputs, outputs, states, and transformation rules.

The operational view focuses on how the system will serve its users. Think of it as the way a user views the system. This view is the key to establishing operating requirements for the system, *how well* and *under what conditions* the system must operate.

An Example of Two Views. Wasserman (1989) has provided us with a good graphic example of the difference between two views of a system. In discussing a revised design strategy that led to the development of the highly successful Xerox 1075 copier, he described the typical engineer's view as machine-centered, a physical view (Figure 2.10). From that perspective, the machine is a combination of gears, wheels, belts, and software. The only external thing the machine has to accommodate is the size and weight of the document that is fed into it, and the size and weight of the paper that comes out. If the user is seen at all, he or she is viewed only as an anonymous, nondescript entity that brings the job to the machine. This is the machine as a black box.

THE SYSTEMS ENGINEERING PROCESS   41

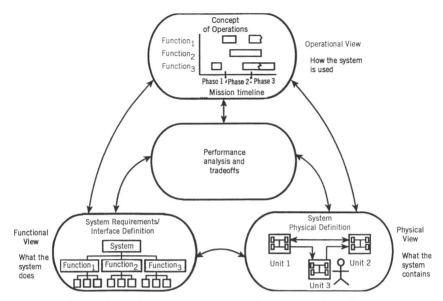

Figure 2.9   Three views of a system. (Adapted with permission from Figure 4/88 TW-2, IBM 1988)

The user-oriented (operational) view of the machine (Figure 2.11) sees it as an information system with the user at the center of the picture. The user is no longer an anonymous, "normative humanoid," but a live individual with distinct needs, abilities, and motivations. One important element that contributed to the eventual success of this machine was its ability to interpret and accommodate the user's requirements. As Wasserman put it,

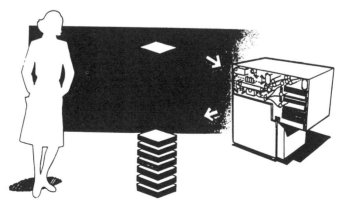

Figure 2.10   A machine-oriented view of a copy machine. (From Wasserman, 1989. Reprinted with permission from Ablex Publishing Corporation)

Figure 2.11 A user-oriented (operational) view of a copy machine. (From Wasserman, 1989. Reprinted with permission from Ablex Publishing Company)

> . . . the job gets processed through the user, and what the user operates has nothing to do with the gears and wheels inside. The user doesn't operate the actual machine; the user operates a *virtual* machine, which stands between her and the actual machine. This virtual machine is the information system which translates the device to the user and the user to the device, and which 'disambiguates' all transactions. It says, 'This is what you lift to put the original in,' 'This is how you position it.' 'This is where you start copying.' 'There is where you put the copy paper in.' 'Here is where you put the toner and how you do it.' (pp. 19–20)

And, as Wasserman observed, the information system had to be designed with the same rigor as any other mechanical, electromechanical, or electronic subsystem in the machine.

Relationships among the Three Views. It's important to emphasize that none of these views is wrong, but that, individually, they are incomplete. Each view provides requirements that are necessary for the design of the system. From the physical view come requirements about such things as the kinds of hardware and software that will be required. From the functional view one derives requirements about inputs, outputs, interfaces, and software that will transform inputs into outputs. The functional view also provides information about the numbers and kinds of personnel the system will require and is the starting point for the allocation of functions to humans, hardware, and software (more about this in Chapter 4). From the operational view come requirements about time-ordered sequences of functions and operating procedures. All must be merged to produce a final and complete list of requirements the system must meet.

The physical, functional, and operational views are compelling ways of describing a system, but they often lead systems engineers and program managers away from using human-factors specialists. As I've indicated, there are human-factors implications in all three, and professionals with human-factors expertise are needed to address all of them.

Application to the Bank Teller Example   From the two documents prepared so far (Tables 2.3 and 2.4), we can see some of the requirements that our automated bank teller must meet. For example:

- The system shall provide some sort of mechanism for obtaining customer personal identification numbers.
- The system shall have logic to establish the validity of personal identification numbers.
- The system shall have some sort of mechanisms for advising customers if the personal identification numbers entered are invalid.

Some requirements are not explicitly apparent in either the operational need or operational concept documents, but may result from logical or other considerations. For example: "The system shall provide feedback to customers within X seconds." Note that, in determining the X in this requirement, the human-factors professional may have to conduct some sort of study or analysis to find out what response time would be acceptable to customers.

As an exercise, prepare a list of ten requirements that must be met by the machine or system for which you prepared an operational need and operational concept statement.

A subtle, but very important point that is often not appreciated by human-factors professionals is that requirements must be phrased in ways the systems engineer can use. To clarify this point, consider requirements about system users. The operational concept will have identified various classes of users, but may not have described their characteristics. At this point human factors would detail the characteristics of each class of users. Table 2.5, for example, lists some characteristics that may apply to bank customers. Similar lists would, of course, also be prepared for maintainers, service personnel, and supervisors or managers.

As an exercise, prepare a list of characteristics that describe potential users of the machine for which you prepared operational need and operational concept statements.

Many human-factors books and articles seem to imply that the human-factor job is done when user characteristics such as those in Table 2.5 have been identified. That is not so. User characteristics do not constitute system requirements. A systems engineer has no control over who will use the system once it has been put into use. User characteristics have to be translated into requirements that can be mean-

TABLE 2.5 Bank Customer Characteristics

Bank customers:
- Will be of either sex;
- Will range in age from 18 to 85;
- May have no prior computer or keyboard experience;
- May have no more than a grammar-school education;
- May have either English or Spanish as their native tongue;
- May cover a wide range of social-economic status;
- May have visual acuity no better than 20/100;
- May have hearing disabilities;
- Will range in height from 58 to 75 inches;
- May have physical handicaps that make it impossible for them to raise their hands above shoulder height;
- May rely upon the use of crutches or wheelchair.

ingfully imposed on the engineer. Table 2.6 illustrates how that is done. If it had been established that users may have no more than a grammar-school education, a system requirement that follows is that instructions should be phrased at the eighth-grade level. An engineer can design to that requirement, but cannot design to the user characteristic from which that requirement was drawn. Similarly, if it had been established that users may have visual acuity no better than 20/100, a requirement that follows is that alphanumeric characters shall be at least 10-point type. This also is a requirement to which an engineer can design. In other words, human-factors professionals must not be content with merely listing user characteristics; they must take the next step of deriving requirements that the system must meet so that it can be used by persons having those characteristics.

As an exercise, extract system requirements from the list of user characteristics you prepared for your machine.

TABLE 2.6 Translating User Characteristics into System Requirements

User Characteristics	System Requirements
May have no more than a grammar school education	Instructions shall be phrased at the eighth-grade level
May have visual acuity no better than 20/100	Alphanumeric characters shall be at least 10-point type
May have hearing disabilities	Both visual and auditory feedback shall follow all user inputs
May arrive on crutches or in a wheelchair	Access shall be at the same level as the adjacent sidewalks; There shall be no stairs. Forty inches of unimpeded access shall be provided.

System Design

Having elaborated requirements the system must meet, the system engineer is now ready to begin system design. This step typically requires a considerable amount of human-factors work. Now human-factors professionals, in concert with other members of the design team, may do some or all of the following:

- Allocate major functions to humans or to machines;
- Prepare functional flow diagrams or operational sequence diagrams of the interchanges between humans and machines;
- Identify and develop user–system interfaces;
- Perform user–hardware–software tradeoffs;
- Prototype and evaluate user interfaces;
- Configure work spaces;
- Address major environmental design concerns.

In the case of our automated bank teller, the operational-need and operational-concept documents have identified four major system scenarios:

1. Transferring funds from savings accounts to checking accounts, or vice versa;
2. Dispensing funds from either savings or checking accounts;
3. Dispensing statements of balances in savings or checking accounts;
4. Accepting deposits into savings or checking accounts.

Functional Analysis and Allocation of Functions Functional analysis and the allocation of functions to humans or to machines (hardware and/or software) is discussed in greater detail in Chapter 4. This is relatively straightforward for our bank teller example, but is not so easily decided in other more complex systems. Some human tasks involved in carrying out banking activities are:

- Activating the system;
- Providing identification and account numbers;
- Choosing a type of transaction;
- Specifying the amount of funds involved in the transactions;
- Confirming selections;
- Interpreting and following error message instructions;
- Removing hard-copy transaction records;
- Completing and entering deposit slips into receptacles; and
- Removing currency.

Some system functions that are involved in carrying out the four major activities are:

46   2: SYSTEMS AND SYSTEMS ENGINEERING

- Verifying that the user is a valid customer;
- Verifying that funds are available to complete the transaction requested by the customer;
- Debiting or crediting accounts, as appropriate, following each transaction;
- Keeping records of all transactions and providing daily summaries;
- Displaying error messages;
- Keeping records of all errors;
- Providing feedback to customers;
- Alerting service personnel when supplies of money and confirmation slips are running low; and
- Keeping records of all system malfunctions.

Developing User–System Interfaces   Operational-sequence diagrams (Chapter 4) are graphic representations of operator or user tasks, as they relate sequentially to both equipment and other operators. They are a useful first step in the development of user–system interfaces. Figure 2.12 is a possible operational sequence for the first scenario, transferring funds. The use of the phrase *possible operational sequence* is deliberate, because you can easily conceive of other variations. For example, two steps that are shown separately in Figure 2.12—entering identification and account number(s)—might be combined in another system. Similarly, another operational sequence might combine the operations of electing to transfer funds and specifying the direction of the transfer.

Decisions about whether steps should or should not be combined and whether functions should be allocated to people or machines are resolved in trade studies (see Chapter 8). For example, what are the advantages and disadvantages of having the user activate the system, versus having the system activate itself through some sort of sensing device as the user approaches it? Should the user be required to enter account number(s), or should the system do this automatically at the same time as it determines the validity of the user's identification? These and other questions are major design decisions that impact the architecture of the system and the way users interact with it. Their resolutions involve human-factors considerations of usability that may have to be addressed through prototyping.

Note, too, that the operational sequence in Figure 2.12 assumes that the user makes no mistakes.

As an exercise, prepare an operational sequence showing human tasks and machine functions to deal with an error the customer might make in performing each of the six tasks in Figure 2.12.

The operational sequence in Figure 2.12 shows a particular allocation of functions to humans and to machines. Conventional banking systems, which our automated banking teller is supposed to replace, have many of the machine functions in Figures 2.12 performed by another human—a live teller.

THE SYSTEMS ENGINEERING PROCESS   47

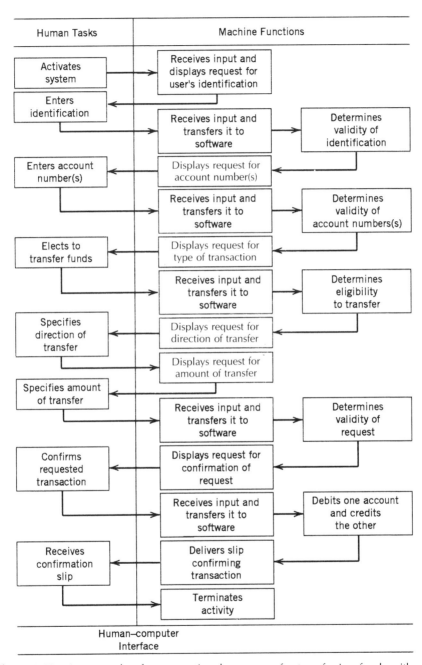

Figure 2.12 An example of an operational sequence for transferring funds with an automated bank teller.

The directional arrows show the way in which human tasks interact with machine functions. The human-factors professional is not normally concerned with machine functions, but is very much concerned with the human–machine interfaces.

Identifying and Describing Human–System Interfaces. A human–system, or human–machine interface is the external boundary of a system through which the human (user, operator, or maintainer) interacts with the system. It contains the controls that people use to make the system do things and the displays that convey information from the system to its human counterparts. On an automobile, the instrument panel and controls, such as the steering wheel, brake, and accelerator, constitute the interface between a driver and vehicle. The keyboard, mouse, and the screen constitute the interface between a user and a computer.

A diagram, such as the one in Figure 2.12, enables one to see quickly what kind of an interface the system will require. In this case, the interface should contain something to deal with every action indicated by a directional arrow that crosses the vertical line between the human tasks and machine functions. Reading from top to bottom, this system requires devices that will allow the customer to:

- activate the system;
- enter some form of identification;
- enter account numbers;
- enter the selection of a type of transaction;
- enter the selection of a direction of transfer;
- enter the amount of funds involved; and
- confirm the requested transaction.

Reading again, from top to bottom, this system will require displays that request:

- the user's identification;
- the user's account number(s)
- the type of transaction requested;
- the direction of transfer;
- the amount of the transfer; and
- confirmation of the request.

One of the useful functions of such an analysis is ensuring that nothing is overlooked in the way of controls and displays.

Keep in mind that Figure 2.12 shows an operational sequence for only one scenario—transferring funds. Operational sequences for all the other scenarios—for example, depositing funds—will also yield lists of input and output devices, many of which may serve multiple functions. For example, the same input device might be used to input a user's identification, account number(s), and amounts of funds to be transferred, deposited, or requested. Similarly, the same output device might serve to display requests for various kinds of customer actions, provide feedback, and display error messages. Note again that these decisions may involve still other trade studies. Should the display that requests various kinds of customer actions also be used to display error messages?

As an exercise, identify some other questions that might require trade studies.

The end result is a description of each interface with a list of the kinds of input devices, output devices, and forms of dialog it will contain. At this stage exact devices have not been specified. This is a description at the *system* level. So, for example, we know that there has to be some sort of a mechanism by which the customer can activate the system, but what that mechanism will be is still to be determined.

Description of the Working Environment. As a part of system design, the human factors professional should prepare a general description of the working environment, paying particular attention to any special environmental conditions, such as vibration, excessive noise, or weightlessness, that might have a serious impact on human performance. Such environmental conditions must be figured into the designs of user–system interfaces.

Personnel Requirements. Finally, a statement of personnel requirements should be prepared at the system-design level. This should include a statement of anticipated staffing requirements, levels of skill that will be required, and training programs that may be necessary to bring skills up to required levels. Once again, at this stage these requirements cannot and should not be stated in great detail because they may, and typically do, change as system development proceeds. Still, at least some preliminary understanding about personnel requirements is necessary, because they may have serious impacts on system design. If, for example, it appears that a system may be so complex that it would require too many operators or maintainers, or that it would require skills not available in the working population (a point further discussed in Chapter 7), it may be necessary to rethink basic system-design concepts to deal with these difficulties.

Hardware and Software Design   The next step in the systems-engineering process is the detailed design of hardware and software. At this stage, the human-factors professional may become involved in one or more of the following:

- Developing user–system interfaces;
- Performing user–system tradeoffs;
- Prototyping and evaluating user interfaces;
- Configuring user interfaces; or
- Developing documentation.

Developing User–System Interfaces. Descriptions of user–system interfaces at the system level merely identified generic kinds of input and output devices. The task now is to get down to more specific details about exactly what those devices should be. For example, in the case of our automated bank teller machine, we noted that there should be some sort of a mechanism whereby a customer may input identification information. What should that mechanism be? A card with a magnetic stripe? A keyboard? A speech-recognition program that enables a customer to say his or her identification? Similar questions can be directed to every one of the input and output devices derived at the system-design stage.

At this stage, the human-factors professional makes use of all the information learned in college courses, the information contained in human-factors guidelines, and recommendations in any of a large set of sources, experience with similar devices in other systems, and the results of any studies that may have been performed with real devices, prototypes, or models.

In arriving at interface requirements it is important for the human-factors professional to ensure that user needs drive system architecture and the choice of input and output devices. In many systems user interfaces are complex displays driven by a particular software architecture. If that architecture is configured without any regard for the user's needs, it is likely that the usability of the system will be constrained. As a simple example, the software architecture may require a tall menu structure when the user needs a flat menu interface, or even an entirely different kind of dialog, for example, form-filling.

Performing User–System Tradeoffs. Although some trade studies are almost certainly required at the system-design level, in the process of specifying detailed requirements for specific input and output devices, the human-factors professional is almost certain to run up against more questions, conflicting requirements, or constraints that require additional trade studies. For example, should visual display terminals on computer systems be monochrome or contain color? Color terminals are more expensive than monochrome ones, and a basic question might be: Does the use of color result in performance that is enhanced enough to justify the additional expense? For a system containing a single terminal, this might not be a serious consideration. But some systems, for example, airline and railroad ticketing systems, may use thousands of terminals and the incremental costs of providing color terminals can be decisive.

Some trade studies involve questions that are exceedingly complex. For example, what level of automation should be provided in the system? Or, what is the optimum balance between skill levels required by selection versus training programs to bring skills up to levels required by the system?

In all cases, the most difficult part of the job for human factors is to provide quantitative data on human performance that can be expected in various alternative designs and to provide those data on time, and in terms that are meaningful and useful to the engineer.

Simulating and Evaluating User Interfaces. Almost without exception, interfaces have to be modeled, mocked up, or simulated in some way to evaluate their adequacy and effectiveness. There is no substitute for actually seeing what an interface will look like and trying it out. Simulations run the gamut from paper-and-pencil drawings, through plywood or plastic mockups, to actual working models and rapid prototyping. The kind of simulation that is most appropriate depends on a number of factors, among them: the stage of design, the complexity of the design, the tentativeness of the human-factors recommendations, and the criticality of those recommendations for system performance.

Configuring User Interfaces. During this phase of design, the human-factors professional should begin *configuring,* or drawing, the way the interface will look. This may be done on paper or with computer-generated models. However it is done, it should include the physical dimensions of operator and maintainer workspaces, doors, and emergency exits, if they are required; the dimensions and placement of equipment in the workspaces; and the physical placement and dimensions of operator and maintainer panels, controls, and displays. Drawings or computer-generated figures are usually not sufficient, and models or mockups of some sort are almost indispensable. Evaluations of the layouts may be made using such techniques as reach and vision envelopes, anthropometric manikins, or link analyses.

Developing Documentation. Even before designs are finalized, it is important to start preparing documentation—the handbooks, manuals, and instructions that will accompany the system. These might be operator manuals, task handbooks, position handbooks, and maintenance manuals. Perhaps the most important consideration is that the documentation, of whatever kind, should be user-oriented. It should be aimed at providing the kinds of information the user needs in words that are useful and meaningful.

One important benefit of starting the preparation of documentation early is that it very often reveals unexpected difficulties with design. The process of trying to explain how to use something or do something very often points up complexities of procedure or features of design that may require modification.

Hardware and Software Development Hardware construction or selection, and writing lines of code for software follow design. These are not human-factors responsibilities and, by now, most of the human-factors work should have been completed. Still, unexpected problems may turn up that require resolution by human-factors professionals. In one case, some hardware design and construction had been contracted out to another firm. After construction was well under way, during a subcontractor review it turned out that additional items were required on a panel for which design had presumably been finalized. This required reconfiguring the original layout. Similar problems are not unusual and should be anticipated.

System Integration and Test At this point the various subsystems are assembled into the system as a whole and tested to determine whether it does, in fact, meet the requirements that had been established earlier. The primary human-factors role at this point is the assessment of user performance with the actual system—the hardware, software, and documentation. This involves the same kinds of considerations that are involved in the design of laboratory experiments, but much more care has to be given to the various factors involved than is the case with experiments conducted in the rather pure environment of a laboratory. Subjects to be tested have to be representative of the using population, tasks have to be representative of those the system was designed to perform, and dependent measures should be meaningful in terms of the operational need. For a good treatment of the special problems of conducting system tests and experiments, see Parsons (1972).

Installation During this phase, the human-factors professional may only be concerned with training prospective customers and users of the system. In an earlier writing, however, I pointed out an instance in which human-factors professionals were involved in the design of installation procedures (Chapanis, 1991). That case involved the installation of large mainframe computers. It was sparked by management concerns about the time required to install these computers, the large expenditures of manpower required for installation, and the much faster installation times achieved by competitors who sold roughly comparable systems. Human-factors professionals did a study of the installation process and were able to effect a number of changes that resulted in substantial savings in manpower and costs. So it is possible that human factors may be able to contribute, even in this phase of the systems-engineering process.

Maintenance and Operations Once a system has been delivered to its customer and put to use, human-factors professionals are often engaged in follow-up studies, to assess how well the system is performing, to find out what difficulties users may be having, and to suggest engineering changes that should be made in later models of the system. Techniques that may be used at this stage are the same ones that were used initially in the study of similar systems: observations of operations, interviews, questionnaires, critical incident studies, and accident studies (see Chapter 4 for a more thorough discussion of some of these methods).

CONCURRENT ENGINEERING

Concurrent engineering, or simultaneous engineering, is a relatively new concept that you are likely to encounter if you work on any development project. Although it is primarily a management plan, it has some important implications for human factors.

Some Consequences of the Waterfall Development Process

To understand what led to this concept, consider a hypothetical sequence of events that would occur if you adhered literally to the waterfall development process (Fig. 2.7). The product, whether it be an appliance such as a television set, a small system like an automobile, or a large system like an aircraft, would be developed through the stage of full-scale engineering development and turned over to manufacturing. When manufacturing had produced the first versions of the product they would be turned over to a support group to organize manuals, repair and maintenance facilities, spare parts, maintenance personnel, and procedures for the return of products. After manufacturing had produced multiple copies of the product, they would be released to sales and marketing, and, for some products, to installers who would fit the product into a customer's facilities.

In many cases, the first time that human-factors personnel, manual writers, training specialists, quality-assurance engineers, sales personnel, and other interested individuals would have been able to actually see what a product looked like was during the testing phase. This is so late that any changes suggested by these and other persons would be strongly resisted, because of the difficulty in making the changes, their cost, or the delays that would result in meeting delivery schedules.

In addition, adhering to a strictly sequential process very often necessitates costly redesign when new constraints are met after full-scale engineering development. The system developed by the design team might turn out to be very difficult to manufacture, given the plant, materials, and manufacturing personnel available. Or, the system would have been much easier to maintain if it had initially incorporated built-in test equipment or line-replaceable units. When the product is turned over to sales and marketing, it may turn out to have missed some important market requirements. Or, the product may have been designed and manufactured in such a way that it is difficult to install.

Case Study A case study by Burns and Vincente (1994a, 1994b) shows how requirements downstream can force the redesign of products. A large U.S. company was designing a control room for a nuclear power plant in a foreign utility. The human-factors aspects of the design met the requirements of IEC-964(1989), an international standard that specifies ergonomic design criteria and the design process for control rooms of nuclear power plants. The design company also agreed to meet the regulatory requirements of the country in which the plant was located.

When the designs were turned over to the customer, it turned out that the panels would not fit through the hallways of the facility, which had already been built. As a result, the panels had to be resegmented. The panels had also been designed initially on the basis of anthropometric data specific to the local population. But a month after the panels had been resegmented and reissued to the customer, the customer revealed that there were minimum height requirements for operators of the plant. The panels were once again redesigned to be taller, to conform to the minimum operator dimensions. Still, a few weeks later, manufacturing decided that the panels were to be constructed from mosaic materials—a series of small modular blocks covered with plastic. The material came in fixed sizes, one of which was slightly larger than the size of the board. To manufacture the panels easier and more cheaply, the panels were redesigned again—three redesigns over a period of several months!

The Concurrent Engineering Approach

Concurrent engineering attempts to avoid these and related difficulties by having marketing personnel, manufacturers, service providers, sales personnel, and engineering specialists work together with designers throughout the entire development cycle, from concept exploration on. The key is teamwork and integration of expertise from all of a company's functional groups. These individuals are formed into teams, for example, a user interface team, system architecture team, system performance team, testing team, manuals and training team, logistic support team, and each person typically participates on two or three teams. By working together on design concepts, the teams focus on developing high-quality products that can be manufactured at low cost, can be installed and maintained efficiently, and that will meet the customer's requirements.

According to Chao (1993), concurrency shortens the development process by shifting from a "redo until right" philosophy to "do it right the first time." In that way, the time and funds spent on redesign are minimized with a concomitant reduction of cost overruns and schedule slippages. There are other benefits to teamwork. Communication barriers are broken down, and everyone, from designers to sales personnel, takes greater pride in the product, by virtue of having contributed to its design.

Implications for Human Factors

When implemented, concurrent engineering does what I have repeatedly been emphasizing: It gets human factors involved early in the design process. It also means that human-factors personnel do not work in isolation, but in concert with other specialists. By working together in teams, other personnel learn to appreciate more fully human-factors techniques and the value that human factors has to offer.

At the same time, human-factors personnel have to become more fully aware of important requirements—for example, ease of production, servicing, and installation; energy efficiency; disposal—that go beyond the performance and ease-of-use requirements to which we typically design. It also means that we have to learn a little about other specialties, such as hardware and software engineering, and to be more sensitive to the difficulties faced by other specialists. In many cases, it also means that we have to be willing to compromise or modify our ideal human-factors requirements when trade studies show that other considerations override usability issues.

SPEAKING PRACTICALLY

Throughout this chapter I have repeatedly stated that human factors *may* do this or *may* do that in system development. That was deliberate, because the tasks that human-factors professionals perform depend, among other things, on the system being developed, the systems engineer's appreciation and acceptance of human factors, the skills human-factors professionals bring to their tasks, the time and resources available for supporting human factors, and whether management adopts a concurrent engineering strategy or follows a more conventional management plan.

Engineers who develop and design automobiles, for example, are not concerned with training drivers. Neither is this a major consideration for human-factors professionals on the design team, although it may be a major concern for human-factors professionals who work with sales, advertising, and marketing. On the other hand, engineers and human-factors professionals are concerned with follow-up studies on the performance of new vehicles. To take another example, human-factors professionals who work with engineers on the design of nuclear power plants are involved in every phase of the system-engineering process, even including the dismantling of nuclear power plants, and the disposal of accumulated waste products.

All too often, engineers in the past have not appreciated the contributions that human factors can make to system development and have called on human-factors professionals only after systems had been designed, put into operation, and found to be deficient in some respect. That is what happened after the Three Mile Island nuclear power plant disaster. A team of human-factors professionals was called in to study representative plants and their operation for the first time after the accident. Although the team was able to make some improvements in plants currently in operation, they were called in much too late to make really substantial improvements. The system was already designed, the architecture already laid out. Doing a really thorough job of design would mean closing down plants and starting from scratch.

In part, the reason why engineers often fail to see the contributions human factors can make to system development is our own fault. We human-factors professionals have not understood how engineers go about their work, how systems are developed, and the kinds of things we can do to assist in that process. This book hopefully will help to rectify that situation.

SUMMARY

In this chapter I have tried to provide a better understanding of what systems are like and how they are developed. I defined systems and systems engineering, described the system life cycle, and discussed the system-engineering process. Throughout, I have emphasized the contributions that human factors can make to system development. Perhaps the most important lesson to be learned from this chapter is that these human-factors contributions are many and varied, and that they are quite different from those given in most human-factors textbooks. Moreover, these contributions must be made early in system development. In fact, the major ones are made long before drawings have been prepared for the construction of a system. As a general rule, the earlier human factors is involved in system development, the better.

The chapters that follow elaborate on the various points that have only been touched on in this chapter.

REFERENCES

Blanchard, B. S., & Fabrycky, W. J. (1990). *Systems Engineering and Analysis.* (Second Edition). Englewood Cliffs, NJ: Prentice-Hall.

Boehm, B. W. (1988). A spiral model of software development and enhancement. *IEEE Computer, 21(5)*, 61–72.

Brennan, L. (1994). Human considerations in emerging manufacturing systems. In *Proceedings of the 12th Triennial Congress of the International Ergonomics Association* (pp. 111–113). Mississauga, Ontario, Canada: Human Factors Association of Canada.

Budurka, W. J. (1984). Developing strong systems engineering skills. *IBM Technical Directions, 10(4)*, 40–48.

Burns, C. M., & Vincente, K. J. (1994a). *Human Factors Design Guidance: Matching the Advice to Designers' Questions. Final Contract Report for Contract XSE93-00010-(303)*. Toronto, Ontario, Canada: University of Toronto, Department of Industrial Engineering, March.

Burns, C. M., & Vincente, K. J. (1994b). Designer evaluations of human factors reference information. In *Proceedings of the 12th Triennial Congress of the International Ergonomics Association: Volume 4, Ergonomics and Design* (pp. 28–31). Mississauga, Ontario, Canada: Human Factors Association of Canada.

Chao, B. P. (1993). Managing user interface design using concurrent engineering. In *Proceedings of the Human Factors and Ergonomics Society 37th Annual Meeting* (pp. 287–290). Santa Monica, CA: Human Factors and Ergonomics Society.

Chapanis, A. (1991). The business case for human factors in informatics. In B. Shackel & S. Richardson (Eds.), *Human Factors for Informatics Usability.* (Chap. 3, pp. 39–71). Cambridge, England: Cambridge University Press.

Clark, D. W., Cramer, M. L., & Hoffman, M. S. (1986). Human Factors and Product Development: Solutions for Success. Retail Systems Division, NCR Corporation: Workshop notes.

Cushman, W. H., & Rosenberg, D. J. (1991). *Human Factors in Product Design.* Amsterdam: Elsevier.

DOD-HDBK-763 (27 Feb. 1987). *Human Engineering Procedures Guide.* Washington, DC: Department of Defense.

Eisner, H. (1988). *Computer-aided Systems Engineering.* Englewood Cliffs, NJ: Prentice-Hall.

IBM (June 1988). *Systems Engineering Principles and Practices Training Course.* Bethesda, MD: Author.

IBM (July 1993). *Human Factors in Systems Engineering Training Course.* Bethesda, MD: Author.

Kirk, F. G. (1973). *Total System Development for Information Systems.* New York: John Wiley Sons.

Kirwan, B., & Ainsworth, L. K. (1992). *A Guide to Task Analysis.* London: Taylor and Francis.

MIL-STD-721C (12 June 1981). *Definitions of Terms for Reliability and Maintainability.* Washington, DC: Department of Defense.

MIL-STD-882B (30 March 1984). *System Safety Program Requirements.* Washington, DC: Department of Defense.

Parsons, H. M. (1972). *Man–Machine System Experiments.* Baltimore: Johns Hopkins University Press.

Royce, W. W. (1970). Managing the development of large software systems. *1970 WESCON Technical Papers, Vol. 14*, pp. 1–9, August.

Scherzinger, P., Willett, B., Kaucharik, D., & Paquette, G. (1994). Development of a catchbasin cover handline tool: A collaborative design process. In *Proceedings of the 12th Triennial Congress of the International Ergonomics Association* (pp. 32–34). Mississauga, Ontario, Canada: Human Factors Association of Canada.

Wasserman, A. S. (1989). Redesigning Xerox: A design strategy based on operability. In E. T. Klemmer (Ed.), *Ergonomics: Harness the Power of Human Factors in Your Business* (Chap. 1, pp. 7–44). Norwood, NJ: Ablex.

Wymore, A. W. (1976). *Systems Engineering Methodology for Interdisciplinary Teams.* New York: John Wiley Sons.

Chapter **3**

Standards, Codes, Specifications, and Other Work Products

Contrary to popular conceptions, many engineers have more to do with words than with material things. You may have had some inkling of that reality from the frequent references in Chapter 2 to operational-need documents, operational-concept documents, requirements, and specifications. This chapter discusses some of the many documents that drive the systems engineer and the human-factors professional as well. Collectively, these documents impose a structure on the development process so that design proceeds in an orderly way. They form two very broad groups: (a) those that are imposed on engineers and human-factors professionals, and have to be taken into account in design; and (b) those that engineers and human-factors professionals have to write as part of the systems-engineering process. Standards and codes fall into the first of these groups.

STANDARDS AND CODES

Although human-factors textbooks seldom discuss or even mention standards and codes, these documents, and not textbooks, are, or should be, the primary sources consulted for design guidelines and requirements. Literally hundreds of published standards and codes constrain engineers and human-factors professionals because they are imposed on designers in some formal way, for example, by legislation, by contract, or by management decree. Indeed, failure to take standards and codes into account in the design process may be disastrous, particularly if a product, machine, or system is ever involved in a product liability law suit.

Standards

The word *standard* has many different meanings, but, in this context, a standard is a set of rules, conditions, or requirements that define terms; classify components; specify components; specify materials, performance, or operations; delineate procedures; or define measurements of the quantity or quality of materials, products, systems, services, or practices.

Standards fall into two broad classes: *safety* standards and *performance* standards. Most standards are of the first type and, although they are certainly of concern to engineers and human-factors professionals, I shall concentrate on the latter—the performance standards. The latter describe what a product does, what one may do with it, or how well one may use it.

Standards Drafting Organizations Standards are prepared by industrial firms, trade associations, technical societies, labor organizations, consumer organizations, and government agencies. For a list of some of the principal organizations involved in the preparation and dissemination of standards, see Pelsma's book (1987, pp. 161–166).

For our purposes, the most important organization is the American National Standards Institute (11 West 42nd Street, New York, NY 10036). This organization is a federation of trade associations, technical societies, professional groups, consumer organizations, and industries that serves as the United States clearinghouse for voluntary standards activity at the national level. It issues a catalog and makes available for sale over 8,000 standards from A [Abbreviations for Use on Drawings and in Text, ANSI Y1.1-1972(R1984)] to Z (Zirconium, ANSI/NFPA 482-1982). Several ANSI standards are referenced throughout this book, and Appendix B is a compilation of additional ANSI standards that are relevant for human factors. There are, however, three important groups of standards that are not listed in the ANSI catalog.

Federal and Military Standards. Hundreds of federal and military standards are invoked in many large systems projects. A few of them are discussed in greater detail later.

Company Standards. Every major industry has its own standards, and human-factors professionals who work in one of those industries have to be familiar with them. Since these standards are generally "For Internal Use Only," I shall not have anything more to say about them.

Foreign Standards. Every country with even a modest amount of technology has its own standards. Prominent among them are the standards of the Deutsches Institut für Normung, the British Standards Institution, and the Association Française de Normalisation. Roughly corresponding to ANSI, the ISO (International Organization for Standardization, 1, rue de Varembé, Genève, Switzerland) prepares and distributes standards that are widely accepted by many countries.

If you work with any systems or products that are likely to be distributed or marketed in foreign countries, it is vital to be informed about the standards of those countries. You can find unpublished, but well-substantiated accounts of American-made equipment that could not be sold abroad because they did not meet local standards. For example, at one time, millions of dollars of American computer keyboards could not be sold in Germany because they did not meet DIN standards for keyboard dimensions. American-made grocery store check-out work stations once could not be sold in Sweden because they were not designed so that operators could stand or sit.

A Basic Set of Standards Among the thousands of standards available, four, in my opinion, belong in the libraries of any human-factors professional who works with systems or even moderately complex machines. They are:

1. *OSHA Standards.* These are standards prepared by the Federal Occupational Safety and Health Administration. They are embodied in five books of the Code of Federal Regulations (Cited as 29 CFR in the list of references at the end of this chapter.) These five documents, totaling several thousand pages, cover a wide range of topics, among them, walking-working surfaces, means of egress, hazardous materials, personal protective equipment, environmental controls, machinery and machine guarding, and standards applicable to a long list of special industries. Although mostly concerned with safety issues, you cannot ignore them in design. All may be purchased from the Superintendent of Documents, U.S. Government Printing Office, Washington, DC 20402.

2. *MIL-STD-1472D.* This military standard, titled *Human Engineering Design Criteria for Military Systems, Equipment and Facilities,* is now in its fourth revision. It was the first standard to try to deal comprehensively with human-factors considerations in the design of equipment. Incidentally, you should not be misled by its military origin. Many of the recommendations in this standard are general ones that apply to any kind of equipment. For example, recommendations about the design of controls apply to controls for power lawnmowers, washing machines, automobiles, and nuclear power plants, as well as to military machines. You will find this standard invoked in all military, as well as many large commercial projects. It is available free from the Navy Publishing and Printing Service Office, Standardization Document Order Desk, 700 Robbins Avenue, Building #4, Section D, Philadelphia, PA 19111-5094.

3. *NASA-STD-3000.* This huge, five-volume compendium, titled *Man–Systems Integration Standards,* is one of the newest human-factors standards to be issued by a federal agency. It covers a slightly different, and much broader range of topics than does MIL-STD-1472D. Some of the topics, for example, extra-vehicular activity, nutrition and health management, apply specifically to space operations, but, as was the case with MIL-STD-1472D, most of the recommendations are general ones that apply to almost all equipment. It is available from the National Aeronautics and Space Administration, Lyndon B. Johnson Space Center, Houston, TX 77058.

4. *ANSI/HFS 100-1988.* This standard, the *American National Standard for Human Factors Engineering of Visual Display Terminal Workstations,* is the first American standard to deal specifically with human-factors principles and practices in the design of visual display terminals (VDTs), associated furniture, and the office environment in which they are placed. It covers applications such as text processing, data entry, and data inquiry. It may be purchased from the Human Factors and Ergonomics Society, P.O. Box 1369, Santa Monica, CA 90406-1369.

The OSHA standards listed are primarily safety standards; the other three are performance standards.

Standards Language   All standards follow important language conventions:

Shall. Statements containing the word *shall* are mandatory and contractually binding. The following from ANSI/HFS 100-1988 is an example:

The specular reflectance, or gloss, of equipment covers and furniture surfaces **shall** be 45 percent or less when measured with a 60-degree gloss instrument or equivalent device.

Should. Statements containing the word *should,* or *is permitted,* or *is preferred,* are guidelines that are desirable and recommended, but not mandatory. The following is an illustration from ANSI/HFS 100-1988:

Dominant wavelengths above 650 nm in displays should be avoided because protanopes are noticeably less sensitive to these wavelengths.

May. Statements containing the word *may* are optional. They generally concern design features for which there are insufficient data to make firm recommendations, for which there is no consensus among experts, or for design features that are not critical for effective performance. The following from ANSI/HFS 100-1988 are examples:

- Font design may impact legibility and readability.
- The keytop shape for keys described in 7.1, "Layout" on page 35 may be of any shape (square, round, rectangular, etc.) provided spacing requirements are not violated.

Some Limitations of Standards   For all of their usefulness, and despite having almost the force of law, standards have some important limitations.

They Establish Minimum Requirements.   Most standards are prepared by representatives of manufacturers, consumers, government agencies, and scientific, technical, and professional organizations, and are arrived at by a consensus among these various individuals. Since they are consensus statements, they represent the lowest common denominators of agreement, the minimum requirements on which often conflicting parties (for example, manufacturers and consumers) could agree. While this does not nullify their importance or usefulness, it means that, in many instances, prudent human-factors professionals should design to exceed the requirements contained in standards. This is especially true of design features that involve safety issues.

They Are Often Too General. Many of the recommendations contained in published standards are precise enough for design. The following from NASA-STD-3000 are examples:

> Normal opening and closing forces for internal doors shall not exceed 22 Newtons (5 lbs.).
>
> Immediate feedback for operator entries shall have not more than a .2 sec. delay.
>
> The dynamic range of a microphone/input device shall be great enough to admit variations in signal input of at least 50 dB.

Although one may question the validity of the numerical values given in these statements, there is no ambiguity about what they mean. Moreover, an engineer can figure out exactly how to meet those requirements.

On the other hand, many recommendations in standards, even though they are "shall" statements, are not precise enough for design. The following, again from NASA-STD-3000, illustrate this point:

> The coding system shall be simple to use, communicate, and memorize.
>
> Workstations shall be designed such that all external distracting stimuli to the operator are minimized.
>
> The relationships of a control to its associated display and the display to the control shall be immediately apparent and unambiguous to the operator.

These statements, and many others like them, are well intentioned. However, they are not precise enough for design. They leave it to the engineer or designer to decide whether, for example, a coding system is "simple to use, communicate, and memorize" or whether control–display relationships are "immediately apparent and unambiguous to the operator."

Equally important is that it is often impossible for an evaluator or inspector, who typically is not trained in human factors, to decide whether a design does, or does not, meet a standard. This ambiguity is often the source of serious disagreements, sometimes leading to breach-of-contract actions between customers and designers. Designers may think they have met a standard that was imposed on them, but, during final tests and evaluations, customers, using a different interpretation of the standard, may claim that the standard has not been met. For that reason, it is important for human-factors professionals to examine standards carefully, to prepare operational definitions of the ones that affect them, and to get agreement about those definitions before work has begun.

Standards Usually Have to be Tailored. Although customers may impose standards *in toto* on the engineer and include them in SOWs, many parts of a standard may not apply to the system being developed. For that reason, it may be necessary to tailor the standard, that is, designate those parts of it that will be followed, and those that will not. This tailoring process usually is accompanied by an explanation about why certain parts of the standard are not applicable.

Standards Do Not Explain the Systems-Engineering Process. As stated in Chapter 2, the systems-engineering process is a disciplined series of activities that build on one another. Standards are usually only concerned with the design of elements of a system, and have nothing to say about other important phases of system development. Understanding the whole process must come from other sources.

Codes

Codes are very closely allied to standards and some codes, such as the National Electrical Safety Code and Life Safety Code, are ANSI standards. Others, such as the BOCA National Building Code and the BOCA National Mechanical Code, both prepared and issued by the Building Officials and Code Administrators International, Inc. (4051 West Flossmoor Road, Country Club Hills, IL 60478-5795), are not. The latter two are model mechanical and building regulations for the protection of public health, safety, and welfare. An appendix to the BOCA National Building Code lists a large number of related codes prepared by other organizations.

Although primarily concerned with safety matters, codes contain many regulations and recommendations that directly or indirectly address human-factors issues.

GUIDELINES

Design guidelines, including checklists, are generally stated suggestions and recommendations for design. They are neither mandatory nor contractually binding. Nor are they cited in RFPs or SOWs. Nonetheless, they do merit some brief discussion because, if for no other reason, there are so many of them.

Literally hundreds of articles, books, and technical reports have been written with advice, suggestions and recommendations for the human-factors design of everything from cooking utensils to nuclear power plants. Although guidelines are often stated in the imperative mood, neither the designer nor the human factors professional is under any obligation to follow them. Indeed, engineers rarely, if ever, seek them out and read them. Moreover, there is no explicit mechanism to determine whether guidelines have been used and followed during system development.

Guidelines suffer from the same faults as standards. One additional difficulty with them is that some are not consensus recommendations, but are rather the opinions of only one person.

Despite their several drawbacks, the better ones are useful for human-factors professionals as reminders of the variables that should be considered in design, as summaries of the findings of human-factors research, and as starting points in the search for acceptable design alternatives. One of the best set of guidelines, although for a limited set of products, is the Smith and Mosier Guidelines for Designing User Interface Software. If you are ever involved in the design of anything associated with computers, you should certainly consult it. Other useful guidelines are in MIL-STD-1472D and NASA-STD-3000.

HUMAN ENGINEERING PROGRAM PLAN (HEPP)*

So far this chapter has been concerned only with documents that are imposed on the systems engineer and the human-factors professional, or with documents that either or both of these persons must consult. I turn now to some of the documents that must be prepared during system development. These are typically called *work products,* although it would be more appropriate to call them *word* products. Most are the responsibility of the system engineer, although they require inputs, and often substantial inputs, from human factors. There is, however, one document for which human factors in primarily responsible, and that is the Human Engineering Program Plan, or HEPP.

A HEPP is the primary vehicle for telling a customer, or management, precisely how human-factors professionals expect to ensure that the human–system interfaces developed will be effective, reliable, and safe. The requirement for a HEPP is given in a military specification, MIL-H-46855B, Human Engineering Requirements for Military Systems, Equipment and Facilities, and in a Department of Defense Data Item Description (DID), DI-HFAC-80740A, Human Engineering Program Plan. In addition, a number of companies have prepared their own instructions for preparing a HEPP. In the following description I have relied heavily on a report prepared by Lockheed's Systems Effectiveness Organization, because its report is much more thorough than either of the two military documents.

HEPPs are required by most military contracts and may also be written for nonmilitary projects, for example, for programs involving the development of large systems such as space systems, air-traffic-control systems and nuclear power plants. If a project that does not formally require a HEPP has a substantial human-factors component, human-factors professionals are certainly going to have to prepare a plan of some sort. In such instances, the outline of a HEPP serves as a convenient reminder of topics that might be included in the plan.

---

*As stated in Chapter 1, human engineering is an alternative term for human factors.

Contents of a Typical HEPP

A table of contents for a typical HEPP is shown in Table 3.1. As you read about the contents of a HEPP, pay particular attention to the human-factors tasks that must be performed in system development.

Preparing a HEPP is not easy. It is usually prepared as part of a proposal, before a contract has actually been awarded. Since the HEPP will be one factor, and sometimes a major one, that the customer uses in arriving at a contract award decision, the HEPP must be sufficiently detailed and thorough to convince the customer that the human-factors professional fully understands the customer's problem and has a sensible plan for helping achieve a solution to that problem. At the same time, the

TABLE 3.1 Example of Table of Contents for a Typical HEPP

1. Introduction
2. Tailoring of Reference Documents
    2.1 Applicable Documents
    2.2 Military Standards
    2.3 Data Item Descriptions
    2.4 Company Standards
3. Organization
    3.1 Internal Organizational Relationships
    3.2 Customer Liaison
4. Human Engineering in Subcontractor Efforts
5. Human Engineering in System Analysis
    5.1 Definition and Allocation of System Functions
    5.2 Task Analyses
    5.3 Preparation of Functional Flow Diagrams
    5.4 Development of User–System Interfaces
    5.5 Work Space Configuration
6. Human Engineering in Equipment Detail Design
    6.1 Detail Design Studies
    6.2 Equipment Detail Design Drawings and Specifications
        6.2.1 Design Review
        6.2.2 Design Support
7. Human Engineering in Procedures Development
8. Derivation of Personnel and Training Requirements
    8.1 Personnel Requirements Plan
    8.2 Training Plan
9. Human Engineering in Test and Evaluation
    9.1 Human Engineering Test Plan
    9.2 Test and Evaluation Implementation
10. Human Engineering Deliverable Products
11. Time-Phased Schedule and Level of Effort

HEPP must be realistic and carefully refrain from promising more than can actually be delivered. This means that the human-factors professional needs to have a thorough understanding of the systems-engineering process and the role that human factors plays in that process and must anticipate the kinds of human-factors activities that will eventually have to be performed in the development of the system.

Introduction The introduction should briefly state the purpose of the HEPP, the broad goals of the human-factors effort, and how the HEPP complies with the RFP. If the customer has identified special human-factors concerns, either in the RFP or during preliminary negotiations, the HEPP should highlight its responsiveness to those concerns.

Tailoring of Reference Documents This section of the HEPP should list all the documents that will be consulted or followed in the development of the system. If the only documents to be used are those called for in the RFP, they would be listed under the heading of "Applicable Documents" and the remaining items (2.2, 2.3, and 2.4) would be omitted. Usually, however, human-factors professionals will want to reference documents in addition to those called for in the RFP. These would be listed under their appropriate headings.

In the discussion of standards I stated that standards have to be tailored for specific projects. Tailoring means recommending that some requirements in standards be omitted and that the wording of other requirements be modified. This is the place where tailoring, if it is to be done, would be detailed. Since the HEPP should show that it is highly responsive to the customer's desires, the preparer of the HEPP must read all the applicable documents with great care, and justify any tailoring convincingly. At the same time, tailoring is almost always required. Some requirements are simply not applicable to some projects. For example, parts of standards for trucks and tanks usually do not apply to aircraft. In addition, as I pointed out in discussing standards, some requirements are too general for design, and so are open to different interpretations. For self-protection, ambiguously worded requirements in standards should be reworded or operationally defined, so that the human-factors professional is not later in disagreement with customers about whether a requirement has, or has not been met. Above all, avoid words such as "optimal," "maximum," and "easy-to-use."

Organization This section of the HEPP should convince the customer that the group responsible for human factors will be so located organizationally that it will actually be able to affect system development. In some developments in the past human factors has worked in almost total isolation and its findings and recommendations were either ignored or overruled. Human factors must have the support of management with sufficient authority to ensure that its recommendations are heeded.

This section of the HEPP identifies the organizational unit that will be responsible for complying with human-factors requirements, and shows the location of that unit, preferably with an organizational diagram, in the company. It also shows the relationship of that unit to other systems-engineering units—for example, software design, safety engineering, integrated logistic support—that are related to human factors. In addition, it identifies the human-factors personnel who will carry on the work, describe their technical qualifications, and provide estimates of the level of effort in person/years that will be required throughout the entire life cycle of the project.

Project personnel invariably need to consult the customer's representatives for advice and information, and the customer needs to monitor the progress of the project from time to time. The nature of the liaison between the customer's representatives and the human-factors staff should be explicitly identified in this section of the HEPP.

Human Engineering in Subcontractor Efforts   In very large projects it is often necessary to subcontract parts of the project. For example, the preparation of computer programs for rapid prototyping may be contracted-out to software companies if the project's own programmers are overloaded. Or, the development of a training package may be subcontracted if the project's human factors staff does not have the skills necessary to do the work.

Although subcontracting relieves the human-factors team of some work, it carries with it managerial responsibilities. The subcontractor needs to have a clear understanding of what it is to do and that understanding must come from the project staff. The subcontractor's efforts need to be monitored and this will almost certainly include periodic visits, participation of the project staff in the subcontractor's design reviews and mockup evaluations, and scrutiny and evaluation of progress reports prepared by the subcontractor. The HEPP should describe all these activities with almost as much detail as if the work were actually to be done in house.

Human Engineering in System Analysis   This section of the HEPP should specify the kinds of system-level tasks human factors will undertake to support the program. In essence, this encompasses the activities that were described in Chapter 2. The HEPP should elaborate each of those activities and explain how they will be carried out. For military projects, requirements for this section are spelled out in a Date Item Description (DI-HFAC-80745A).

Human Engineering in Equipment Detail Design   The human-factors activities here include those that were discussed in Chapter 2 under the heading of Detail Design. In addition, human-factors professionals may participate in reviews of specifications and detail drawings, tests, and mockup evaluations and provide support to software design if such is required in the system.

Human Engineering in Procedures Development   This section should describe human-factors efforts in developing operational or maintenance procedures that apply to equipment and software under development. The function and task analyses (see Chapter 4) that were performed for equipment design also serve as basic information for the design of operational procedures.

There are wide variations in the amount of responsibility that human factors may bear for this activity. In small companies, human factors may have full responsibility for the development of procedures, but in companies that have separate training and technical publications departments, human-factors professionals may have only an advisory role. In any case, the HEPP should state the extent of human-factors participation in this process.

Derivation of Personnel and Training Requirements   For some projects, the customer may specify the kinds of personnel who will operate or use the system. In those cases, the human-factors professional has no role in personnel selection. As I emphasize in Chapter 7, however, the way a system is designed determines the number of people who may be required to operate it and the skills they will need to do so. Even if human factors has no direct role in selection and training, the customer will certainly want to consult with human factors for assistance in the development of training procedures. Whichever situation applies, this section of the HEPP should state exactly what role human factors will play in delineating personnel and training requirements for both operators and maintainers.

Human Engineering in Test and Evaluation   This section of the HEPP should detail how the tests specified in the section on System Integration and Test will be carried out. For military projects two Data Item Descriptions (DI-HFAC-80743A and DI-HFAC-80744A) are relevant.

Human Engineering Deliverable Products   The products referred to here are essentially reports, although they may be such things as computer programs, drawings, or mockups. The number of them that will be required, and the amount of detail contained in them will vary greatly. For some small projects, there may be only a few reports of no more than a page or two. For very large projects, the number of reports may be specified in RFPs and they may be substantial documents. Some typical reports might include:

- Human Factors System Function Analysis;
- Task Analysis Report;
- Human Factors Simulation Plan;
- Human Factors Design Approach;
- Procedures Documents;
- Personnel and Training Plans; and
- Human Factors Test Plan.

The HEPP should identify and describe briefly each report that will be prepared for the project.

Time-Phased Schedule and Level of Effort   This section of the report should provide a milestone chart identifying each separate human-factors activity to be undertaken, and show the anticipated start date and duration of each activity. The chart should be accompanied with a level-of-effort statement, indicating the number of person-months that will be required for each activity.

A Final Note

A mistake that students, especially engineering students, commonly make in preparing a HEPP is to give solutions: "The system will look like this," "The hardware will consist of X," "Software will perform this function," and so on. Sometimes a student will even append a rough drawing of a system or its components. A HEPP does *not* provide solutions. It is a program of activities to study a customer's problem and to try to find an acceptable solution to that problem.

As an exercise, prepare a HEPP that is to be part of a proposal for developing an automated bank teller.

SYSTEMS-ENGINEERING WORK PRODUCTS

I turn now to work products the systems engineer may have to prepare during the development of a project. Some of these, such as a Technical Management Plan, Technical Risk and Performance Plan, and Quality and Productivity Measurement Plan, do not concern human factors directly. That doesn't mean that these and other documents are unimportant. They are just not strictly germane to this book. I mention these other documents only to impress upon you the enormous amount of paperwork the engineer may have to produce in the development of some systems.

Figure 3.1 summarizes some typical systems-engineering work products that require human-factors inputs and the times in the system life cycle when they are drafted, revised, and produced in final form. This figure is a convenient way of showing the relationship of these documents to one another, and it serves as an outline for the discussion that follows. Since I have already discussed the operational-need and operational-concept documents in sufficient detail in Chapter 2, I shall start with specifications. I shall discuss specifications in a commercial, or nonmilitary, context, but my treatment corresponds approximately to requirements set forth in MIL-STD-1521B and MIL-STD-490A.

Specifications

In some ways specifications are like standards. One difference between them, however, is that a standard provides specifications for generic systems, whereas a specification is written for a specific system. In addition, standards are prepared by people external to a project, whereas specifications are prepared by persons engaged in the development of a system. Both standards and specifications contain "shall" statements that indicate mandatory and binding requirements.

A specification is a detailed and precise presentation of the technical and operational requirements for an entity. It is a contractual document and, in addition to

Major system life cycle phases		Operational-need determination	Concept exploration	Concept demonstration and validation	Full-scale engineering developement
Operational-need document		Final			
Operational-concept document		Draft	Final		
Specifications	System, sub-systems		Draft	Revised	Final
	Hardware, software			Draft	Final
Requirements rationale reports	System, sub-systems		Draft	Revised	Final
	Hardware, software			Draft	Final
Design descriptions	System, sub-systems		Draft	Revised	Final
	Hardware, software			Draft	Final
Design rationale reports	System, sub-systems		Draft	Revised	Final
	Hardware, software			Draft	Final
Reviews		ORR	SRR	SDR	SSR, PDR, CDR, TRR

Figure 3.1 Some work products a systems engineer and human-factors professional may have to prepare or prepare for, and their timing in relation to the system life cycle. (Adapted with permission from Figure SS&R-22, IBM 1993).

stating requirements, it typically itemizes functions and defines parameters, values, and characteristics; spells-out design constraints; indicates how requirements will be verified; and provides a delivery date.

Kinds of Specifications Engineers sometimes refer to a specification tree in talking about specifications. A specification tree is a hierarchy of specifications and the best way to understand that is to refer to the generic four-level system hierarchy (Figure 2.3) in Chapter 2. There is, first of all, a specification for the system as a whole (the uppermost box in Figure 2.3). Then there is a specification for each of the subsystems, another specification for each of the sub-subsystems, and so on, until we arrive at a specification for each of the hardware and software items at the bottom of the chart. There is no specification dedicated to operators and maintainers. The closest approximation to such a specification would be a personnel-requirements document. The specifications in a specification tree define system requirements in successive levels of detail.

Engineers make a distinction between design-to and build-to specifications. A system specification is a design-to document that states system capabilities and performance in solving an operational or customer's problem. It translates a customer's operational need into system functions and requirements, allocates those requirements to subsystems, and includes requirements that are assigned to operators and maintainers. It does not, however, contain implementation detail. A build-to specification gives that detail. It is a faithful rendering of the detailed physical and logical structures that implement the requirements given in a design-to specification.

The distinction between the two kinds of specifications is important for the human-factors professional because the amount of human-factors detail to be supplied depends on the kind of specification for which that information is supplied. For example, a design-to specification for the automated bank teller we have been talking about might contain a specification for a ten-button keyboard, but would not give details on what this piece of hardware would look like. A build-to specification would show the exact size and shape of the keyboard and keys, show the arrangement of the keys, give their colors, and identify the captions on each of the keys.

Human-Factors Inputs to a System Specification Some of the general kinds of human-factors inputs required for a system specification are:

- Requirements related to staffing, operating, maintaining, and supporting the system;
- Dimensional and volume requirements
  - crew spaces;
  - operator station layouts;
  - ingresses;
  - egresses; and
  - accesses for maintenance.
- Human-performance requirements
  - requirements for human–machine interfaces;
  - identification of areas in which human errors would be particularly serious and
  - methods of operation
- Health and safety requirements
  - environmental stresses
- Maintainability requirements;

- Personnel requirements; and
- Training requirements.

Human-Factors Inputs to Lower-Level Specifications Successive lower levels of specification require the same categories of human-factors inputs, but the inputs become progressively more detailed as you go down the specification tree.

Requirements Rationale Reports (R^3)

Specifications contain requirements, and every specification is generally supplemented with a requirements rationale report. These reports explain and justify the various requirements contained in specifications. The human-factors contributions to an R^3 report are explanations of why the operator and maintainer requirements in various specifications are necessary to meet the system's objectives. These reports also document the results of human-factors analyses and trade studies that may have been performed to support operator and maintainer requirements.

Design Descriptions

Specifications state requirements. R^3 reports say why those requirements are necessary. Design descriptions say how requirements will be met. For these reports the human-factors professional may have to provide information on design features that will meet operator and maintainer requirements, such as:

- Human–system interfaces;
- Methods of operation;
- Selection of personnel; and
- Training of personnel

For military projects, two relevant DIDs are the so-called HEDAD-O and HEDAD-M (DI-HFAC-80746A and DI-HFAC-80747A).

Design Rationales

Design rationales say why particular designs were selected to meet requirements. Human-factors professionals are now called on to justify their selection of the hardware and software designs that interface with operators and maintainers. This justification must be supported by appropriate human-factors analyses, such as task analyses, link analyses, and trade studies. Human factors also has to justify the skill levels and training requirements for operators and maintainers that were given in design descriptions.

Reviews

When I described system life cycles in Chapter 2, I stated that engineers have to review and defend their actions and decisions from time to time. (On military projects these reviews are detailed in MIL-STD-1521B.) In these reviews the engineer has to, figuratively or literally, stand up in front of the customer's representatives or management, and describe what progress has been made, where the project stands, what decisions have been made, and what obstacles have been encountered. These reviews are sometimes decision points—points at which a customer, or management, decides whether to let the project continue on to the next phase, revise it, or terminate it. For small projects, these reviews may be more or less informal, but, for large projects, they are sometimes agonizing experiences. Some typical reviews and the times at which they take place are shown in Figures 2.5, 2.8 and 3.1.

As I summarize the human-factors contributions made to these reviews, keep in mind that these contributions are, at first, rather general, but that they become more and more detailed in successive reviews as system development proceeds.

Operational Readiness Review (ORR) This review determines whether the operational need and operational concept are sufficiently well understood to allow the project to progress to the next phase—concept exploration. It would review the operational-concept document prepared by the contractor to determine whether it is responsive to the customer's needs.

System Requirements Review (SRR) A system requirements review is normally conducted after a functional analysis has been completed and decisions have been made about the allocations of functions to equipment, software, and personnel. It defines human activities in broad terms. Trade studies, interface studies, and system safety studies should all have been completed before this review. Note once again that this represents a considerable amount of human-factors work that has to be done early in the development of a system, before any design work has been undertaken.

System Design Review (SDR) A system design review is normally conducted after the completion of a system specification and is intended to evaluate how well the integrated system will meet its requirements. This review elaborates and provides more detail on the human-factors inputs provided to the system requirements review, and covers the following items (still at a system or general level) insofar as they are relevant to the particular project under development:

- Hardware items and interfaces for:
  - operations;
  - maintenance;
  - test; and
  - training
- Human–computer interfaces, if any;
- Facilities;
- Personnel requirements;
- Environmental conditions;

- Updated design requirements for:
  - operations and maintenance functions, items, and facilities; and/or
  - operations and maintenance personnel and training.

Trade studies should also have been completed on automated versus manual operations and on functional breakdowns between hardware, software, and personnel procedures.

Software Specification Review (SSR) A software specification review is normally conducted after a draft of a software component specification has been completed. It is a review of requirements for a software component and, from a human-factors standpoint, is a review to determine whether interface requirements (for example, keyboards, display formats) and usability requirements (for example, dialogues, help facilities, symbology) have been met. A successful review approves the design-to specifications of individual software items to be developed.

Preliminary Design Review (PDR) A preliminary design review is normally conducted after the completion of a hardware component specification and is intended to assure the customer that the design implementation in the design description and design rationale reports is on the right path and that it will meet the customer's requirements. Human-factors engineers will have used rapid prototyping, simulators, mockups, and perhaps other techniques to evaluate the technical adequacy of the hardware component or groups of them. Human-factors inputs now cover:

- Detailed hardware and software interface requirements including such things as:
  - operating modes and functions performed at each station;
  - displays and controls used at each station;
  - exact formats and contents of each display, for example, data locations, spaces, abbreviations, message lengths, special symbols;
  - formats of all operator inputs;
  - control and data-entry devices, for example, cranks, levers, pedals, keyboards, special function keys, cursor controls;
  - status, error, and data printouts;
- Detailed requirements for tests and evaluations of the above;
- Verification of allocation of functions to operators to ensure that their capabilities are utilized and that their limitations are not exceeded; and
- Technical manual and documentation coverage.

A successful review approves the "design-to" specifications and preliminary designs of individual hardware and software items to be developed.

Critical Design Review (CDR) This review is normally the final review before the system goes into production. Its purpose is to review all of the hardware and software items for compliance with all system and component specifications. One reason for such a review is that changes may have been made during the course of system development when, for example, "off the shelf" items are selected to minimize costs. No other new human-factors inputs should be required at this time because they should all have been made prior to this point. At this stage, changes required to make equipment meet human-factors requirements would be extremely, perhaps prohibitively, costly.

At this stage human factors reviews:

- Detailed designs or drawings, schematics, mockups, or actual hardware and evaluates by checklists or other formal means the adequacy of designs with regard to:
  - operator displays;
  - operator controls;
  - maintenance features;
  - anthropometry;
  - safety features and emergency equipment;
  - work space layout;
  - internal environmental conditions;
  - training equipment; and
  - personnel accommodations.
- Time/cost/effectiveness considerations and forced trade-offs of human-factors design features.

A successful review approves "build-to" specifications and related documents and releases them to production. Programmers begin to code and the production line becomes a reality.

Test Readiness Review (TRR) After hardware and software prototypes have been built and are available, the next step is a demonstration that the design task has been completed and that the design will meet contractual requirements. The TRR is a vehicle for assuring, well ahead of time, that the operational tests planned and that the operational definitions of Proof-of-Compliance will be accepted by the customer.

Large, complex systems, such as a nuclear power plant or an air-traffic-control system, will have hundreds of human-interface issues that need to be tested. Since these cannot all be done at the same time, human factors has to show the customer how its plan will sequence tests over a period of time and how much of what kind of resources (people, hardware, software, facilities) the tests will require of the customer.

As a minimum, the TRR requires the following human-factors inputs to those plans:

76   3: STANDARDS, CODES, SPECIFICATIONS

- Operational definitions of operability and usability;
- Definition of test procedures, in accordance with those definitions;
- Definition of test personnel to match the user population; and
- Procedures for the conduct of the test and evaluation of the test data.

Some Final Words

By now you should have been impressed by the enormous amount of paperwork that must be produced in the development of most large systems. Thorough documentation of all analyses, studies, and decisions at the time they are made is also vital because of the necessity to justify those activities later in rationale reports and in reviews. An additional reason for meticulous documentation is that many large systems take years, sometimes almost decades, to produce, and the development process may outlive the engineers or human-factors professionals who started work on them. Complete documentation is important to instruct new personnel who may join a project after it is well under way, or even to refresh the memories of those who may find it difficult to recall why things were done, in certain ways, years ago.

You should also have noticed that a substantial amount of human-factors work has to be done early in system development, long before any hardware or software may actually have been produced.

Although I have used generally accepted names for the various work products discussed in this chapter, they may not exactly be the names used in other organizations, or in the industry where you work. By whatever name they are called, however, work products like those I have described are required in large development projects.

SUMMARY

This chapter has reviewed some of the practices, standards, and specifications that may be imposed on systems engineers and human-factors professionals by customers, by government agencies, or the company for which those individuals work. A work product, the human engineering program plan (HEPP), primarily a human factors responsibility, was described in detail. Several other work products—specifications, rationale reports, and design descriptions—that the systems engineering team may need to prepare were then covered. Finally, the chapter described various reviews that the systems engineer may have to prepare for. All these work products incorporate or require human-factors inputs, and the chapter summarized the kinds of inputs that the human-factors professional can or should make to each of these work products. These inputs result from the use of a broad spectrum of methods, which I address in the next chapter.

REFERENCES

ANSI/HFS 100-1988. (Feb. 4, 1988). *American National Standard for Human Factors Engineering of Visual Display Terminal Workstations.* Santa Monica, CA: Human Factors and Ergonomics Society, Inc.

Building Officials and Code Administrators International, Incorporated (1993). *The BOCA National Building Code/1993.* Country Club Hills, IL: Author.

DI-HFAC-80740A. (26 May 1994). *Human Engineering Program Plan.* Washington, DC: Department of Defense.

DI-HFAC-80743A. (26 May 1994). *Human Engineering Test Plan.* Washington, DC: Department of Defense.

DI-HFAC-80744A. (26 May 1994). *Human Engineering Test Report.* Washington, DC: Department of Defense.

DI-HFAC-80745A. (26 May 1994). *Human Engineering System Analysis Report.* Washington, DC: Department of Defense.

DI-HFAC-80746A. (26 May 1994). *Human Engineering Design Approach Document-Operator.* Washington, DC: Department of Defense.

DI-HFAC-80747A. (26 May 1994). *Human Engineering Design Approach Document-Maintainer.* Washington, DC: Department of Defense.

IBM (July 1993). *Human Factors in Systems Engineering Training Course.* Bethesda, MD: Author.

MIL-H-46855B (5 Apr. 1984). *Human Engineering Requirements for Military Systems, Equipment and Facilities.* Redstone Arsenal, AL: US Army Missile R&D Command.

MIL-STD-490A (4 June 1985). *Specification Practices.* Washington, DC: Department of Defense.

MIL-STD-1521B (4 June 1985). *Technical Reviews and Audits for Systems, Equipments, and Computer Software.* Washington, DC: Department of Defense.

MIL-STD-1472D (14 March 1989). *Human Engineering Design Criteria for Military Systems, Equipment and Facilities.* Washington, DC: Department of Defense.

NASA-STD-3000 (Oct., 1989). *Man–Systems Integration Standards.* Houston: National Aeronautics and Space Administration, Lyndon B. Johnson Space Center.

Pelsma, K. H. (Ed.) (1987). *Ergonomics Sourcebook: A Guide to Human Factors Information.* Lawrence, KS: The Report Store, A Division of Ergosyst Associates, Inc.

Smith, S. L., and Mosier, J. N. (Aug. 1986). *Guidelines for Designing User Interface Software.* Report No. ESD-TR-86-278, MTR 10090. Bedford, MA: The MITRE Corporation.

Systems Effectiveness Organization 62-91 (Sept. 1984). *Developing a Human Engineering Program Plan (HEPP).* Report Number LMSC/D889095. Sunnyvale, CA: Lockheed Missiles & Space Company, Inc.

29 CFR Parts 1900 to 1910 (§§1901.1 to 1910.999). (July 1, 1993). *Labor.* Washington, DC: Office of the Federal Register, National Archives and Records Administration.

29 CFR Part 1910 (§1910.1000 to End). (July 1, 1993). *Labor.* Washington, DC: Office of the Federal Register, National Archives and Records Administration.

29 CFR Parts 1911 to 1925. (July 1, 1993). *Labor.* Washington, DC: Office of the Federal Register, National Archives and Records Administration.

29 CFR Part 1926. (July 1, 1993). *Labor.* Washington, DC: Office of the Federal Register, National Archives and Records Administration.

29 CFR Part 1927 to End. (July 1, 1993). *Labor.* Washington, DC: Office of the Federal Register, National Archives and Records Administration.

Note: The dates given for the standards, DIDs, specifications and codes here and throughout this book were the current ones available to me at the time I was writing the manuscript for this book. All, however, are revised from time to time. If you use any of them in product development be sure you have the most recent version.

Chapter **4**

Human Factors Methods

In Chapters 2 and 3 I repeatedly pointed out places where human-factors inputs were required during system development, and I mentioned briefly some methods that could be used to provide those inputs. In this chapter I describe in greater detail some of the major methods used by human-factors professionals to:

- analyze systems;
- provide data about human performance;
- make predictions about human–system performance;
- evaluate whether the performance of a human–machine system meets design criteria.

The chapter also shows the interrelationships among the various methods, and provides examples of how they have been applied. I have chosen examples from a wide range of applications. Some are from simple systems; others from enormously complex ones. Some are biomechanical in nature; some deal with interface problems; and others with cognitive issues. My intent in selecting these examples has been (1) to illustrate the scope and usefulness of these methods in answering a variety of questions that frequently arise during system development, and (2) to provide you with sample solutions that may, by analogy, suggest ways you could tackle problems that might be encountered in your work.

I devote a considerable amount of attention to methods because human-factors specialists engaged in system development cannot rely solely on recommendations, guidelines, checklists, or standards to do their work for them. Guidelines don't analyze systems, nor do they design systems, or test and evaluate them. Instead, the human-factors specialist has to depend on methods like those in this chapter to support the three basic activities of system development—*analyze, design,* and *test.*

BASIC VERSUS APPLIED RESEARCH

A great deal of the basic human-factors research you read about in journals and textbooks produces generalizations that are often designed to answer some theoretical question or to fit into some theoretical framework. Applied research, the kind I write about in this book, is highly specific and theory-independent, often intended to answer some deceptively simple question: Can people do this exact task under these special conditions? Or, Can people do A better than B, C, or D?

You will find an example of the former question in my description of a maintenance problem on pages 124 and 125. The only question involved was: Can maintenance personnel exert the force required to do a certain task under a set of special conditions? On pages 106—109 I discuss research undertaken to answer the same simple kind of question but with a very much more complex system. In this case the question was: Can operators perform a complex set of tasks to prepare a space satellite within a certain time period before it begins its next pass around the earth? Then on pages 129—131 I describe some research to answer another kind of simple question: Can sonar operators monitor sonobuoys from a helicopter better with a 4- or an 8-channel receiver?

No theory is involved in any of these examples, nor do I know of any theory that would have provided the answers. These and a great many other questions you will read about in this book only require performance data. Note, too, that generalizations will not suffice. What is required are specific and dependable answers. That doesn't mean that this kind of research is any less important. In fact, the performance of complex systems often hinges on getting dependable answers from applied research. For example, the maintainability of nuclear submarines depended heavily on the answer to the first question and the performance of space satellites depended on the answer to the second.

I said that applied questions are often deceptively simple because you will also find that doing the research to answer those questions is often very difficult, much more difficult than doing basic laboratory studies. Applied research usually means using elaborate equipment and facilities that match some operational situation, careful selection of subjects to match some specific population, tasks and methods that simulate the operational ones, and dependent measures that are operationally relevant.

So, in describing and discussing human-factors methods I do so from a systems engineer's point of view. As an engineer, you are not interested in these methods for their ability to provide data about human performance that merely add to a store of basic knowledge or settle some theoretical issue. An engineer's interest is intensely practical: What information, what work product, does this method provide that I can use in the design of some specific thing? In other words, an engineer wants to know: What does this method do for me? Moreover, generalities are of little or no use; an engineer needs specific answers.

Description versus Prediction

Another difference between the basic, or theoretical, and practical approaches to methods is the difference between description and prediction. In conventional research you observe what happened, describe what happened, analyze what happened, and try to explain what *happened*. (Notice the past tense.) In system development the goal is prediction: Will this system meet its design objectives when it is built and put to use? How well will people be able to use it if it is designed this way or that? How can we design it so that maintainers *will be able* to service it easily and keep it operating? (Notice the use of the future tense.) Unfortunately, as Meister (1993) argues, our methods have given us many good descriptive databases, but have failed to produce equally useful predictive databases. A further complication is that every development project is for a system that has never been built before and, in the beginning, there is often some uncertainty about whether the system can actually be built.

The application of methods in system development is much harder than doing basic research for another reason. In basic research, if you make a mistake in designing a study, carrying it out, collecting and analyzing data, or formulating a theoretical explanation, no great harm is done, because science tends to be self-correcting. In fact, the investigator who errs is almost certain to gain a kind of prominence, perverse though it may seem, because he or she will almost certainly be cited by other investigators who will follow-up and try to rectify the deficiencies in the original study. By contrast, in research applied to system development, you have only one chance. If your predictions are wrong and if you're lucky, the system will merely not do as well as expected. If your predictions are wrong and you're not lucky, though, your system might fail or result in disaster.

Keep these thoughts in mind, because they are one key to the successful integration of human factors into system development.

SOME ADDITIONAL PRELIMINARY REMARKS

Although I have called them human-factors methods, they do not belong exclusively to the human-factors profession. Some have been borrowed from other professions, for example, industrial engineering and safety engineering, and adapted to our purposes. Moreover, the methods are not used exclusively by human-factors professionals but rather are often used by or in conjunction with other professionals. The methods I discuss do not, by any means, cover all the techniques available to the human-factors professional. For much more thorough treatments from somewhat different perspectives, see the excellent books by Meister (1985) and Kirwan and Ainsworth (1992), or DOD-HDBK-763. What follows is a selection of those methods I have found most useful in the practical business of developing systems.

My treatment of each method is highly abbreviated but follows a standard format. I first say something about the method and define it. Then I describe the inputs—the information needed to use the method. Next, I describe how the method is used, and finally, I describe the work products—what the method tells us that can be used in the design process. In most instances I give examples of how the method was used in some practical situation(s) and provide additional commentary.

Most of these methods are the distillation of a considerable amount of practical experience, born of the necessity for finding ways of getting useful answers to difficult questions. They usually lack the rigorous and formal underpinning that you find, for example, in experimental designs built around the analysis of variance. Many of them also require the exercise of a considerable amount of judgment in their application. For those reasons the methods described in this chapter are not necessarily ideal models to be followed faithfully, nor are the instructions I give for using them inviolate. You should not feel constrained to adhere meticulously to the methods as I describe them if they seem to require modification. You will find variations in the way these methods have been applied in some of the examples I give later in this chapter. On the other hand, keep in mind that the methods and procedures described in this chapter did not come about by accident. They were developed to meet specific needs and they have all been used successfully. Think of them collectively as a process of problem solving—solving the problem of how to arrive at a set of requirements that will result in a usable system.

A Sequence of Human-Factors Methods

The human-factors methods used in system development are sequential, in the sense that the products of one method are generally required for the application of the next one. Figure 4.1 shows such a sequence. For example, you cannot do a task analysis until after you have decided what functions will be allocated to humans. That is also the reason why controlled experimentation is so far along in the sequence. To do a controlled experiment you have to know or have some idea about the tasks people will perform in the system. That means that experiments can be done only after a task analysis, informal or formal, has been completed.

Methods Are Used Iteratively Figure 4.1 shows a linear sequence of methods, but that is only to simplify the illustration. In actual practice, these methods may be used several times during system development, with increasing detail and precision as development progresses. For example, I pointed out, in Chapter 2, that in preparing an operational concept in the early stages of system development, you may perform a functional analysis and task analysis. Since no hardware or software may exist at this early stage, and since no designs may have been committed to paper, function and task analyses performed at that time can only be high-level descriptions of gross functions and tasks that will be performed. Details, complete with times and other supplementary information, can only be made later as decisions about hardware and software have been made and more exact information about them becomes available.

Sometimes, too, the results of analyses performed later in the sequence may require a reexamination of earlier steps. For example, the results of a workload analysis may reveal that operators have, or will have, excessive workloads. This

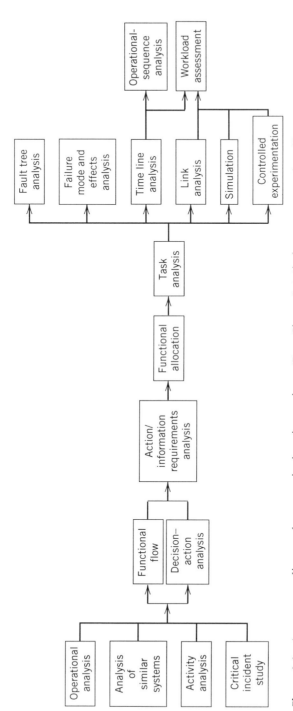

Figure 4.1 A sequence of human-factors methods and procedures. (From Chapanis & Shafer, 1986. Reproduced by permission of IBM.)

4: HUMAN FACTORS METHODS

may then require going back to the stage of functional allocation, perhaps with a view to reallocating some human functions to software, or to other operators.

Because the methods in Figure 4.1 are used iteratively, you cannot say unequivocally that some methods are used only at certain stages of system development. It is possible that all these methods may be used in every stage of system development.

A final point is that not all methods may be used on all projects and in all stages of development. For example, in developing an updated version of a system, some functional requirements may be identical to those in the earlier one. In such a case, there is no point in repeating analyses that have already been done for a previously designed and proven system. You have to select and use those methods that are appropriate for providing the answers needed at any time.

THE METHODS

Operational Analysis

I have already had quite a bit to say about operational analysis (called mission analysis on military projects) in Chapter 2, when I discussed operational-need and operational-concept documents. Here I summarize the method more formally.

Definition An operational analysis is an analysis of projected operations to obtain information about situations or events that will confront operators and maintainers in using a new system. It typically contains scenarios, verbal descriptions of events, and summarizes anticipated operations, assumptions, and environments. Scenarios should be sufficiently detailed to convey an understanding of all anticipated operations and should cover all essential system functions, such as failures, both hard and soft, and emergencies. An operational analysis may also include an operational profile, a graphical or pictorial model of the operations that a system is expected to perform (see Figure 2.6).

Inputs Primary inputs for an operational analysis come from information contained in Requests for Proposals (RFPs), planning documents, and system requirements documents provided by the customer, and from knowledgeable experts.

Procedure Analysts use their expertise and experience, and consultations with systems engineers and representatives of users of the projected system, to extract implications for operators and maintainers.

Products An operational analysis will yield:

- Descriptions of the situations or events that will confront operators and maintainers;
- A list of system operational and maintenance requirements;
- Descriptions of assumed operations;

- A list of operations that appear feasible and those that may overstress the system;
- A list of environmental factors that may affect system performance;
- A list of constraints, e.g., personnel and training, that may affect system performance; and
- A list of foreseeable failures and evaluations of the consequences of those failures.

Additional Commentary   Although in the past, operational analyses have typically been done by systems engineers or operations analysts, ideally human-factors professionals should be members of the analysis team. A user-oriented operational analysis is critical to all later stages of development and it is important that operator and maintainer activities not be minimized or overlooked.

Analysis of Similar Systems

This method, and the following two—activity analysis and critical incident study—are typically done with existing systems to obtain information useful in planning for a new one. As I pointed out in my preliminary remarks, however, they, like all the methods in Figure 4.1, are not necessarily confined to any specific phase of system development. Activity analysis and critical incident studies could be performed on simulations or prototypes of a new system or, for that matter, on systems that have just been developed and put into operation. As a general rule, however, analyzing similar systems is a good way to begin any project.

It is difficult to think of any kind of machine or system that does not have a predecessor. The first nuclear power plants were designed partly on the basis of experience gained from the operation of their fossil-fueled predecessors. New models of automobiles generally try to eliminate complaints and difficulties discovered in existing ones. And it is understandable that the first space vehicles were designed with the help of experienced jet-aircraft pilots. Experiences gained from systems in use are valuable sources of information that should be capitalized on.

Definition   In the analysis of similar systems, one or more observational methods may be used to discover salient features of systems that are similar to the one under consideration.

Inputs   The data for the analysis come from one or more of the following:

- Structured observations;
- Interviews;
- Questionnaires;
- Activity analyses;
- Critical incident studies; and/or
- Accident investigations.

Although there are no hard and fast rules about how antecedent systems should be studied, the kinds of things the analyst might look for are:

- Data on the operability of the older system;
- The maintainability of the older system;
- Numbers of people required to staff the system;
- Skills required to operate and maintain the system;
- Training required to bring operators to proficiency;
- Historical data on human factors design problems encountered in the design of the system; and
- Problems encountered by personnel in using the system.

Procedure Depending on the availability of systems, facilities, records, and personnel, the human-factors professional applies one or more of the observational methods above and uses analysis methods appropriate to each one to extract relevant information. Some of the kinds of records that often contain useful information are:

- Productivity records;
- Maintenance records;
- Training records; and
- Accident or incident reports.

Products Some of the useful products that may result from such an analysis are:

- Identification of environmental factors that may affect personnel;
- Preliminary assessments of workloads and stress levels;
- Assessments of skills required and their impact on selection, training, and design;
- Estimates of future staffing and manpower requirements;
- Identification of operator and maintainer problems to be avoided in the new system; and
- Assessments of the desirability of and consequences of reallocating system functions.

Two Examples I mentioned in Chapter 2 some work that I had done on a project to design an improved production line for the manufacture of large items. In Phase I of the project, I studied the existing production line, interviewed workers on the line, and studied accident records. The results revealed a number of classic human-factors design faults that I recommended should be eliminated in the new line. A simple example is that gauges that had to be read were, in some instances, located in the production line. In order to check these gauges, a production worker had to stop the production line. My recommendation to remote gauges to positions outside the production line actually speeded up the production process, because the line did not have to be stopped when the gauge indicated normal readings.

The second example relates to the use of automatic teller machines (ATMs). Rogers and colleagues (1994) conducted telephone interviews with 100 older adults about their use of ATMs and discovered a number of difficulties users have with these machines. You might want to consult their article to help you with the ATM design problems I assign at various places throughout this book.

Activity Analysis

Definition Activity analysis is a method for measuring and quantifying how operators spend their time. It is especially adapted to situations in which operators do many different things in no fixed order.

Inputs The inputs are the activities of operators going about their normal duties in an operational system or in a simulation of one.

Procedure From direct observation, or from film or video recordings, the analyst periodically or aperiodically samples activities being performed and classifies them into a set of categories, for example:

- Entering data into computer;
- Using telephone;
- Using copier.

These data are then aggregated over some appropriate time period, for example, a day, and activity-frequency tables or graphs are constructed, showing the percentage of time spent in various activities. It is also possible to prepare state-transition diagrams showing the probabilities of specific activities following other activities. For example, if the operator is performing activity A, what is the probability that he or she will next engage in activity B, C, or D?

Products Among the useful products of activity analyses are the following:

- Estimates of, or verification of, staffing and personnel requirements
- Assessments of skill levels required and their impact on selection, training, and design;
- Assessments of workloads and stress levels;
- Assessments of the desirability of, and consequences of, reallocating systems functions; and
- Indications where changes in procedure or system design would improve performance.

Examples Figure 4.2 is a classic example of an analysis performed by Christensen (1949) on the activities of navigators in long flights. The example is taken from my book (Chapanis, 1959). The table is fictitious, but the categories are the ones that Christensen actually observed and the percentages are those that he actually found.

4: HUMAN FACTORS METHODS

Activity: First Navigator - DC-6B				
Operator: J. Hodgkins		Recorder: G. Wendell		
Time: 0130	Date: 9/7/57	Sampling interval: 5 Sec.		
Flight: Reconnaissance squadron A/c No. 456-F				
Remarks: Departed Juneau, Alaska, 0055. Weather cloudy. Enroute to Los Angeles.				

Activity	Tally	Sum	% of Grand Total
Log work	ÏÅÆ ÏÅÆ ÏÅÆ ÏÅÆ ÏÅÆ ÏÅÆ ÏÅÆ ÏÅÆ	40	16.7%
Interphone	ÏÅÆ ÏÅÆ ÏÅÆ ÏÅÆ ÏÅÆ ÏÅÆ ÏÅÆ //	37	15.4%
Chart work	ÏÅÆ ÏÅÆ ÏÅÆ ÏÅÆ ÏÅÆ ÏÅÆ ÏÅÆ /	36	15.0%
Inactive	ÏÅÆ ÏÅÆ ÏÅÆ ÏÅÆ ÏÅÆ ///	28	11.7%
Transition	ÏÅÆ ÏÅÆ ÏÅÆ ÏÅÆ ÏÅÆ /	26	10.8%
Sextant work	ÏÅÆ ÏÅÆ ÏÅÆ ///	18	7.5%
Eating	ÏÅÆ ÏÅÆ ///	13	5.4%
E-6B computer	ÏÅÆ ////	9	3.8%
Map reading	ÏÅÆ ////	9	3.8%
Astrocompass	ÏÅÆ	5	2.1%
Auxiliary radar	////	4	1.7%
Radio	///	3	1.2%
Altimeter	//	2	0.8%
Drift reading	/	1	0.4%
Other activity	ÏÅÆ ////	9	3.8%
Grand total		240	100.1%

Figure 4.2 A type of data sheet that could be used in activity analyses (Reproduced by permission of Johns Hopkins University. From Chapanis, 1959). The data sheet is fictitious, but the categories in the left-hand column are the ones studied by Christensen (1949) and the percentages in the right-hand column are those found by him in his study.

One interesting thing revealed by this analysis is that nearly a third of the navigator's time was spent in paper work (log work and chart work), but there was no place for the navigator to do this work. He typically did it on his knee. This pointed up a serious mismatch between what was required and what was provided.

The second example (Figure 4.3) is a more contemporary analysis performed by Pew and colleagues for Bolt, Baranek, and Newman, Inc. (1979). This was a study of the times spent in change-of-address interviews conducted in several Social Security Administration district offices. The purpose of the study was to compare the times required to conduct these interviews in the conventional way, or with either of two computer systems (A or B). Note that the total interview time with either of the two computer systems was greater than the time required by the current process. The more detailed analysis shows some of the reasons for that. For example, with both computer systems interviewers spent a considerable amount of time using the keyboard, which, of course, was not used in the face-to-face interviews, and spent more time waiting (for computer feedback).

Critical Incident Study

Definition This is a method of identifying sources of operator/maintainer–system difficulties in operational systems, or in simulations of them. It is especially useful

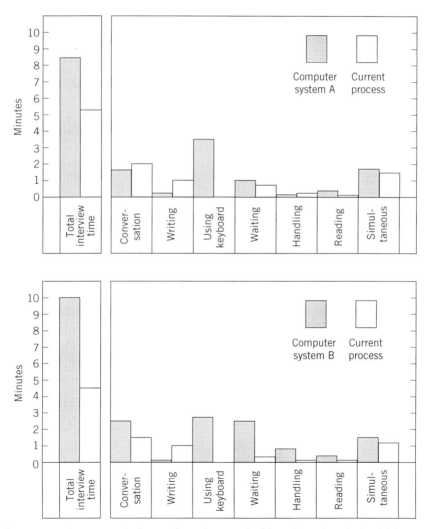

Figure 4.3 Times spent in face-to-face change-of-address interviews in Social Security Administration district offices compared with times obtained with either of two computer systems. (Constructed from data in Pew et al., 1979)

when a system has been in operation and difficulties are observed or suspected, but the nature and severity of those difficulties are not known.

Inputs Operators supply first-hand accounts of critical incidents, that is, accidents, near-accidents, mistakes, and near-mistakes they have made in using a display, operating a piece of equipment, or in carrying out some operation. In essence, the investigator interviews or otherwise questions a large number of people who have used a particular instrument, machine, or system, and asks them to "Tell me about some error, near-error, mistake, or near-mistake you have made, or some accident or near-accident you have had in using this particular device."

One reason for collecting incidents involving near-errors, near-mistakes, and near-accidents, as well as real errors, mistakes, and accidents is that we are interested in situations that cause difficulties. Often it is just a matter of chance whether a particular operation results in an accident or a near-accident. For example, during a cutting operation on a shearing machine, whether the end of an operator's glove or finger is cut off depends on where the operator's hand happened to have been positioned. Our interest is in the situation that would allow such events to take place whether or not there has been any injury or accident.

To be useful, the incidents must be detailed enough (a) to allow the investigator to make inferences and predictions about the behavior of the person involved, and (b) leave little doubt about the consequences of the behavior and the effects of the incident.

Procedure After collecting a large number of incidents, the analyst reads through them and groups them into categories, for example:

- Mistakes and near-mistakes in reading an indicator;
- Mistakes and near-mistakes in using a control;
- Mistakes and near-mistakes in interpreting an instruction.

The analyst then uses human-factors knowledge and experience to hypothesize sources of difficulty and how each one could be further studied, attacked, or redesigned to eliminate it.

Products The products are sources of serious human–system difficulties in the use or operation of a system with suggested solutions to those difficulties.

Examples A classic human-factors study using this method was conducted by Fitts and Jones (1947) in analyzing errors made in operating aircraft controls. They asked pilots to:

> Describe in detail an error in the operation of a cockpit control (flight control, engine control, toggle switch, selector switch, trim tab, etc.) which was made by yourself or by another person whom you were watching at the time.

Altogether they collected 460 "pilot error" experiences, some of them frightening. Following are just three accounts:

> A B-25C with full bomb and gas load was taking off from a 3,500-foot strip with trees at both ends. We crossed the end of the runway at an altitude of two feet and were pulling up over the trees ahead when I gave the 'wheels up' signal. The airplane continued to brush the tree tops at a constant 125-mph speed with T.O. [take-off] power. The copilot had pulled up the flaps instead of the wheels.

> On take-off in a C-47 with approximately fifty persons on board, the *right* engine quit. The pilot and copilot both reached for the *left* feathering switch and finally got the *left* engine feathered. No one was close enough, or realized in time, to prevent this mistake. It was very fortunate that several open fields were straight ahead—no one was killed but several persons were bruised.

> On a routine combat mission, a recently checked-out copilot, flying as first pilot, had been given a position as wing man. Going over the target, an engine was damaged by flak and heated up and lost considerable oil. Therefore, he had his copilot feather the engine. He remained in formation over the target and all the way home. On arriving at the field, he requested the copilot to unfeather this damaged engine. The copilot . . . told the pilot that there was no reason to unfeather the engine, but to go ahead and land with only three engines . . . [An argument developed] . . . and finally the first pilot became angry and reached for the feathering button. He hit the wrong one, feathering the second engine. At this time, both pilots put their heads in the cockpit trying to get their engines unfeathered. All this time, they were still in formation. As a result of concentrating on the unfeathering procedure, they skidded into the flight leader, cutting off his fuselage just in front of the vertical stabilizer and his plane plummeted to earth 1,000 feet below killing all eleven occupants.

Table 4.1 summarizes the results of their study. Perhaps the most interesting aspect of these data is that, although they were collected over 40 years ago, they are just as valid today as they were when they were collected. We find exactly the same kinds of mistakes being made with controls on all sorts of machines and systems—in automobiles, in nuclear power plants, on computers, and on VCRs.

An example of a different sort is a critical incident study that Safren and I (1960) made, of errors in the administration of medications in a large hospital. Over a period of seven months we collected 178 incidents, of which 80 percent (143) were real errors and 20 percent (35) were near-errors. These were then classified into seven major categories. A kind of error reported most frequently was that a patient received or almost received a wrong dosage of a drug. These, incidentally, were not small errors, since in some instances a patient might receive as much as 24 times the required dosage of a drug!

The "common sense" approach would probably be to blame the nurse, pharmacist, or physician for making such "stupid blunders." Our emphasis was the reverse. It was to ask, "What is there about the medication system that will allow such errors to occur?" In answering that question we were able to identify ten categories of immediate causes of these errors.

TABLE 4.1 Sources of 460 Control Errors Identified from a Critical Incident Study*

Source of Error	Percentage
Substitution: Confusing one control with another	50
Adjustment: Operating a control too slowly, too rapidly, or to an incorrect position	18
Forgetting: Failing to check, unlock, or use a control at the proper time	18
Reversal: Moving a control in a direction opposite to that required	6
Unintentional activation: Accidentally activating a control	5
Inability to reach	3

*Adapted from Fitts & Jones (1947).

One cause could be traced to misread abbreviations. At that time a physician might write a prescription for a certain medication to be administered q.n., which is an abbreviation for the Latin term *quaque nocte,* or "once every night." Another prescription might be for a medication to be administered q.h., which is an abbreviation for *quaque hora,* or "once every hour." Whether a patient received a medication once a day (at night), or 24 times a day, depended on the distinctiveness of the letters n and h. If you have ever seen how many physicians write, you can understand why I have called this an error-provocative situation, a situation that almost literally invites people to make mistakes. For an illustration of how the label on a pharmacy bottle resulted in patients receiving ten times the required dosage of a drug, see Chapanis (1980, pp. 112 and 113).

As an exercise, collect some critical incident accounts from your friends or classmates about mistakes or near-mistakes they have made in operating controls for some common machines or appliances. See if you can classify them according to the categories in Table 4.1.

Additional Commentary The percentages of errors found in a critical incident study do not necessarily reflect their true proportions in operational situations, because the incidents are dependent on human memory and some incidents may be more impressive, or more likely to be remembered than others.

Note, too, that the critical incident technique only identifies problems, not solutions to those problems. Occasionally, solutions seem obvious, as in the case of the medical abbreviations described above—change the system of ordering dosages. More typically, further studies are usually necessary to find ways of mitigating or eliminating those problems. The Fitts and Jones (1947) study, for example, motivated a whole series of studies on control coding, ways of preventing accidental activation, reach envelopes, and the design of controls resulting in guidelines and recommendations that can routinely be found in many textbooks and standards.

Functional Flow Analysis

We have now arrived at one of the most useful techniques for determining system requirements. It is also more highly structured, that is, with more definite rules for its application, than the methods I have discussed so far.

Definition Functional flow analysis is a procedure for decomposing or identifying the sequence of functions or actions that must be performed by a system.

Inputs The analysis uses information from operational analyses, analyses of similar systems, or activity analyses.

Procedure Starting with system objectives, functions are identified and described iteratively with higher, top-level functions being progressively expanded into lower levels containing more and more detailed information. Functions or actions are described by verb-noun phrases, for example, activate system, plan route, check temperature, or verb-adjective-noun phrases, for example, record subsystem status, select menu item. Functions are also numbered in a way that clarifies their relationship to one another and permits traceability of functions through the whole system. For example:

- *Zero-order functions* are the initial gross functions that meet system objectives and are numbered 1.0, 2.0, 3.0, etc.
- *First-level functions* analyze each zero-level function into functionally related tasks and are identified by numbers such as 3.1, 3.2, 3.3, etc.
- *Second-level functions* analyze each first-level function, or task, if it can be further subdivided, into subtasks and are identified by numbers such as 3.2.1, 3.2.2, 3.2.3, etc.
- *Third-level functions* analyze each second-level subtask that can be subdivided into actions and are identified by numbers such as 3.2.4.1., 3.2.4.2, 3.2.4.3, etc.
- *Higher level functions* are carried out to such additional levels of detail as may be necessary, and are identified by extensions of the numbering system above.

The entire analysis is organized so that one can easily find the input and follow the flow through functions to the resulting output. Each individual function is contained in a rectangular block and numbered for reference, generally according to its sequence. Some analysts identify on the output line the output product, for example, motor running, directory file, the inputs to the next function. Once a number has been assigned to a function, it retains that number no matter where the function may be used or repeated later. Each functional flow diagram is linked to its next higher functional flow by means of a reference block broken in the middle (see Figures 4.6 and 4.7). Arrows enter function blocks from the left, and exit from the right, and so show the normal sequence of system functions, left to right, and, if necessary, down. Whenever arrows join or split, their junctions are shown with "and," "or," and "and/or" gates in circles. An "and" gate means that all the following or preceding functions must be performed; an "or" gate means that one or more of the following or preceding functions may be performed.

Products The products are *functional flow diagrams,* also referred to as *functional block diagrams, functional flows,* or *functional flow block diagrams,* that provide a sequential ordering of functions that will achieve system requirements, and a detailed checklist of system functions that must be considered in assuring that the system will be able to perform its intended mission. The system functions so identified are needed for the solution of trade studies and determinations of their allocation among operators, equipment, software, or some combination of them.

Examples Figure 4.4 shows a simple functional flow chart for providing meals. One of the interesting things about this top-level flow chart, and many others as well, is that it is timeless. Essentially the same basic functions have been carried out for thousands of years. The major change that has occurred over the years is the technology. Instead of obtaining food by killing it in a hunt or grubbing it from the soil, as our ancestors did, most of us obtain our food by going to a supermarket. Instead of cooking our foods on sticks or in pots over an open fire, we cook our food in microwave or conventional ovens. Instead of eating our food with our hands, as our ancestors undoubtedly did, we use dishes and implements such as knives, forks, and spoons.

Another important thing to notice is that there is a kind of similarity about the top-level functions in all systems: you prepare to do something, do it, and clean-up and put things away afterward. This is also illustrated in Figure 4.5, which shows the zero-order functions for a helicopter on a military mission. Preparations are made in the pre-launch phase, the helicopter is launched in the next phase, it flies out to where it is to do its business (the transit out phase), it does its business (the tactical phase), flies back (the transit back phase), and finally returns to the place from which it was launched (the recovery phase). Notice that the functions flow, or follow one another, in a linear manner.

Each zero-level function can then be further analyzed into subfunctions. Figure 4.6, for example, shows that zero-level function 1.0, "prepare and present flight briefs," can be analyzed into three first-order functions, numbered 1.1, 1.2, and 1.3. The starting point is the reference block labeled "initiate mission." These first-order functions flow directly into the next zero-order function, 2.0 "assume pre-launch posture." Coincident with the three functions associated with preparing and presenting flight briefs, there is another function, numbered 1.4, preparing the helicopter, going on at the same time, indicated by the "and" gate.

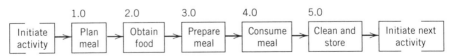

Figure 4.4 A zero-order functional-flow chart for providing meals.

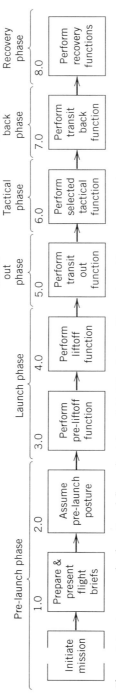

Figure 4.5 A zero-order functional-flow chart for a LAMPS mission. (Reproduced by permission from Figure HFM-19, IBM 1993)

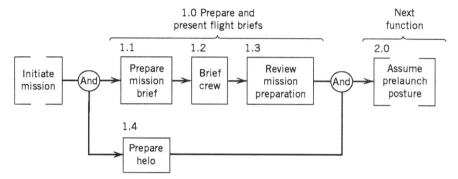

Figure 4.6 A first-order functional-flow chart for one zero-order function in Figure 4.5. (Reproduced by permission from Figure HFM-20, IBM 1993)

Most first-level functions can be divided into second-level ones. In this illustration, the first-level function 1.1, "brief crew," can be subdivided into two second-level functions (Figure 4.7). The starting point is the previous first-level function 1.1, "prepare mission brief," and the two second-level functions flow into the next first-level function, 1.3, "review mission preparation." Notice the numbering system, a feature of functional flow diagrams. This is especially important because it helps to identify and trace functions easily, even in very complex systems. Notice too that the two higher-level functions at the beginning and end, the reference functions, are shown in broken rectangles, the lower-level ones in unbroken rectangles.

Second-level functions may or may not, depending on the complexity of the system, be analyzed into still higher-level functions. There is no hard and fast rule about how far one carries out these analyses. This is where experience and judgment of the analyst come into play.

An important point to observe is that, in this analysis, nothing has been said about implementation. We have merely identified general functions that the system is to perform. How these functions are to be carried is not specified. For example, consider function 1.2.1, "obtain briefing information." This function might be carried out by having an observer say what he or she observed, by having information obtained via telephone, by having information presented on a display board, or

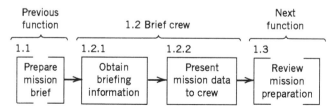

Figure 4.7 A second-order functional flow for one first-order function in Figure 4.6. (Reproduced by permission from Figure HFM-21, IBM 1993)

by still other means. The same applies to function 1.2.2, "present mission data to crew." This might be done by someone communicating information to the crew orally, by handing out hard-copy data from a FAX machine, by presenting data on a CRT, or by still other means. A functional flow diagram identifies functions, but leaves open the exact means by which functions will be implemented.

Incidentally, don't be fooled by the tidy little illustrations given here. A complete functional flow diagram for a complex system can very well cover the wall of a large room!

As an exercise, prepare a functional flow diagram showing the steps involved in preparing for and taking a vacation trip by automobile. Carry out your diagram at least to second-level functions.

Decision-Action Analysis

Decision-action analysis is so similar to functional flow analysis that usually only one or the other would be used. The principal difference between the two methods is that decision-action analysis deals with systems in which binary decisions have to be made and, for that reason, it is generally used when the system is software oriented.

Definition Decision-action analysis is a method of identifying and depicting the sequence of functions or actions that must be performed by a system containing decisions that can be phrased as questions with binary (yes/no) choice alternatives.

Inputs The inputs are the same as those for functional flow analysis: the results of operational analyses, analyses of similar systems, and activity analyses, supplemented with information provided by knowledgeable experts.

Procedure Functions or actions are identified and described as in functional flow analyses, with the addition of decisions phrased as questions with binary choices. Each function is a short verb-noun combination with occasional adjectives or other modifiers. Each decision is placed in a diamond symbol and is phrased as a question that may be answered with a binary, yes/no response. Function blocks and decision diamonds are given reference numbers as is done with functional flow diagrams.

Products The products are decision-action diagrams, also referred to as information flow charts, decision-logic diagrams, or operation-decision diagrams, that show a sequential ordering of functions and decisions that will meet system requirements. These constitute an extensive checklist of system functions and decisions that must be considered to ensure that the system will be able to perform as intended. The diagrams are easily translated into logic flow charts for computerized sections of the system.

Supplementary Details Decision-action analyses, like functional flow analyses, may be made at progressively detailed levels. A zero-level analysis may diagram only gross system functions and decisions, each of which may then be analyzed into lower-level ones. For example, one decision in driving might be "Engine status OK?" This can be decomposed into lower-level decisions such as "Fuel level OK?", "Engine oil OK?", and "Temperature OK?"

As with functional flow diagrams, the functions and decisions in decision-action diagrams are, at first, general ones with no indication of who or what will accomplish them. Once they have been allocated to people, hardware, or software (in functional allocation, to be discussed later), single-line symbols are usually used to indicate that a function or decision will be made by people; double-line symbols that they will be performed by machine or software.

Example In Figure 4.8 I have modified and recast some information given by Phillips and Melville (1987) into a decision–action diagram. It illustrates a small set of functions and decisions involved in dealing with potential airspace conflicts in the air-traffic-control Initial Sector Suite Subsystem. Although their first function is wordier than usual and could have been abbreviated simply to "Review situation display", the additional words seem justified in this case, since they clarify exactly what the controller should be reviewing the situation display for. The two decisions are self-explanatory.

As an exercise, prepare a decision-action diagram showing the functions and decisions involved in carrying out transactions with an automated bank teller machine. Be sure to include such decisions as verifying (1) the customer's ID and (2) the legality of the transactions requested.

Action-Information Analysis

As I have said several times above, functions and decisions identified up to this point are listed without reference to the agent (person or machine) that will execute them. Allocating those functions to people, hardware or software is the next logical step and it is sometimes taken following the completion of functional flow or decision–action analyses. Often, however, it is useful to have even more detail before attempting to make those allocations and an action-information analysis is a method for providing that detail.

Definition Action-information analysis elaborates each function or action in functional flow or decision–action diagrams by identifying the information that is needed for each action or decision to occur. This analysis is often supplemented with sources of data, potential problems, and error- or accident-inducing features associated with each function or action.

Inputs The inputs are data from the analysis of similar systems, activity analyses, critical incident studies, functional flow and decision—action analyses, supplemented with comments and data from knowledgeable experts.

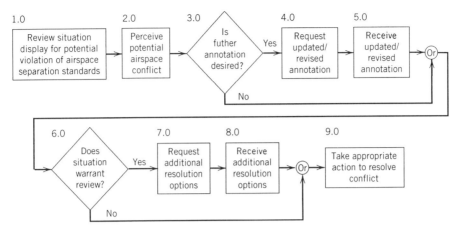

Figure 4.8 A decision–action diagram constructed from data in Phillips & Melville (1987).

Procedure Each function or action identified in functional flow or decision-action analyses is studied by the analyst who uses his expertise and all available information to identify and describe information requirements, sources of information, potential problems, error-inducing features and any other relevant commentary.

Products The products are detailed lists of information requirements for operator–system interfaces, early estimates of special personnel provisions likely to be needed, support requirements, potential problems, and probable solutions. The analysis may also produce suggestions for improvements in the design of hardware, software, and procedures.

Example Table 4.2 is an example of an action-information analysis for a very simple and familiar function, preparing to fill the gas tank on an automobile. The function (action) is listed in the left-hand column with its appropriate identifying number from a complete functional flow diagram. The next column shows three kinds of information associated with that function (action). The third column shows some of the difficulties associated with getting that information. The fourth column lists some source of information.

If you drive only one car, all this may seem commonsense and even trivial. But, if you're driving a borrowed or rented car you could have some problems. Let me illustrate. A car I rented recently had, on the floor next to the driver's seat, a lever for releasing the cover over the gas tank filler. I drove up to a gas station, got out of the car and pulled on the lever to release the cover. It didn't work even after repeated, increasingly frustrated tugs on the lever. Not only that, but a gas station attendant couldn't get it to work. When I became thoroughly annoyed and got into the car to cool off I discovered the secret: the gas tank cover could not be released if the car door was open!

While driving another new car I pulled into a gas station, lowered the window,

TABLE 4.2 Example of an Action-Information Analysis: Filling an Automobile Gas Tank

Function (action)	Information Requirements	Difficulties	Sources of Information
7.2.3 Prepare to fill gas tank	(1) Kind of gas required: (leaded, unleaded, octane rating)	On old cars may not be apparent	(1) Instruction on instrument panel next to gas gage (2) Owner's manual (3) Gas station attendant
	(2) Location of gas tank cover (right side, left side, behind rear license plate)	May not be immediately apparent and so may drive up to wrong side of gas pump.	
	(3) Method of unlocking gas tank cover	Special anti-theft features may not be apparent	

and handed out the ignition key to an attendant so that he could unlock the gas tank cover. Try as he might, he couldn't pry the cover open. When I got out of the car to try to help him, the source of the difficulty struck me: The gas tank cover on that car could not be opened if the car doors were locked!

I use these examples because they show that there may be hidden difficulties and sources of frustration in what may, at first glance, appear to be straightforward activities. Don't overlook potential problems that may be lurking in "simple" and "obvious" things.

In contrast to the almost trivial information requirements for the example I have just given, the analysis of complex systems may yield a very large number of them. Endsley and Rodgers (1994), for example, list 200 or so (depending on how you count them) specific items of information needed by air traffic controllers for what is termed "situation awareness," assessing (1) the rapidly changing location of each aircraft in three-dimensional space, (2) the future locations of aircraft relative to one another, and (3) pertinent parameters (destination, fuel, communications, etc.) associated with each aircraft. Study this list and try to imagine how you could display or convey all that information to a controller.

As an exercise, prepare an action-information analysis of one of the functions you identified in the decision-action diagram of transactions with an automated bank teller machine. For example, the function "customer selects a transaction to be executed."

Functional Allocation

In discussing functional flow, decision-action, and action-information analysis I said that the functions, decisions, and actions identified in those analyses were general ones with no indication of who or what would be performing them. That is not strictly correct because, in many cases, such decisions are almost literally forced on the designer by state-of-the-art capabilities or limitations, cost, or other considerations. But, in the development of most systems, some functions may not have been allocated up to this point. This is the point at which those allocation decisions are made.

There are some significant differences of opinion about this procedure. Whereas everyone, human-factors professionals and engineers (see, for example, Blanchard & Fabrycky, 1990) alike, acknowledge the importance and usefulness of techniques such as functional analysis, functional allocation as a process is much more controversial. On the one hand, Meister (1985) and DOD-HDBK-763 (1987) give models for allocating functions quantitatively, whereas Fuld (1993) argues that a formal function allocation process is an abstract notion that provides no unique methods or results. I think the practical approach lies somewhere between those two extreme positions because, one way or another, functions *are* allocated to people, hardware, software, or some combination of them. In my discussion of this method I have relied partly on the moderate position taken by Price (1985).

Definition Functional allocation is a procedure for assigning each system function, action, and decision to hardware, software, human operators, or maintainers or some combination of them.

Inputs The data needed for this procedure come from (a) functional flow analyses, decision-action analyses and action-information analyses; (b) past engineering experience with similar systems; (c) state-of-the-art performance capabilities or machines and software; and (d) store of known human capabilities and limitations.

Procedure Identify all those functions that must be allocated to personnel or equipment for reasons of safety, limitations of engineering technology, human limitations, or system requirements, and remove them from further consideration for the time being. List the remaining functions, those that could be performed either manually or by some combination of personnel and equipment. Prepare descriptions of the various ways in which each of these functions could be implemented. Establish weighting criteria by means of which each design alternative can be compared with the others. Compare alternative configurations in terms of their effectiveness in performing the given function according to those criteria, and select the alternative that is likely to perform best and be most cost-effective.

Products The products are allocations of system functions to hardware, software, human operators, or maintainers, or some combination of them. The functions allocated to human operators or maintainers are the ones for which task descriptions are prepared and on which task analyses are performed. The functions so allocated are also useful in identifying user skills and information needs and in providing preliminary estimates of staffing, training, and procedures requirements and workload assessments.

Amplifying Information Functions may be dedicated, that is, automatically allocated, for a number of reasons. A system requirement for a million computations per second can be met only by a machine. Safety considerations may dictate that maintenance be performed by robots, rather than by live technicians. On the other hand, a requirement that the amounts of checks be entered into customers' accounts can today be met only by human operators. Sometimes, too, functions are allocated to human operators because of a requirement that present personnel tasks be preserved.

Many functions, however, are not dedicated. Verifying the validity of a bank customer's identification number, or ID, could be done by having someone—some person—check the ID against a printed list or a list stored in a computer. Alternatively, the verification could be done entirely by computer. Note, however, that the choice between these two methods would interact with other functions, for example, how the customer expressed his or her ID. If the ID were spoken, that is, delivered orally, a human operator would have to receive that information because speech recognition technology is not yet sufficiently advanced to recognize numbers correctly when they are spoken by anyone with just one repetition. As a result, if the verification function were entirely allocated to a computer, it would mean that the bank customer's task would be different. The code could not be spoken, it would have to be entered into a computer in some way.

Criteria in Making Allocation Decisions. A number of criteria are used in making allocating decisions:

- Performance capability, including predicted reliability;
- Costs involved in implementing the function, including hidden costs, for example, the costs required for selecting and training personnel;
- Technological feasibility;
- Producibility, that is, the time required to implement the function;
- Maintainability;
- Number of personnel required;
- Power requirements;
- Safety;
- Support requirements; and
- Political considerations (for example, the decision to send humans into space was partly a political decision).

Of these, performance is generally the most important.

Criteria such as these are assigned weights, and there is no single hard-and-fast rule about how those weights are assigned. Meister (1985) suggests one method. DOD-HDBK-763 (1987) suggests another. In essence, however, the criteria are compared against each other, perhaps in a simple ranking, and assigned numerical values according to judgements of their relative importance. Candidate alternatives are then rated on each of these criteria, again with some simple rating scheme. Ratings are multiplied by their weights and summed. The design alternative with the highest score is the preferred solution.

As you can see, the procedure for allocating functions is not very precise. Judgment is involved in selecting criteria, determining the weights to be assigned to those criteria, and rating how well each of those criteria is met by the alternative design configurations. Imprecise though it may be, one of the principal values of this procedure is that it forces designers to recognize explicitly the criteria they use in making allocations and the relative importance of those criteria. That is a considerable step above making allocations by intuition alone.

Functional allocation is not a one-time matter, but may be revised and refined as design progresses. Changes in hardware and software, due to unexpected technological breakthroughs or unforeseen technical difficulties in design or programming, may require reconsideration of human involvement in the performance of some functions. Or, task analyses and workload analyses performed later may indicate excessive workloads or tasks that exceed human capabilities, perhaps requiring automating some tasks that had been previously assigned to human operators.

Who Makes the Allocations? Human-factors specialists almost never make allocation decisions by themselves. In fact, most are made by engineers and designers. Ideally, however, human-factors professionals should participate in this process because they can contribute important information about human capabilities. For example: Can human operators perform certain functions? With what accuracy can human operators perform these functions? How rigorously must operators be selected and how thoroughly must they be trained to operate? And so on. Such considerations may be critical in arriving at allocations that will maximize system performance.

An Example Many years ago postal sorting in large post offices was done by delivering trays of letters to sorters. A sorter would take up a batch of letters in his or her hand, read the address on letters one by one, and insert them into one of a large number of compartments according to their destinations (see Figure 4.9). Planners forecast that the volume of mail anticipated in years to come would overwhelm this method of sorting mail and that the process had to be automated. I was fortunate to have participated in some of the early discussions about the design of such an automated system.

Some functions that were identified in these discussions were:

- Incoming sacks of mail had to be emptied.
- Mail had to be sorted into groups by size.
- Letter mail had to be arranged in trays with the postage stamp in the upper-right-hand corner.
- A decision had to be made about whether the amount of postage was correct.
- Addresses had to be read.
- Letters had to be distributed according to their zip codes.

Figure 4.9 "Pigeon hole" method of sorting incoming mail in the General Post Office, New York City. This manual method of sorting mail was used for about 200 years prior to the Postal Reorganization in 1970. (*Photo courtesy of the U.S. Postal Service*)

Some of these functions, sorting into groups by size, orienting letters in trays with the postage stamps in the upper right, were tedious for human operators. Machine methods could perform them faster and more economically. But consider the determination of whether the amount of postage was correct. Theoretically, one could devise an automated system that would weigh each letter and, by suitable coding of postage stamps, for example, bar codes or phosphorescent stripes, determine the amount of postage on the letter. But there's more to it than that. Applicable rates for letters vary according to a number of other criteria, for example, whether they originate from tax-exempt organizations, and whether they are destined for domestic or foreign addresses. To determine the amount of postage required one would also have to perform the next function, that is, read the address. No one at the time, now roughly 50 years ago, could think of any automated way of reading addresses in all of the ways that they could be printed, typewritten, or handwritten. In fact, that still cannot be done by machine or computer. So these two functions, determining the correctness of the postage and reading addresses, had to be assigned to human operators, because the functions were beyond the state of engineering art.

Now in large automated post offices, trays of letters are delivered on a conveyor to sorters, who sit in front of a simple keyboard. A machine arm picks up a letter, places it in front of the sorter, who then has six-tenths of a second to read the address and four-tenths of a second to key in the zip code (Figure 4.10). Once that is done, the letter is whisked-off and sorted automatically. Since the operation is

Figure 4.10 A mechanized multiple letter-sorting machine. (*Photo courtesy of the U.S. Postal Service*)

machine-paced, it is not ergonomically ideal and it is a dull and monotonous job. But it was one of the decisions that followed from the trade-offs that were considered during functional allocation.

Task Analysis

Task analysis is another important method in the human-factors repertoire, because it provides inputs to virtually all the remaining methods in this chapter.

Definition Task analysis is a method for producing an ordered list of all the things people will do in a system, with details on:

- information requirements;
- evaluations and decisions that must be made;
- task times;
- operator actions; and
- environmental conditions.

Inputs The data needed for task analysis come from any or all of the methods discussed so far, supplemented with information provided by experts who have had experience with similar systems.

106   4: HUMAN FACTORS METHODS

Procedure   The analyst, together with subject matter experts, if they are available, uses his experience and knowledge to list and describe all tasks, subdividing tasks into subtasks, and amplifying each with relevant supplementary information.

Products   The principal product of a task analysis is an ordered list of all the tasks that people will do in a system. This list can be used later to (a) estimate the time and effort required to perform tasks; (b) determine staffing, skill, and training requirements; (c) determine human–system interface requirements; and (d) provide inputs to reviews and specifications.

Supplementary Information   Tasks are where hardware and software come together to accomplish functions. In fact, you cannot have systems integration without tasks. Despite the importance of task analysis, there is no single or best way to perform it. Meister (1985), for example, summarizes and compares 13 variations of the procedure and, as he says, it is not relevant to say that any one is bad or good. Task analysis is a design tool that can be modified and adapted to whatever the specialist finds useful for a specific project. The minimum requirements of a task analysis are those listed in the definition above. Table 4.3, however, is a longer list of some kinds of data usually collected in task analyses. Note, for example, that some pairs of questions go together: "What information is required by the operator?" and "What information is available to the operator?" If information required by the operator is not available, this identifies an important mismatch that should be rectified. Another such pair is "How much workspace is required by the action?" and "How much workspace is available for the action?"

Figure 4.11 shows one kind of form that has been used successfully in conducting task analyses on several large projects. An important feature of this particular form are the numbers in parentheses at the bottom of the form. They refer to specific paragraphs of MIL-H-46855B and served as one way of satisfying the customer (the government, in this case) that the contractor was complying with the requirements of that standard.

An Example   This example illustrates the application of task analysis to predict performance for a system that had not yet been built. The system, the Data System Modernization (DSM) system, was designed to serve as the means of communicating with and controlling space vehicles or space satellites. During any one pass of a space vehicle there was an eight-minute window available for communicating with the satellite. The question asked by the designers was: Can all the activities necessary to configure and test the Remote Tracking Station (RTS), Range Control Complex (RCC), and Mission Control Complex (MCC) resources be done in eight minutes?

Since this was a unique system and had not yet been built, no answer was available from the literature or from experience with earlier systems. The approach taken by the human-factors professional was to perform a task analysis and time

TABLE 4.3 Data Collected in Task Analyses

Information requirements
    What information is required by operators?
    What information is available to operators?
    How do operators evaluate the information?
    How do operators reach a decision about the action to take?
Action requirements
    What action do operators take?
    What body movements are required by the action?
    How often must the action be taken?
    How precisely must the action be taken?
    How fast must the action be taken?
    What feedback do operators get about the action?
Working conditions
    How much workspace is required by the action?
    How much workspace is available for the action?
    What tools or equipment do operators need?
    What job aids are required?
    What special hazards are involved?
Staffing requirements
    How many operators are required?
    How will multiple operators interact?
Training requirements
    What training or experience do operators need?
Environmental considerations
    Where is the work environment located?
    What is the nature of the work environment?

analysis to predict the answer. The first step was to identify the major RCC and MCC activities from the DSM operations concept. Figure 4.12 shows the five major activities identified for the range control complex. Next, each of the major activities was further subdivided into subtasks from data contained in the DSM specifications, computer program configuration item (CPCI) documents, and the DSM operations concept. A task analysis was then performed of the tasks and subtasks on the basis of information contained in available documents and interviews with CPCI and systems designers. Figure 4.13 shows such a task analysis for Task Number 5, "Transfer control of range resources to MCC," in Figure 4.12. Time estimates from the American Institutes for Research (AIR) data store (Munger et al., 1962) were then entered into the table to yield a total time of 42 seconds for the task.

Task analyses similar to that in Figure 4.13 were performed on the other four major tasks in Figure 4.12 and the results are shown in Figure 4.14, which gives the estimated times for each of the five tasks and the cumulative time (4.35 minutes) for all five tasks. These data were then combined with similar analyses performed on

Start time	Task or step	Description	Start cue	Information provided to operator	Evaluations and decisions	Operator actions	Systems feedback	Duration and frequency	Communication link(s)	Concurrent tasks	Other data and problems
T − A or T + B seconds, minutes, or hours.	No. and/or name of task or separately analyzed task steps.	1. Brief descriptive overview of listed task or steps. 2. Include task steps in description if not listed separately. 3. Details of equipment & procedures likely to indicate required knowledge and skills of personnel 4. General equipment (like consoles) if not covered elsewhere.	Cue, event or condition causing task to start.	1. Types of information required to perform task. 2. How provided: by CRT, printed sheets, voice, etc. 3. Types of equipment, forms, displays, etc. 4. Approx. format and amounts of data presented. 5. Requirements to find or detect data having certain characteristics or changes. 6. Content and coding characteristics likely to indicate required knowledge and skills of personnel.	1. All mental performances of operator. 2. Seaches and identification of relevant data. 3. Interpretations of what data shows about the system. 4. Determinations of what needs to be done. 5. Selections and planning of specific actions. 6. Extent and complexity of mental tasks. 7. Performance requirements likely to indicate required knowledge and skills of personnel.	1. Types of actions and data inputs required to complete task. 2. How performed: by controls, keyboards, voice, body actions etc. 3. Types of equipment, controls, keyboards, etc. 4. Approx. forms and amounts of work performed or data entered. 5. Content and coding characteristics likely to indicate required knowledge and skills of personnel.	Cue, event or condition verifying operator action.	Task or step duration in seconds, minutes or hours. Any times or frequencies at which tasks must be repeated.	1. Nature and amount of any coordination and communication with other people and activities. 2. Number of personnel required. 3. Specialties and experience.	1. Summary of other simultaneous or interleaved monitoring, manual performances, talking, etc. 2. Approx. extent and demand of concurrent performances. 3. Their approx. frequencies and durations.	1. Other relevant conditions in equipment, environment, procedures, etc. affecting operator performances. 2. Any special problems. 3. Suggested solutions.
(11)			(1)	(2)	(3) (4)	(5) (6)	(12)	(10)	(14) (16) (18)		(7) (8) (9) (13) (15) (19) (20)

Note: Numbers in parentheses refer to descriptions listed in MIL-M-46866B, Para. 3.2.1.3.2

Figure 4.11 A task analysis data form used in IBM's SUBACS project. (Reproduced by permission from Figure HFM-36, IBM 1993)

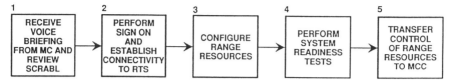

Figure 4.12 Functional flow of the five major activities in the Range Control Complex for the DSM system. (Reproduced by permission from Figure HFM-38, IBM 1993)

the MCC activities with the results shown in Figure 4.15. The final answer was that all activities could be performed within eight minutes, but with very little spare time for handling equipment malfunctions or other contingencies. But that's not the end of the story. As a result of this analysis some hardware and software components were redesigned with changes in procedures and a subsequent reduction in the total time required for all activities.

Fault Tree Analysis

This method and the next one, failure modes and effects analysis, are both specifically concerned with errors. Predicting human errors and evaluating their consequences is obviously no simple matter, and attempts to quantify human reliability have been severely criticized from theoretical and practical points of view (see Adams, 1982, for an example). Nevertheless high-risk systems are being designed, built, and operated and, however imperfect our methods may be, some attempts have to be made to anticipate mistakes that operators or maintainers may make and to try to design against those mistakes.

Safety engineers have devised a number of methods for evaluating system errors primarily from equipment or operational points of view. Some of those methods have been adapted by human factors to deal with human errors. Two well-known methods I shall describe are, in a manner of speaking, the reverse of each other. Fault tree analysis starts with an undesirable event or failure (for example, a train or aircraft collision, injury to personnel) and attempts to determine those combinations of events and circumstances that could lead to it. It is, in short, a method of accident investigation. In failure modes and effects analysis, one starts with components—and for human-factors purposes one of the components is a human operator or maintainer—and deduces the consequences of a failure in one or more of those components.

Definition Fault tree analysis determines those combinations of events that could cause specific system failures, faults, or catastrophes.

Inputs Data are used from virtually all the products of human-factors methods earlier in the sequence, but especially from functional flow analyses, decision-action analyses, action-information requirements analyses, and task analyses. These data are supplemented with information from data stores on human reliability.

Start time	Task or step	Description	Start cue	Operator actions	Information provided to operator	Evaluations and decisions	Systems feedback	Duration and frequency	Communication link(s)	Concurrent tasks	Other data and problems
	1. Initialize transfer of resources	RC alerts GC to stand by for resources transfer		Speech communication		Verify readiness of MCC to receive resources	Audio	5.0	Voice link		
	2. Call up DRG connectivity frame	RC display on R.S. frame needed for entry of connectivity data		Using lightpen on 3278 select "display connectivity" transaction – Shift – Selection – Shift loc – Review display on R.S. – Speech communication	MCC DRG connectivity destination information		ECSP53 frame updated and DRG connectivity to MCC(s) (ECRR50)	1.0 6.0 2.0 (MR) 2.0 3.0			
	3. Start transfer process	After verification with GC RC initializes transfer resource request processing		Using lightpen on 3279 select "transfer from RCC to MCC(s)" transaction – Shift – Selection – Verify display	MCC DRG connectivity destination information	Verify with GC for proper connectivity ports	ECRR51 frame	5.0 0.8 6.0 (MR) 2.0 (MR) 2.0	Voice link		
	4. Complete transfer process	RC presses enter PFK to complete transfer and then verifies with GC that range resources are under MCC control		Depress enter PFK on 3279 KBD – Shift – PFK – Speech communication	Confirmation of connectivity between RTS and MCC	Verify with GC completion of transfer	ECRR51 frame updated and composite RTS status appears on the same screen	0.6 1.5 2.0 (MR) 4.0	Voice link		
	End of RC's prepass activity							42			

Figure 4.13 Task analysis for one of the tasks in Figure 4.12. (Reproduced by permission from Figure HFM-39, IBM 1993)

THE METHODS **111**

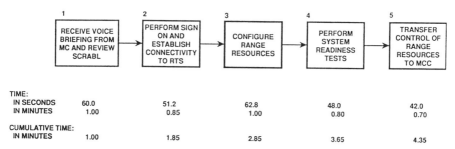

Figure 4.14 Estimated times for the five major tasks in Figure 4.13. (Reproduced by permission from Figure HFM-40, IBM 1993)

Products The products are probabilities of various undesirable events, the probable sequences that would produce those undesirable events, and the identification of sensitive elements that could reduce the possibility or probability of a mishap.

Supplementary Information Fault trees are constructed with symbols to represent events and to describe the logical relationships between events. Symbols called *logic gates* show how events at one level of a tree combine to produce an event at the next higher level. The most commonly used logic gates are the AND and the OR (Figure 4.16). An AND gate represents a condition in which an event above the gate

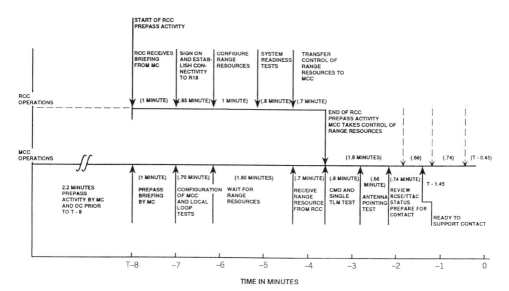

Figure 4.15 A timeline analysis combining estimated times for activities in the Range Control Complex and Mission Control Complex for the DSM system. The solid lines are time estimates for monitoring one channel; the dashed lines for monitoring five channels. (Reproduced by permission from Figure HFM-41, IBM 1993)

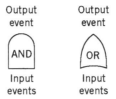

Figure 4.16 AND and OR gates used in fault tree analysis.

(the output) will occur if and only if all the events below the gate occur at the same time. For example, in a punch-card verification system an error will occur only if the person punching the cards and the person verifying the punching each make exactly the *same* error at the *same* place. An OR gate represents a condition in which any of the events below the gate (the inputs) will lead to the event above the gate (the output). For example, a control error will result if an operator uses the wrong control, if the operator uses the correct control improperly, or if the control is not functioning properly.

Figure 4.17 is a simple example of the way probabilities combine in fault trees. B will occur if either D or E occur and the probability of that happening is

$$p_B = 1 - (1 - p_D)(1 - p_E).$$

C will occur only if both F and G occur simultaneously and the probability of that happening is

$$p_C = p_F p_G.$$

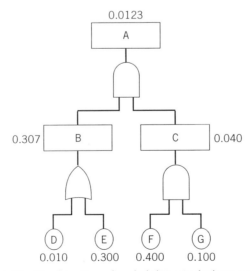

Figure 4.17 Combination of probabilities in fault tree analysis.

Likewise, A will occur only if B and C occur simultaneously and the probability of that happening is

$$p_A = p_B p_C$$

As an exercise, for the fictitious data in Figure 4.17, what strategy would reduce A by the greatest amount: Reduce D, E, F, or G by 50%?

Two important limitations of fault tree analysis for human factors work are that: (a) each event must be described in terms of only two possible conditions and no more; and (b) it is extremely difficult to attach exact probabilities to human activities. Although we do have data stores in human reliability (see, for example, Munger et al., 1962, and Swain & Guttmann, 1983) they are, by no means, as good, as complete, or as valid as we would like.

Failure Modes and Effects Analysis

Definition This is a method for deducing the consequences for system performance of a failure in one or more components and the probabilities of those consequences occurring. For human factors, of course, the components of primary interest are operators and maintainers.

Inputs Data from virtually all the products of the human-factors methods discussed earlier in the sequence in Figure 4.1 provide inputs to this method, but the most useful ones are functional flow analyses, decision/action analyses, action/ information requirements analyses, and task analyses. These inputs are supplemented with information from data stores on human reliability.

Procedure The analyst starts with components—in this case, operators or maintainers—and identifies the various kinds of errors humans could make in carrying out subtasks, or functions. Estimates are made of the probabilities or frequencies of making each kind of error and the consequences of each kind of error are deduced by tracing its effects through a functional flow diagram to its final outcome.

Products The products are a list of those human failures that would have critical effects on system operation, the probabilities of subsystem or system failures due to human errors, and identification of those human tasks or actions that should be modified or replaced to reduce the probability of serious system failures.

Supplementary Information Although the rationale underlying this method is straightforward, carrying it out is difficult for two reasons: (a) it is almost impossible to anticipate, or foresee, all the kinds of errors people can make; and (b) our data on human reliability are tenuous. Despite these difficulties, the procedure usually pays off handsomely in reducing or eliminating error-inducing features in a system.

Example Kirwan (1987) has provided us with an excellent example of the use of both fault tree and failure modes and effects analysis, using first one then the other. It is a particularly important study because it is a rare example of a thoroughly documented error analysis published in the open literature. Three other significant aspects of this study are: (1) the system had an unusual feature—operators had to deliberately *delay* taking immediate action during an emergency; (2) the study was done during the detailed engineering phase, that is, before the system was built; and (3) the analyses led to several recommendations for reducing the likelihood of human error that were incorporated into the final design of the system.

The study concerned some operations on a North Sea Offshore Oil Platform. The system and the study were complex, and my description gives only bare-bones essentials. (For details, consult the original [Kirwan, 1987] article). The operations involved compressors and separators containing flammable hydrocarbons that were essential to the oil-production process. In the event of an emergency (fire, explosion, gas leak, power failure) the compressors and separators had to be depressurized and vented (blown down) to reduce the risk of fire or explosion. The unusual feature was that, in order to reduce the risk of explosions, pipe ruptures, fires, and radiation hazards, the depressurization had to be done in stages: depressurization of either the separators or compressors could begin only after the other one had been fully vented. This meant that, when an emergency occurred and depressurization of separators, for example, had begun, operators at each of two different locations had to wait at least four minutes before starting depressurization of the compressors. You can appreciate that this would be an extremely stressful situation. Operators would know an emergency had occurred from, among other things, alarm signals and noise, but they had to deliberately wait what would probably seem like an eternity before taking final action.

Using a fault tree type of analysis, Kirwan prepared scenarios (Remember the discussion of scenarios in operational analysis?) of five possible accidents: (1) a failure of the two computer control systems due, for example, to a power failure; (2) a compressor leak or fire; (3) a platform emergency (fire, gas leak) in the area of the compressors; (4) a platform emergency near the compressors; and (5) a platform emergency remote from the compressors. Keep in mind that, at this point, the system has not yet been built, so these scenarios are predictions.

For each scenario Kirwan prepared what he called operator action event trees (essentially fault trees) to identify where human errors could contribute to the probability of system failure (flare overload). Task descriptions were then prepared of the actions that operators could, or should take, and the consequences of those actions, that is, which ones would result in a safe outcome and which would lead to failure (flare overload). Using data primarily from U.S. nuclear power accidents, he then estimated the probabilities that operators would commit any one of nine kinds of errors, for example: that an A operator would unintentionally hit the wrong buttons on their panel; that an A operator would realize he had unintentionally hit the wrong buttons and would try to recover from that error; that the A operators, due to stress, would forget to wait four minutes; that the B operators (in another location) might be unable to communicate with the A operators and so take action on their own before the four-minute waiting period had expired; and so on. The error probabilities so estimated were then compounded as in failure modes and effects analysis to predict the probabilities that each of the five accident scenarios would occur. The most dangerous scenario, the one with the highest predicted probability of occurrence (.074), was a platform emergency near the compressor modules. These analyses then led to recommendations for changes in the design of the emergency depressurization panel, the incorporation of timing interlock devices, improvement in communication and emergency procedures, and the use of high-fidelity simulators for training.

As an exercise, prepare a list of all the foreseeable errors you could make in programming a VCR to record a television program.

Timeline Analysis

Timeline analysis follows easily and naturally from task analysis. Although the times required for various activities were certainly determined during a task analysis, they were not necessarily ordered, or arranged in sequence. A timeline does that. It may also integrate times from separate task analyses to show overlapping activities of two or more persons.

Definition In timeline analysis, charts are produced showing sequences of operator or maintainer actions, the times required for each action, and the time at which each action should occur. It is a relatively minor and easily performed extension of task analysis.

Inputs The main inputs are data from task analyses.

Procedure The durations of tasks identified in task analyses, and the times at which these tasks occur, are plotted in charts or graphs.

Products The method produces plots of the temporal relationships among tasks, the durations of individual tasks, and the times at which each task is, or should be performed. These plots (1) permit the analyst to verify that necessary tasks can be performed, and that there are no incompatible tasks; (2) serve as inputs for workload evaluation; and (3) provide early estimates of the number of personnel that may be required.

Examples Although not plotted in a chart, the cumulative times at the bottom of Figure 4.14 constitute a timeline, and the data in Figure 4.15 are a time line of the simultaneous activities of range control and mission control operations. Notice that the latter chart integrates data from two separate task analyses. I have already commented on the value of this analysis.

Figure 4.18 is an example of a time chart of a segment of pilot and copilot activities during helicopter flight. Note, for example, that the plot shows the copilot

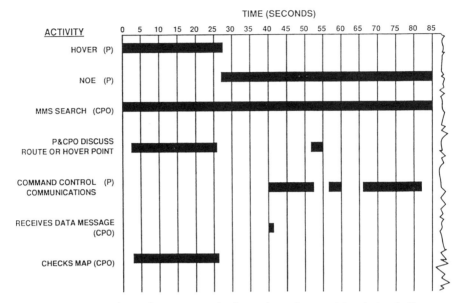

Figure 4.18 Time chart of a segment of pilot and copilot activities during helicopter flight (P = pilot, CPO = copilot observer, NOE = nap-of-the-earth flight, MMS = mast-mounted sight. (From Shaffer, Shafer, & Kutche, 1986. Copyright by the Human Factors and Ergonomics Society, Inc., and reproduced by permission.)

checking a map and searching a multi-purpose display (MPD) associated with the MMS equipment. Since these tasks cannot be done simultaneously, it means that the copilot is overloaded at this point and must time-share those tasks. On the other hand, it is possible to fly the helicopter and talk at the same time as is shown for the pilot.

Link Analysis

We come now to a method that is quite different from those I have discussed so far. It is concerned with the physical arrangement of items and so is a design activity rather than an analysis procedure. It is located so far along in the sequence of methods (Figure 4.1) because it presumes that decisions, or at least tentative ones, have been made about items of equipment, personnel, and procedures.

Definition Link analysis is a method for arranging the physical layout of instrument panels, control panels, workstations, or work areas to meet certain objectives, for example, reduce total amount of movement, increase accessibility.

Inputs The primary inputs required for a link analysis are data from activity analyses and task analyses and observations of functional or simulated systems.

Procedure The steps involved in a link analysis are:

- List all personnel and items to be linked.
- Measure preferably, or otherwise estimate, frequencies of linkages between items, operators, or operators and items.
- Measure, or estimate, the importance of each link.
- Compute frequency-importance values for each link.
- Starting with the highest link values, successively add items with lower link values, and readjust to minimize linkages.
- Fit the layout into the allocated space.
- Evaluate the new layout against the original objectives.

Products The products are recommended layouts of panels, workstations, or work areas that minimize linkages or meet other objectives.

On the Definition of Links A link is any sequence of use of two instruments or any sequence of action. It is a connection between (a) a person and a machine or part of a machine, (b) two persons, or (c) two parts of a machine. For example, if an operator has to use a telephone, this is a person–machine link. When operator A goes to talk to operator B (or give B a piece of paper), this establishes a person–person link. When an operator first twists knob C and then knob D, this identifies a linkage between C and D.

Examples The procedure described above was used to reconfigure the Combat Information Center (CIC) aboard the *U.S.S. Louisville*. The original layout of this CIC is shown in Figure 4.19, and, as you can see, it was a very complex assemblage of men and machines. To follow the procedure it is not really necessary to know what all these equipments were or what all these men did. You will notice from a study of the figure, however, that a number of men (plotters, radar operators, and radio men) remained at their posts during air attacks. A few other men (six in all) had to move about to carry out their duties. An important thing to emphasize about these paths of movement is that there was no fixed order in which they occurred. The actions of any particular officer depended, to a large extent, on the way the battle went. About all that can be said about these paths of movement is that they occurred with certain relative frequencies. Figure 4.20 shows link values between the critical men and equipments in this system. The link values in this application are *use-importance* links, that is, they give some indication of both the *frequency* with which each link was used and its *importance*. The use values were obtained by recording the number of times each man used each linkage and then condensing those numbers to integer values of 1, 2, or 3. The importance values were obtained by having experienced officers assign a value of 1, 2, or 3 to each link, according to their judgments of its importance, averaging those ratings, and rounding the averages to 1, 2, or 3. The final link values in Figure 4.20 are the products of those two numbers. There is nothing sacred about having link values span a range of 1 to 9 or

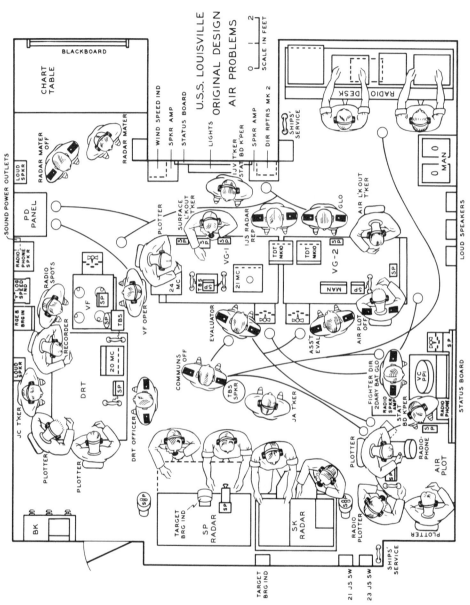

Figure 4.19 Original layout of the combat information center (CIC) aboard the *U.S.S. Louisville*. (From Chapanis, 1959. Reproduced by permission of Johns Hopkins Univ.)

		Men					
		Comm. off.	Eval.	Asst. eval.	GLO	Fight. dir.	VF oper.
Men	Communications officer (Com. off.)		6			6	
	Evaluator (Eval.)	6		5		7	
	Assistant evaluator (Asst. eval.)		5				
	Gunnery liaison officer (GLO)						
	Fighter director (Fight. dir.)	6	7				
	VF radar operator (VF oper.)						
Machines	VC radar					8	
	Air plot	1	9				
	Radio desk	1					
	PD panel				2		1
	VG radar no.1		9				
	VG radar no.2			8	7	4	
	VF radar				3		9
	Director repeaters				7		

Figure 4.20 Linkages between men and men, and between men and machines in the CIC of the *U.S.S. Louisville*. (From Chapanis, 1959. Reproduced by permission of Johns Hopkins Univ.)

about this method of determining them. Other ways of quantifying linkages could have been used, but, in the absence of any theoretical rationale, the simple ranks used in this case produced satisfactory results.

The Schematic Link Diagram. Figure 4.21 is a schematic representation of the same data that appear in Figures 4.19 and 4.20, except that only the critical items of equipment and critical men are shown in the diagram in approximately their correct spatial relationships. A diagram like Figure 4.21 shows clearly why this is not the most efficient arrangement of men and equipments. Notice, for example, that, to get to the radio desk, the communications officer had to (a) travel a long way, and (b) cross the paths between the evaluator and air plot, between the evaluator and fighter director, and between the fighter director and the *VG*-2 radar. Notice also that the gunnery liaison officer had a long way to go to the *VF* radar and to the *PD* panel, both of which he had to consult from time to time. Notice, too, that paths crossed, that is, men literally bumped into each other. The objectives of the analysis were to shorten the paths of movement and to increase accessibility.

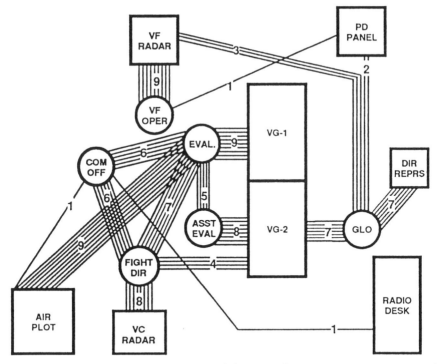

Figure 4.21 A schematic representation of the same data given in Figures 4.19 and 4.20. (From Chapanis, 1959. Reproduced by permission of Johns Hopkins Univ.)

An Improved Layout. A relatively simple analysis of this problem yielded the solution illustrated in Figure 4.22. There you can see that the paths between men and men and between men and machines never cross, and that the total lengths of the paths have been reduced.

Installing the Revised Layout. When the analyst has arrived at a solution such as that in Figure 4.22, it must be tested to see whether it will fit into the space actually available. Men and equipment must be positioned to avoid pillars and other fixed architectural features; to allow free access to doors, elevators and corridors; and to make full use of recessed areas, corners, and cubicles. This fitting job can be done on a floor plan of the space with cut-outs representing the men and various items of equipment.

During this fitting procedure, think of the layout in Figure 4.22 as a loose-jointed arrangement with flexible linkages between men and men and between men and equipment. Men and equipment can be moved about, so long as the approximate relationships in Figure 4.22 are maintained and as long as none of the paths are made to cross one another. You can see that this gives you considerable leeway in making up the final arrangement. For example, the radio desk could be rotated 90° and placed along the same wall as the air plot. The *VC* radar could be moved to a number of alternative positions without destroying the basic pattern. Still other modifications are possible.

Evaluating the Layout. The final step in a link analysis is an evaluation of the old and new arrangements to ensure that there has been an improvement. In this example, three assessment criteria were used: (1) *an index of walking*—the total distance walked by all the men during typical operations; (2) *an index of crowding*—a simple scoring system to determine the amount of interference introduced by crowding; and (3) *an accessibility score*—a measure of the passage space available for each person to enter the room and reach his station. In the original layout the men had to walk a total of 62 feet during an air attack. The revised layout reduced that to exactly half—31 feet. In the original layout the general accessibility rating was only fair; in the revised layout it was excellent.

An Office Example. Figure 4.23 shows another way of representing a link table. The data come from an article by Rhodes (1984) and show the interrelationships between personnel and between personnel and equipment in an office. Rhodes called this an activity relationship chart, but you will recognize it as a frequency link

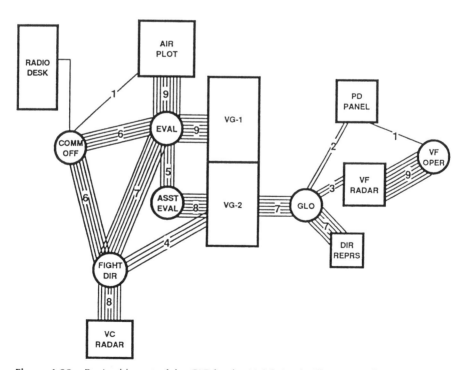

Figure 4.22 Revised layout of the CIC for the *U.S.S. Louisville*. (From Chapanis, 1959. Reproduced by permission of Johns Hopkins Univ.)

4: HUMAN FACTORS METHODS

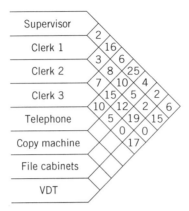

Figure 4.23 Interrelationships between personnel and between personnel and equipment in an office. (Constructed from data in Rhodes, 1984)

chart. Incidentally, DOD-HDBK-763 (1987) refers to this way of representing linkages as a correlation matrix, a term sometimes used by other authors.

Rhodes also published an association chart (Figure 4.24) as a companion to the activity chart. You will recognize his association chart as an importance link chart.

As an exercise, prepare a diagram of an office layout that will minimize the frequency linkages in Figure 4.23 and make use of the importance linkages in Figure 4.24. Assume that the office is square in shape, and that one wall has a door in the center and a large window on each side of the door. A door in the center of the opposite wall opens into a storage and utility room.

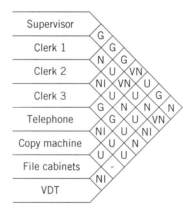

Figure 4.24 Association chart to accompany Figure 4.24. VN = very necessary, N = necessary, U = useful, G = general use, NI = not important. (Constructed from data in Rhodes, 1984)

Other Applications. Link analysis has been applied to somewhat simpler problems, like arranging the controls on a control panel to minimize hand movements and arranging displays on a panel to minimize eye movements. For example, Yamanoi and colleagues (1982) show link analyses of hand movements made in the operation of four types of electrocardiographs on the market at that time. They then devised mockups of revised panels that greatly reduced the number of movements and their flow from one to another. So the technique, you see, can be applied to diverse layout problems.

A Link Analysis Tool. The examples given here are relatively small and simple ones. Despite the apparent complexity of the CIC shown in Figure 4.19, only a small number of men and equipments were involved. The revised layout was achieved with paper and pencil. For large systems with many elements and links, paper-and-pencil solutions become prohibitively time consuming and expensive. A microcomputer-aided link analysis tool described by Glass, Zaloom, and Gates (1991) enables one to perform link analyses for a wide variety of systems in a desktop computing environment. The tool runs on an IBM-XT and compatible microcomputers; considers wage rates of personnel, if they are relevant; permits the rapid comparison of alternative solutions; and produces link analyses to include in a report.

Simulation

Definition Simulation is a basic engineering and human-factors method used to predict the performance of systems, or parts of systems that do not exist, or to allow users to experience and receive training on systems, or parts of systems, that are complex, dangerous, or expensive.

Inputs Simulations involve some combination of (a) hardware, (b) software, (c) functions and tasks elucidated in task analyses, and (d) operating procedures.

Procedure Models or mockups are prepared incorporating some or all of the inputs, and operators, maintainers, or users perform typical tasks on the models or mockups.

Outputs Simulations allow you to: (a) make predictions about system performance, (b) assess workloads, (c) evaluate alternative configurations, (d) evaluate operating procedures, (e) provide training, and (e) identify accident or error-provocative situations and mismatches between personnel and equipment.

Elaboration Simulations cover a wide range of techniques and devices from simple representations on paper to extremely complex devices such as aircraft training devices and simulators for training nuclear power operators. They may be reduced-scale or full-scale constructions of plywood and plastic, computer simulations of the human body used in anthropometry, or interactive computer programs allowing for rapid prototyping.

Simulation may be used in all phases of a system life cycle. In the concept-formulation stage, simulations help to develop and portray concepts of equipment and layouts, and identify potential problem areas and additional study areas. During preliminary design, simulations may be used to identify design requirements for ease of maintenance, and to develop preliminary specifications for equipment operability and maintainability. During engineering development, they aid in the detailed design of equipment panels, cables and duct routing, and in developing preliminary installation, operating, and maintenance procedures. In the production-and-operation phase, simulations may be used to refine installation procedures, familiarize installation personnel with procedures, and familiarize operational and maintenance personnel with the system.

For a good discussion on the use of simulation and simulators, and the behavioral issues and problems common to their use, see the 147-page report by Jones, Hennessy and Deutsch (1985). A 21-page appendix to that report describes and illustrates a number of simulators in five categories: (1) engineering simulators, (2) training simulators, (3) research simulators, (4) battle simulators, and (5) embedded simulation.

Examples The first example I shall use illustrates the use of simulation to solve a biomechanical problem. Although the simulation was a simple one, the application for which it was done had some unusual features (Winters & Chapanis, 1986). In a large submarine a maintainer must frequently crawl through a large tube to reach a spherical enclosure in the nose of the vessel. This enclosure contains a number of electronic cabinets like the one shown in Figure 4.25. Each cabinet consists of three tiers of drawers, with three drawers in each tier and three tiers of standard electronic modules, or SEMs. The SEMs are metal-framed cards containing microelectronic circuits. The rear edge of each SEM contains numerous connector pins, each pin requiring a certain nominal force to make firm electrical contact with a rear pin array. During maintenance, defective SEMs must be extracted and replaced with new ones.

In this instance, engineers proposed that each SEM was to have 250 connector pins that would require a combined minimum force of 15.9 kg to be sure the SEM was seated properly. Opposing these purely engineering considerations were human-factors concerns about whether maintainers could exert the required amounts of force under the conditions in which they had to work. The electronic cabinets were so hot that maintainers could neither brace themselves against cabinets at their backs nor rest their hands on the surface of the cabinets. Hence, the normal procedure for inserting SEMs is by use of two thumbs on the SEM's push area, from a free-standing position, and without contacting surrounding, already seated SEMs in the drawer, the cabinet drawer surfaces, or other surrounding enclosures. The question raised by the engineers was: Can Navy maintainers seat SEMs requiring 15.9 kg of force under these conditions?

Since the answer was not available in the human-factors literature, the solution

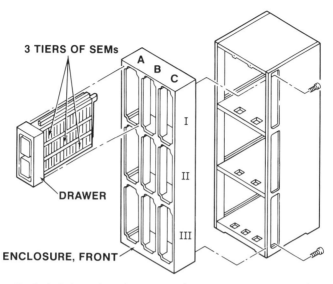

Figure 4.25 Exploded view of an electronic cabinet. (From Winters & Chapanis, 1986. Reprinted by permission.)

was to conduct tests on a simulated SEM and cabinet, with 102 subjects who matched the Navy population in age, height and weight, and with an enclosure and procedures matching the operational conditions. A force gauge registered the amounts of force that the men could exert with their thumbs on simulated SEMs in 27 different locations. Figure 4.26 shows a part of the data for only three locations. The criterion for acceptance was that 95 percent or more of the men should be able to exert the design force. In some locations, for example, in the upper tier of the center drawer (Figure 4.25) the task was easily accomplished. In the bottom-left drawer about 90 percent of men could exert the required force, but in the top-center location only about 50 percent of the men could do so. In fact, in more than 50 percent of the locations (14 of 27), fewer than 95 percent of men could exert the design force.

This was one of those satisfying instances in which a human-factors question was asked *before* design was finalized. The simulation was constructed, tests were run, the results analyzed, and recommendations for redesign made in only a few months time.

Other Examples. You can find in the literature examples of studies that are conceptually similar to the one I have just described, but that have used different methods of simulation and of evaluation. I shall cite only one. Using subjects drawn from a working population, Klemmer and Haig (1988) tested models of telephone handsets to find the weight and balance that was most acceptable. Their evaluation methods were psychophysical: subjects made paired comparisons and absolute judgments of 11 different handsets. They concluded that the optimum handset should weigh about 176 g and should be balanced between transmitter and receiver.

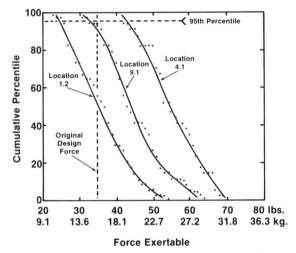

Figure 4.26 Forces exertable by 102 men at three positions of a simulated cabinet. (Constructed from unpublished data in Winters & Chapanis, 1986)

As an exercise, prepare on paper a scaled-down representation of an automated bank teller machine. Be sure to show dimensions that would accommodate 90 percent of the potential user population. Identify all displays, controls, and supplementary features, such as apertures for the receipt of deposits and delivery of funds. Exercise the simulation by "walking through" the actions a customer would make in carrying out typical transactions.

Usability Testing   Simulation is also involved in the much more sophisticated tests conducted in usability laboratories. These tests are typically used in software and computer system development, but they have also been used in the evaluation and development of other products, for example, hand-held controllers for CD players, steam irons, and telephone sets (de Vries et al., 1994). Usability laboratories have made a relatively recent appearance on the human-factors scene; almost all have been constructed since 1980. They consist of two or more rooms, at least one of which is a test room and another an observation and control room. The rooms typically have one-way mirrors between them and are elaborately instrumented with equipment such as audio equipment, video cameras, scan converters, monitors, and computer programs that allow the evaluator to observe the person undergoing the test and to record everything subjects say or do—even every keystroke that is made and the timing of keystrokes.

Evaluations are typically made after requirements have been defined, a task analysis has been performed, and an initial design has been specified. Based on the task analysis, realistic tasks are brought into the laboratory for subjects to work on. Environments—for example, offices, living rooms, medical laboratories—are also sometimes simulated. The simulations are, of course, never complete because, among other things, working conditions can never be duplicated exactly. The presence of monitoring and recording equipment serves as a constant reminder to subjects that they are being observed, and the requirement to "think aloud" is not a normal part of working.

Despite their shortcomings, usability evaluations are almost uniformly productive. They can be used to determine:

- Which of two or more specific interface designs is better or best;
- How people navigate through the product and how frequently they get "lost" in doing so;
- The familiarity of words used in the interface;
- The effectiveness of icons used in the interface;
- The effectiveness of the documentation and on-line help;
- Whether users can "walk-up-and-use" the interface;
- Nominal times for learning tasks; and
- Nominal times for performing tasks.

Recommendations for design or redesign follow from the results of the tests. In some laboratories, facilities for rapid prototyping allow the evaluator to change interfaces and to evaluate alternatives rapidly.

A joint special issue (1994, Volume 13, January and April, Numbers 1 and 2) of *Behaviour and Information Technology* is devoted to usability laboratories. The collection of articles in this issue describe and illustrate laboratories in 13 different organizations. The articles show many different layouts, describe the equipment contained in them, elaborate on the procedures for conducting usability evaluations, and discuss the uses and limitations of usability testing.

Prototyping   Prototyping, or rapid prototyping, merits some discussion because it is used so commonly in designing and testing computer interfaces these days. It is essentially a method of simulating the functions and behavior of user–system interfaces with the capability of rapidly changing interface features. The methods range from simple slide presentations to complex interactive systems. Prototyping can be used to (a) verify or validate requirements, concepts, and functional flows or tasks; (b) evaluate alternative designs and workplace configurations; (c) determine compliance with requirements for user or system performance; (d) identify problems of usability or functionality; and (e) generate specifications of requirements for human–computer interfaces.

The benefits of prototyping are that it can (a) reduce the number of design changes; (b) sensitize system designers to problems by virtue of working directly with users; (c) reduce costs and impacts on schedules because users are less likely to reject the system; and (d) improve system performance due to better system architecture and panel design.

Some Pitfalls of Prototyping.   Despite its several potential advantages, prototyping suffers from a number of difficulties. These difficulties are shared by all forms of simulation. First and foremost is that prototypes are simulations. Like all simulations, they are never complete. They never have all the features and characteristics of the systems they simulate. Prototype hardware and software may not be representative of what is, or what will be, in the final system. Environmental conditions may be impossible to simulate, or may be overlooked. In addition, there may not be any experienced users to test until a new system has been in place for a while.

For these and other reasons, it may be difficult to generalize from (1) the controlled laboratory conditions in which prototypes are tested to real life; (2) the prototype to the rest of the system; and (3) prototype users to the ultimate user population.

Prototyping Tools. There are currently available a number of so-called prototyping tools, computer programs that can be used to automate panel design and chain panels into an interactive scenario. Examples are *Demo II, Generalized Application Simulator (GAS) 2.0, Instant Replay, Matrix Layout, The PROTOtyping System (PROTO),* and *Smalltalk/V 286*. All have some usefulness, but no one tool serves all purposes. Some have a specific facility for designing HELP; others do not; some have a facility for error messages; others do not; some permit the creation of user-defined symbols and icons; others do not.

In selecting a tool, there are a few factors that should be considered:

- Can the tool prototype anticipated dialog styles?
- Does the tool support a graphical or textual interface?
- Does the tool have a special facility for error messages?
- Does the tool have the ability to use screens as templates for new screens?
- What hardware is required to run the tool, for example, PC or mainframe?
- Can the tool convert the prototype into reusable code?
- What level of expertise does the tool require, for example, nonprogrammer, programmer?
- Does the tool fit the job that is to be done?

Conducting Prototype Evaluations. The steps in conducting prototype evaluations are similar to those in planning a controlled experiment, a topic considered next. Examples of objective data that may be collected are:

- Times to complete tasks;
- Percentages of users able to complete tasks within a set time;
- Error rates; and
- Numbers of calls for help.

These objective measures are usually supplemented with subjective complaints, ratings of satisfaction, and ratings of ease of use.

Controlled Experimentation

If you have had a course in any of the sciences, you are already familiar with controlled experimentation. You know that it is a powerful method of scientific inquiry and that it takes time, resources, and careful planning. Unfortunately, in the practical business of system development, you rarely have the luxury of conducting elaborate controlled experiments. Keep in mind, too, that controlled experiments are not the best way of collecting many kinds of information that are needed in system development.

Definition Experimentation is a highly controlled and structured version of simulation using controlled observation with the deliberate manipulation of some variables to answer one or more hypotheses.

Inputs Variables or factors of interest may be used from any of the human-factors methods discussed up to this point.

Procedure

- Identify independent, dependent and controlled variables.
- Set-up conditions of test, apparatus, facilities.
- Select tasks for the subjects to perform.
- Prepare instructions and test protocols.
- Select subjects to match some desired population.
- Select an appropriate form of experimental design.
- Determine the observational methods to be used.
- Run the tests.
- Analyze the results statistically.

Products Controlled experiments yield quantitative statements of:

- the effect(s) of some variable(s) or factor(s), the independent variable(s), on others, the dependent variables, in a controlled setting;
- the relationships or correlations between variables or factors; and/or
- differences between alternative configurations, procedures, or environments.

Supplementary Comment I shall not add any further details about controlled experimentation because they are so generally well known and have been discussed in a large number of excellent and comprehensive textbooks.

An Example The example I shall discuss has a couple of interesting features. It was a controlled experiment done not in a laboratory, but in the field with operational equipment, and was done to answer a design question before design was finalized.

Airborne tactical officers and sonar operators on a certain type of anti-submarine warfare (ASW) helicopter monitor up to 20 or so sonobuoys in the water through receivers in the aircraft. At the time this experiment was done, four-channel receivers were used, and the question was: Can mission effectiveness be improved by providing receivers with eight channels instead of four? Common sense would suggest that eight channels would be better, but there was a price to be paid in weight and space for double the equipment. The experiment was done to provide data for a tradeoff analysis, since an answer was not available in the human-factors literature and there was no way to predict an answer by analytic methods.

It so happened that there was a P-3 weapon systems trainer in a Lockheed Electra aircraft with two sonar operator stations side-by-side. Each station had four channels and detection functions similar to those planned for the system under development. Moreover, there was a chair on rails so that an operator could quickly move from one receiver panel to the other. This was the apparatus around which the experiment was designed.

Two independent variables were tested: (a) reception with either four or eight channels; (b) tests made with and without a job aid, a worksheet to help the air tactical officer keep track of which sonobuoys had been monitored and which had not.

Procedures that were standardized (controlled) were: (a) sonobuoys were dropped in a set pattern two minutes apart from an altitude of 5,000 feet with the aircraft flying at 150 knots; (b) targets were introduced suddenly on randomly selected targets; and (c) clear target acoustics and signatures were used, since the purpose of the tests was to test time-sharing in channel assignment rather than acoustic detection.

The experimental design was technically a multivariate, repeated-measures design with one replication. Four P-3 tactical officer and sonar operator teams each ran two missions under each of four conditions: (a) four channels, without job aid; (b) four channels, with job aid; (c) eight channels, without job aid; (d) eight channels, with job aid.

Figure 4.27 shows the results obtained with three dependent measures chosen for their operational significance: mean time to detection, mean errors per mission, and deviation from optimum monitoring time per sonobuoy. The first two measures were self-evident, the third requires a few words of explanation. Sonobuoys are analog devices, that is, they fill up with information over time and the best time to retrieve that information is just after a sonobuoy has filled. Once information has been retrieved (emptied) from a sonobuoy, it is available to start accumulating new information.

Considering the small numbers involved in this field study, the results have a satisfactory degree of consistency and support two recommendations: (1) Design for eight-channel operation, and (2) provide a job aid.

The sequel to this story is that, after the recommendations had been made, a vendor discovered how to use new technology to make lighter, eight-channel receivers at no extra cost, and with no increase in physical dimensions. This was a satisfying conclusion.

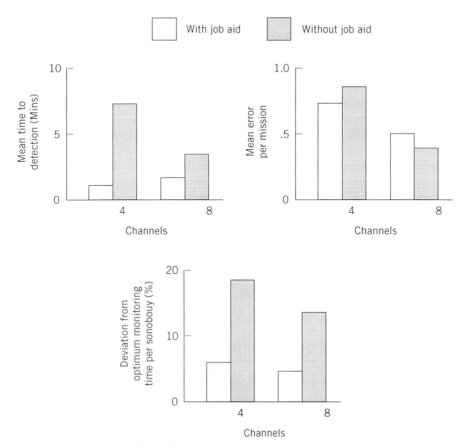

Figure 4.27 Results obtained in an experiment to test two systems of monitoring sonobuoys in IBM's LAMPS project. (Reproduced by permission from Figure HFM-63, IBM 1993)

As an exercise, design an experiment to test whether to use a CRT, digital readout, or voice output to instruct customers in the use of an automated bank teller machine. Be sure to specify such things as: the characteristics of each of the devices that are the independent variables, the variables to be controlled, tasks to be performed, dependent measures to be used, and subjects to be tested. Justify each of the decisions you make in devising your experimental plan.

Operational Sequence Analysis

Operational sequence analysis was derived from methods engineering and adapted to the analysis and description of systems (Kurke, 1961). It is one of the most powerful techniques available to the human-factors specialist. Although sometimes cumbersome and expensive to construct because of the amount of data that must be collected and integrated to construct them, operational sequence diagrams (OSDs) are nonetheless one of the cheapest and quickest ways to simulate systems. They are less expensive than mockups, prototypes, and computer programs that attempt to serve the same purpose. Moreover, OSDs allow one to visualize easily interrelationships between operators and between operators and equipment, spot interface problems and explicitly identify decisions that might otherwise go unrecognized.

Definition Operational sequence analysis is a method that combines events, information, actions, decisions, and data to provide a graphic presentation of operator tasks as they relate sequentially to both equipment and other operators. It is particularly appropriate for the analysis and description of highly complex systems requiring many time-critical information–decision–action functions among several operators and items of equipment.

Inputs Data from all prior analyses are used in the construction of OSDs.

Procedure The leftmost column of the diagram is reserved for a time scale and successive columns show external inputs, operators, machines, and external outputs. It is generally convenient to place in adjacent columns operators and machines with which they interface. The flow of events (actions, functions, decisions) is then plotted from top to bottom against the time scale on the left using special symbology (Figures 4.28 and 4.29). Interactions among events are identified and plotted.

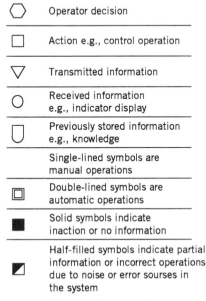

Figure 4.28 Symbology proposed by Kurke (1961) for operational-sequence diagrams (OSDs). (Copyright by the Human Factors and Ergonomics Society, Inc. and reproduced by permission)

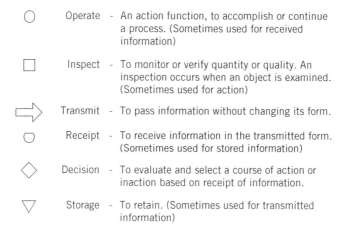

Mode of transmission and receipt may be indicated by a code letter within the symbols.

V— Visual
E— Electrical/electronic
S— Sound (verbal)
IC— Internal communication
EX— External communication

T— Touch
M— Mechanically
W— Walking
H— Hand deliver

Figure 4.29 Symbology for OSDs in DOD-HDBK-763 (1987).

Product The product is a time-based chart showing the functional relationships among system elements, the flow of materials and/or information, the physical and sequential distribution of operations, the inputs and outputs of subsystems, the consequences of alternative design configurations, and potential sources of human–system difficulties.

Supplementary Comment Once again there is nothing sacred about either of the two sets of symbols in Figures 4.28 and 4.29 and, as you see, the symbols in 4.29 refer to some of those in 4.28. I like some features of each of the two. The important point is to be sure to indicate clearly what symbology you are using and then be consistent in your use of it.

An Example Kurke (1961) has used OSDs to compare two ship collision avoidance systems, a manual system and a system that uses a computer to do most of the calculations (Figure 4.30). Let's consider the manual system first (on the left). When the watch officer discovers another ship in the area, he must judge whether it represents a threat (is on a collision course with his own ship). If it appears that the other ship might be a threat, the officer plots its course relative to that of his own ship to determine whether the two ships are on a collision course. If the officer decides to maneuver his ship to avoid the other one, a new course is plotted and the

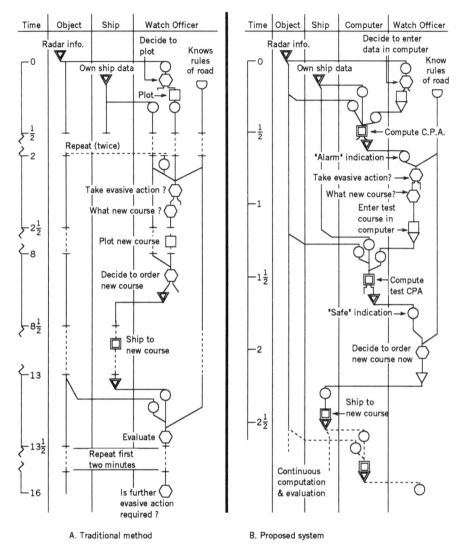

Figure 4.30 OSDs for two ship collision-avoidance systems. (From Kurke, 1961. Copyright by the Human Factors and Ergonomics Society, Inc. and reproduced by permission)

officer's own ship is ordered to a new course and heading. After the change the relative course of the other ship is replotted and the collision threat reevaluated to determine if any further evasive action is necessary.

With the computer system (on the right in Figure 4.30), the course of the other ship is entered into a computer which continually evaluates the possibility of a collision. When the evaluation indicates a collision course, the computer triggers an alarm which alerts the watch officer who enters a new test course into the computer.

It figures out the closest point of approach (CPA) and decides whether that is safe. If the computer decides that it is safe, it so informs the watch officer who then orders the ship to the new course. Throughout all this activity the computer continually computes and evaluates the situation.

As you can see from the illustration, this scenario is completed much more quickly with the computerized system than with the manual and it appears on the diagram as though it were accomplished very quickly. Consider, though, that if the watch officer had to enter a number of items of information (latitude, longitude, present compass heading, new compass heading, present speed, new speed) and if the information had to be entered by means of a menu dialog, this transaction could conceivably take considerably longer than a few seconds and could be the source of inadvertent errors. This points to one place where particular attention should be given to the way this information is entered into the computer. Can you see some other places where system problems could arise?

Workload Assessment

We all, at one time or another, have experienced varying degrees of workload. If you are an experienced driver, cruising along a divided interstate highway on a clear day with moderate traffic constitutes a light workload. Let a thunderstorm appear with such a heavy downpour that your windshield wipers can hardly keep up and your workload is greater. Change the locale to the center of a strange city at rush hour, with cars and pedestrians crowding you from all directions and the workload is dramatically increased.

One human-factors goal in system development is to keep operator workloads at reasonable levels and to ensure that workloads are distributed equitably among operators. Unfortunately, like fatigue, workload is hard to measure quantitatively because, as a report from the National Research Council (Huey & Wickens, 1993, p. 55) says, workload has no "generally accepted definition, standardized procedures and units of measurement, or an absolute standard against which to compare a particular task or candidate measures." Nonetheless, as a practical matter, the human-factors professional engaged in system development is almost certain to be required to get some sort of an estimate of workloads that operators and maintainers will experience.

Generally speaking, workload increases (a) the more things you have to do; (b) the shorter the time you have in which to do them; (c) the greater the difficulty of the tasks; and (d) the more stressful the conditions under which they have to be done. The many attempts to measure workload fall into three major classes: (a) performance measures, measurements of performance in primary or secondary tasks; (b) physiological measures, for example, heart rate, respiration rate, muscle tension; and (c) subjective estimates. Most are descriptive, that is, they measure, or attempt to measure, workloads experienced by operators while they are engaged in some activity or have just completed some activity. As I said earlier, however, in system development the aim is to *predict* performance before a system has been developed and is available. What follows are some suggestions about ways of approaching this problem.

Definition Workload assessment, or workload analysis, is a procedure for appraising operator and crew task loadings or the ability of personnel to carry out all assigned tasks in the time allotted or available.

Inputs Task time, frequency, and precision data are used from any or all of the following: studies of similar systems, task analyses, timeline analyses, controlled experiments, observations of operational systems, or simulations of them, and human-factors data stores. These sources of information are supplemented with judgments and estimates by knowledgeable experts.

Procedure One method of predicting workloads (recommended in DOD-HDBK-763) is to estimate the time required to perform a task or tasks, divided by the time available or allotted to perform it or them. The times required to do tasks may be estimated from data obtained in one or more of the preceding analyses, for example, tasks performed in similar systems, task analyses, timeline analyses. The ratios so obtained are plotted on a chart of time on the job.

An extremely simple measure that is the reverse of the above was used with apparent success by McLeod and Sherwood-Jones (1992). For successive time samples they computed the proportion of the time during which the operator was not busy. This kind of measure could presumably be used predictively by using data from the inputs above.

As a practical predictive approach, Shafer (1987) has proposed that workload be defined as the number of things an operator has to do within any particular time period modified by their levels of difficulty, that is:

$$\text{Workload} = f(\text{things to do})(\text{level of difficulty})$$

The things to do are physical (control activations), mental (things in memory, mental tasks, information load), visual (things to see, visual inputs), vocal (voice commands, communication), or auditory (communications or signals received), and they can be counted with more or less precision from task analyses, for example. Levels of difficulty are physical (precision of movement), mental (precision of recall, precision of computation, level of ambiguity), visual (ambiguity of presentations), vocal (precision of speaking), and auditory (signal-to-noise ratios). These levels of difficulty may be estimated by knowledgeable experts or by human-factors professionals.

Sometimes expert ratings of workloads in actual systems or simulations of them may be sufficient. Some justification for this simple approach comes from Wierwille and Eggemeier (1993, p. 276) who reviewed a number of workload measurements techniques and concluded that "Research has demonstrated clearly that rating instruments are among the most sensitive, most transferable, and least intrusive techniques for workload estimation."

Products However measured, the products are quantitative assessments of estimated workloads for particular tasks and specific times. If workload estimates are within acceptable limits, hardware and software design requirements can be more precisely stated. If the workload estimates indicate workload overloads, alternative allocations of functions and allocations of some tasks to other operators are required. Major underloads suggest the need for alternative allocations of functions or assignment of additional tasks.

An Example The example I have chosen illustrates the use of experts to provide estimates of workloads using a real system and a simulation of one. Figure 4.31 shows workload estimates made by helicopter pilots who were given a functional flow for a typical mission and asked to rate how much effort they thought it would take to fly the mission as a single pilot in their present helicopter. A simple numerical rating scale was used and the workload designations—high, medium and low—are arbitrary. These same pilots then viewed a video tape of the same mission in which many of the functions had been automated. Following the viewing they again rated the workload they thought they would experience with the new helicopter. As you see, the ratings were much lower. After the helicopter had been built, the human-factors team had the rare opportunity to validate their predictions. Workload measures made during several phases of flight with the new system agreed surprisingly well with the subjective estimates that had been made earlier (Shafer, 1987).

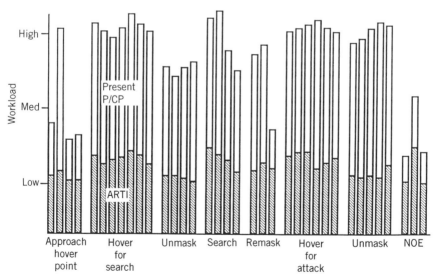

Figure 4.31 Pilot ratings of how difficult it would be to fly their present helicopters as a single pilot (*clear bars*) and an automated Advanced Rotorcraft Technology Integration (ARTI) helicopter (*striped bars*). The major phases of flight are identified at the bottom (*NOE* = nap-of-the-earth flight); each bar above major phases is for a separate subactivity. (Reproduced by permission from Figure HFM-67, IBM 1993)

Supplementary Commentary As you may infer, without a suitable metric for measuring workload, it is difficult to say exactly what constitutes acceptable limits, and what is overload or underload. Once again, you have to fall back on the judgements of experts. A final note: Workload estimates are only relative. All we can say with certainty is that A is a heavier or lighter workload than B. Though some measures may have the appearance of precision, they are nonetheless very rough, and subject to large errors of estimation.

For more thorough discussions of workload and its applications, consult Eggemeier and Wilson (1991), Gopher and Donchin (1986), Hart and Wickens (1990), or O'Donnell and Eggemeier (1986).

SUMMARY

In this chapter I have discussed in some detail a number of human-factors methods and procedures that are used to (1) analyze systems; (2) provide data about human performance for the evaluation of alternative designs, allocation decisions, trade-off studies, and design; (3) make predictions about human–system performance; and (4) evaluate whether the performance of human–machine systems meets design criteria. These methods may be used throughout all phases of system development but with increasing amounts of specificity and detail as development progresses. Although I did not say so earlier in the chapter you should recognize that these methods take time and that they need to be used early enough to affect design.

The methods in this chapter have certain features that need emphasizing:

1. They are not interchangeable or substitutes for one another. Activity analyses and critical incident studies, for example, each yield different kinds of information and neither is like task analyses or link analyses.
2. They build on one another. OSDs, simulations, and controlled experiments use information from task analyses and task analyses, in turn, are done on tasks that have been allocated to humans during functional allocation.
3. With few exceptions (functional flow analyses, controlled experimentation) the methods do not follow precisely definable procedures. There is a considerable amount of flexibility in exactly how they may be applied.
4. Most of them involve judgment and, for that reason, they are highly dependent on the skill and expertise of the analyst.
5. You will have noticed the frequent mention of inputs from "knowledgeable experts" in this chapter. This is another way of using a principle frequently mentioned in the human-factors literature: involve users in the design process. Experienced, and in some cases inexperienced, operators are a source of valuable information about the strong points and deficiencies of old systems and about ways in which new systems might be used. They can frequently provide insights that even experienced human-factors professionals might not see.

6. The products of these methods are inputs to the specifications and reviews discussed in Chapter 3 and to the system requirements to be discussed in Chapter 8. It's probably safe to say that no system involving people has ever been designed using only guidelines, recommendations, and standards. Despite the weaknesses inherent in many of these methods, the examples I have provided show that they are useful and, indeed, indispensable tools for the development of usable machines and systems.

REFERENCES

Adams, J. A. (1982). Issues in human reliability. *Human Factors, 24,* 1–10.

Blanchard, B. S., & Fabrycky, W. J. (1990). *Systems Engineering and Analysis.* Englewood Cliffs, NJ: Prentice-Hall.

Chapanis, A. (1959). *Research Techniques in Human Engineering.* Baltimore: The Johns Hopkins University Press.

Chapanis, A. (1980). The error-provocative situation: A central measurement problem in human factors engineering. In W. E. Tarrants (Ed.), *The Measurement of Safety Performance* (pp. 99–128). New York: Garland STPM Press.

Chapanis, A., & Shafer, J. B. (1986). Factoring humans into FSD systems. *IBM Technical Directions, 12(1),* 15–22.

Christensen, J. M. (1949). Arctic aerial navigation: A method for the analysis of complex activities and its application to the job of the arctic aerial navigator. *Mechanical Engineering, 71(1),* 11–16, 22.

de Vries, G., van Gelderen, T., & Brigham, F. (1994). Usability laboratories at Philips: Supporting research, development, and design for consumer and professional products. *Behaviour and Information Technology, 13,* 119–127.

DOD-HDBK-763. (27 Feb. 1987). *Human Engineering Procedures Guide.* Washington, DC: Department of Defense.

Eggemeier, F. T., & Wilson, G. F. (1991). Performance and subjective measures of workload in multitask environments. In D. Damos (Ed.), *Multiple-task Performance* (pp. 217–278). London: Taylor and Francis.

Endsley, M. R., & Rodgers, M. D. (1994). Situation awareness information requirements analysis for en route air traffic control. In *Proceedings of the Human Factors and Ergonomics Society 38th Annual Meeting* (pp. 71–75). Santa Monica, CA: Human Factors and Ergonomics Society.

Fitts, P. M., & Jones, R. E. (July 1, 1947). *Analysis of Factors Contributing to 460 "Pilot Error" Experiences in Operating Aircraft Controls.* Army Air Forces Air Materiel Command, Engineering Division, Aero Medical Laboratory, Wright-Patterson Air Force Base, Ohio, Report No. TSEAA-694-12. (Reprinted as Paper 8, Pages 332–358, in H. W. Sinaiko (Ed.), *Selected Papers on Human Factors in the Design and Use of Control Systems.* New York: Dover Publications, 1961.)

Fuld, R. B. (1993). The fiction of function allocation. *Ergonomics in Design, January,* 20–24.

Glass, J. T., Zaloom, V., & Gates, D. (1991). A micro-computer-aided link analysis tool. *Computers in Industry, 16,* 179–187.

Gopher, D., & Donchin, E. (1986). Workload—an examination of the concept. In K. R. Boff, L. Kaufman, and J. P. Thomas (Eds.), *Handbook of Perception and Human Performance: Volume II. Cognitive Processes and Performance* (pp. 41/1–41/49). New York: John Wiley & Sons.

Hart, S. G., & Wickens, C. D. (1990). Workload assessment and prediction. In H. R. Booher (Ed.), *MANPRINT: An Approach to Systems Integration* (pp. 257–296). New York: Van Nostrand Reinhold.

Huey, B. M., & Wickens, C. D. (Eds.) (1993). *Workload Transition*. Washington, DC: National Academy Press.

IBM (July 1993). *Human Factors in Systems Engineering Training Course*. Bethesda, MD: Author.

Jones, E. R., Hennessy, R. T., & Deutsch, S. (Eds.) (1985). *Human Factors Aspects of Simulation*. Washington, DC: National Academy Press.

Kirwan, B. (1987). Human reliability analysis of an offshore emergency blowdown system. *Applied Ergonomics, 18*, 23–33.

Kirwan, B., & Ainsworth, L. K. (1992). *A Guide to Task Analysis*. London: Taylor and Francis.

Klemmer, E. T., & Haig, K. A. (1988). Weight and balance of a new telephone handset. *Applied Ergonomics, 19*, 271–274.

Kurke, M. I. (1961). Operational sequence diagrams in system design. *Human Factors, 3*, 66–73.

McLeod, R. W., & Sherwood-Jones, B. M. (1992). Simulation to predict operator workload in a command system. In B. Kirwan & L. K. Ainsworth (Eds.), *A Guide to Task Analysis* (pp. 301–310). London: Taylor and Francis.

Meister, D. (1985). *Behavioral Analysis and Measurement Methods*. New York: John Wiley & Sons.

Meister, D. (1993). Non-technical influences on human factors. *CSERIAC GATEWAY, 4(1)*, 1–3.

Munger, S. J., Smith, R. W., & Payne, D. (Jan. 1962). *An Index of Electronic Equipment Operability: Data Store*. Report AIR-C43-1/62-RP(1). Philadelphia: American Institutes for Research.

O'Donnell, R. D., & Eggemeier, F. T. (1986). Workload assessment methodology. In K. R. Boff, L. Kaufman, & J. P. Thomas (Eds.), *Handbook of Perception and Human Performance. Volume II: Cognitive Processes and Performance* (pp. 42/1–42/49). New York: John Wiley & Sons.

Pew, R. W., Hoecker, D. G., Miller, D. C., & Walker, J. H. (June, 1979). *User Interface Requirements for Computer-Assisted Service Delivery*. Report Number 4045. Cambridge, MA: Bolt, Beranek and Newman.

Phillips, M. D., & Melville, B. E. (1987). Analyzing controller tasks to define air traffic control system automation requirements. In *Human Error Avoidance Techniques Conference Proceedings* (pp. 37–44). Warrendale, PA: Society of Automotive Engineers.

Price, H. E. (1985). The allocation of functions in systems. *Human Factors, 27*, 33–45.

Rhodes, W. (1984). Human factors in office design. *Canadian Office, September*, 24–25.

Rogers, W. A., Gilbert, D. K., & Cabrera, E. F. (1994). An in-depth analysis of automatic teller machine usage by older adults. In *Proceedings of the Human Factors and*

Ergonomics Society 38th Annual Meeting (pp. 142–146). Santa Monica, CA: Human Factors and Ergonomics Society.

Safren, M. A., & Chapanis, A. (1960). A critical incident study of hospital medication errors—Part 1. *Hospitals, 34(9),* 32–34.

Safren, M. A., & Chapanis, A., (1960). A critical incident study of hospital medication errors—Part 2. *Hospitals, 34(10),* 53.

Shafer, J. B. (1987). Practical workload assessment in the development process. In *Proceedings of the Human Factors Society 31st Annual Meeting* (pp. 1408–1410). Santa Monica, CA: Human Factors Society.

Shaffer, M. T., Shafer, J. B., & Kutche, G. B. (1986). Empirical workload and communications analysis of scout helicopter exercises. In *Proceedings of the Human Factors Society 30th Annual Meeting* (pp. 628–632). Santa Monica, CA: Human Factors Society.

Swain, A. D., & Guttman, H. E. (Aug, 1983). *Handbook of Human Reliability Analysis with Emphasis on Nuclear Power Plant Applications.* Report NUREG/CR-1278, SAND80-0200, RX, AN. Washington, DC: Nuclear Regulatory Commission.

Wierwille, W. W., & Eggemeier, F. T. (1993). Recommendations for mental workload measurement in a test and evaluation environment. *Human Factors, 35,* 263–281.

Winters, J. A., & Chapanis, A. (1986). Thumb push forces exertable by free-standing subjects. *Ergonomics, 29,* 893–902.

Yamanoi, N., Yajima, K., Aoki, T., Aoki, K., Kinoshita, S., Tanaka, H., & Furukawa, T. (1982). Link analysis of electrocardiograph manipulation and application for design of electrocardiograph. In K. Noro (Ed.), *IEA'82: The 8th Congress of the International Ergonomics Association* (pp. 180–181). Tokyo: Japan Ergonomics Research Society.

Chapter 5

Human Physical Characteristics

In Chapter 1 I stated that human factors applies information about human abilities, human limitations, and other human characteristics to design. In other words, human abilities, limitations, and other characteristics are the raw materials with which human factors deals. What are those characteristics?

Thousands of books have been written about human anatomy, physiology, psychology, anthropology, and sociology and no single book, much less a couple of chapters, can even begin to summarize all that material adequately. This chapter and the next discuss only a few of the most important human attributes that interest and often perplex the engineer, and make the job of the human-factors professional so difficult and challenging. The material in this chapter is highly selective, concentrating on those characteristics that have direct relevance to design.

My approach is different from what you will find in many other textbooks because I first describe some very general characteristics that distinguish people from the hardware and software with which engineers work. The remainder of the chapter is concerned with the human being as a biological and structural entity, and deals with the following questions: How do we describe human beings dimensionally? What are they made of? What are the mechanisms of the body that keep it alive and functioning? What are some of the major environmental determinants that affect human performance?

THE STARTING POINT

I start with some very basic, but fundamental characteristics about people—characteristics so basic that we tend to overlook them. Yet they pervade all human-factors activities and thus can never be ignored.

People Are Complicated

Perhaps the most important single thing that can be said about people is that they are complicated. In fact, no one is likely to challenge the assertion that people are more complicated than any other physical system in existence. Modern computers perform complex arithmetic and logical operations at speeds that stagger the imagination. Yet, compared with the human body, even the most complex computer is a simple device. Here is what the eminent mathematician, John von Neumann, once had to say about this subject:

> In comparing living organisms, and, in particular, that most complicated organism, the human nervous system, with artificial automata, the following limitation should be kept in mind. The natural systems are of enormous complexity . . . With any reasonable definition of what constitutes an element, the natural organisms are very highly complex aggregations of these elements. The number of cells in the human body is somewhat of the general order of 10^{15} or 10^{16}. The number of neurons in the central nervous system is somewhere of the order of 10^{10}. We have absolutely no past experience with systems of this degree of complexity. All artificial automata made by man have numbers of parts which by any comparable schematic count are of the order of 10^3 to 10^6. In addition, those artificial systems which function with that type of logical flexibility and autonomy that we find in the natural organisms do not lie at the peak of this scale. The prototypes for these systems are the modern computing machines, and here a reasonable definition of what constitutes an element will lead to counts of a few times 10^3 or 10^4 elements. (von Neumann, pp. 2–3)

Though this was written several decades ago, during which time computers have become vastly more complex than those von Neumann knew, the validity of his basic comparison remains unchallenged.

One troublesome aspect of human complexity are the paradoxical motivations that sometimes cause people to act in unexpected or perverse ways. They not only use, but also misuse, and sometimes abuse equipment. This is particularly important these days, because one of the arguments used successfully in many product-liability law suits is that designers should have anticipated the ways in which equipment could be used improperly and should have designed against those contingencies.

Those same complex motivations and behavior patterns also make it difficult to study people, because they sometimes confound experimenters by doing completely unexpected things. Indeed, they may even deliberately thwart the purposes of a study by their actions. Although we don't have foolproof ways of circumventing these problems, by being constantly aware of them we can sometimes mitigate or compensate for their effects.

There Are No Design Drawings for People

In the discussion that followed the paper from which von Neumann's quotation was taken, the psychiatrist Warren McCulloch had this to say:

> I confess that there is nothing I envy Dr. von Neumann more than the fact that the machines with which he has to cope are those for which he has, from the beginning, a blueprint of what the machine is supposed to do and how it is supposed to do it. Unfortunately for us in the biological sciences . . . we are presented with an alien, or enemy's machine. We do not know exactly what the machine is supposed to do and certainly have no blueprint of it. (von Neumann, p. 32)

Engineers and designers work with design drawings. The drawings for a large computer are very complicated and may run to thousands of pages, but there *are* drawings. There are no comparable design drawings for people. To be sure, we have *models* of one or another feature of human beings, their characteristics and behavior, and one can find a number of such models in the human-factors literature and in this book. But models are only approximations, usually crude approximations, and are always incomplete. They do not approach the precision of engineering design drawings. As a result, in human factors we are faced with the problems of trying to arrive at generalizations about organisms whose structures are only approximately delineated and whose mechanisms are imperfectly understood.

People Differ

Not only do we not have a blueprint of a human being, but also consider that, such a blueprint, if one were ever fortunate enough to devise one, would have to be different for each and every person on earth, meaning something on the order of 6 to 7 billion blueprints.

Many of the ways in which people differ are easy to observe and measure. It's obvious that people differ in anthropometric dimensions—in height, weight, arm reach, and in the hundreds of different ways in which bodies can be measured—and that the range of differences is large. You only have to look around you to be impressed by how much we differ. Not so readily apparent, but no less important, are large differences among people in their sensory capacities, motor abilities, mental abilities, personalities, and attitudes. Strange as it may seem, some differences, such as color-vision defects, may be undetected for long periods of time, even by persons possessing the deficiency.

The Range of Human Capacities A systematic study of the range of human capacities was made a number of years ago (Wechsler, 1952), but the data are still helpful. Wechsler collected data from a number of sources in which large numbers of people had been tested. Since it is virtually impossible to determine an absolute range, for example, the shortest to tallest person alive, he defined the range as the distance between the 0.2% and 99.9% points in a distribution of measurements.

Using his criterion of range and his definitions of traits, capacities, and abilities, the ranges of 18 linear traits, for example, sitting height, stature, interpupillary distance, vary from 1.16:1 to 1.44:1. The ranges of 17 motor capacities, for example, tapping, card-sorting, simple reaction time, vary from 1.65:1 to 3.43:1. The ranges of 17 perceptual and intellectual abilities, memory span for digits, CFF (critical flicker frequency), vocabulary, general intelligence, vary from 1.45:1 to 3.87:1. The ranges of 16 productive capacities, for example, loom operation, typing, checking grocery orders, polishing spoons, vary from 1.40:1 to 5.00:1. More recent measurements show that these ranges are, in some cases, conservative.

Some Examples of Individual Differences Figure 5.1 shows absolute threshold measurements, the faintest light that a dark-adapted eye can see, measured in different parts of the eye. The data show a number of interesting things about the eye and its capacities, for example, the blind spot—the region where nerve endings leave the eye; the relative insensitivity of the fovea—the central part of the eye; and the region of greatest sensitivity, about 20° to 40° from the fovea; but look at the range of sensitivities. The mean plus- and minus-two standard deviations includes roughly the middle 95 percent of people. Since the ordinate of the chart is in logarithmic units, the difference between any two units represents a factor of 10 to 1. In the fovea, then, measurements for the middle 95 percent of people indicate a range of about 100 to 1 in sensitivity. The range is even greater in the periphery of the eye.

Figure 5.2 shows data on some individual differences in muscle strength. Under the conditions in which these measurements were made (subjects seated and re-

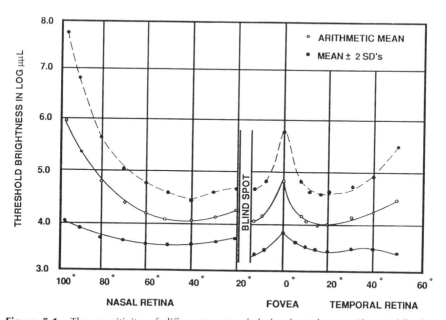

Figure 5.1 The sensitivity of different parts of dark-adapted eyes. The middle line represents average values obtained on 101 subjects; the upper and lower lines are average values plus and minus two standard deviations. (Data of Sloan, 1947, after Chapanis et al., 1949. Published with permission from the *American Journal of Ophthalmology*. Copyright by the Ophthalmic Publishing Company.)

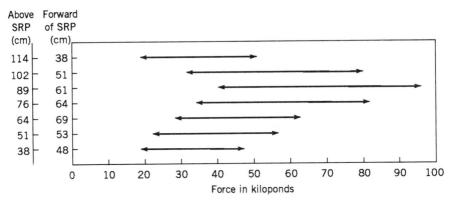

Figure 5.2 The ranges of maximum static push forces (5th to 95th percentiles) exerted by 55 young male subjects. Measurements were made with a handle assembly in the center line of the seat but at various locations above and forward of the Seat Reference Point (SRP). (Constructed from data of Laubach in NASA Reference Publication 1024)

strained with a lap belt), greatest forces could be exerted at about 89 cm above the seat reference point. Once again, however, notice the ranges. They are very large. Keep in mind, too, that these are ranges from the 5th to 95th percentiles and so do not include 5 percent of the weakest and 5 percent of the strongest men. In addition, these measurements were made on young, healthy males. The ranges would have been much greater if they had included female subjects and persons of various ages.

Figure 5.3 shows individual differences in short-term memory. In this study, 52 high school students were given unlimited time to reproduce from memory 1,000 eight-digit numbers, one at a time, after seeing each number for five seconds. Fewest errors were made in the first digit position, errors increased in successive digit positions, reaching a maximum at position seven and decreased at the last digit position—a general finding called the *serial position effect*, important in the design of codes involving short-term memory. Notice, however, the very large range of errors committed. At digit position seven, for example, the best subject made only 17 errors, the poorest 530, a ratio of over 31:1!

One more example. Figure 5.4 is a distribution of intelligence quotients measured with the Wechsler Adult Intelligence Scale (WAIS) on a carefully selected sample of 1,880 adults. We don't have any test that really measures intelligence. Indeed, psychologists are not even sure what "intelligence" really is. However, the WAIS test is one of the best available measures of mental ability. It is a difficult test to administer, because it takes several hours and is done one-on-one, that is, a single examiner tests a single examinee. That means that, in administering the test and scoring results, a trained examiner can take into account special difficulties an examinee might have, for example, language handicaps or emotional problems.

The most important point I would like to make about these data is that engineers and designers typically fall into the *superior* and *very superior* categories of mental ability. Among the various professions, engineers rank among the highest in general

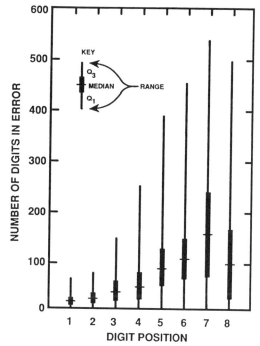

Figure 5.3 Errors made by 52 high-school students who each reproduced 1,000 clearly legible eight-digit numbers, one at a time, after viewing each for five seconds. (Unpublished data from the study by Chapanis & Moulden, 1990)

mental ability. In addition, engineers are facile with numbers and understand mechanisms. Yet they must design for large numbers of people who do not have these abilities.

To illustrate, adults at the lower limit of the "average intelligence" range should be able to do problems like these in their heads*:

1. How many inches are there in two-and-a-half feet?
2. How many oranges can you buy for 36 cents if one orange costs six cents?

but not these:

3. How many hours will it take you to drive 120 miles at the rate of 40 miles per hour?
4. You buy $8.50 worth of groceries and hand the clerk a $20 bill. How much change will you get?

---

*For obvious reasons, these are not exactly the questions used in the test. They are unvalidated approximations used only for illustrative purposes.

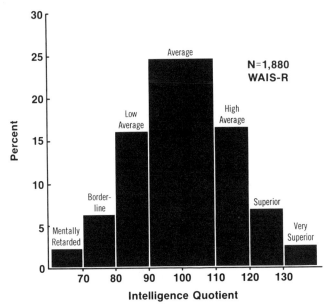

Figure 5.4 Intelligence quotients of 1,880 randomly selected persons on the Wechsler Adult Intelligence Scale. (Based on data in Wechsler, 1981)

At first glance, you might wonder: what is so hard about questions 3 and 4? The difficulties, though they undoubtedly seem trivial to you, are the concept of "rate" introduced in question 3, and the use of decimals in question 4.

As impressive as that simple exercise might be, it still does not cover the full range of mental ability. Keep in mind, too, that half the population is below average in mental ability and designing equipment for people with limited abilities is a challenge that cannot be ignored. For engineers it may be one of the hardest design challenges they face.

The foregoing are sufficient to illustrate one of the most disconcerting things about people—we differ in every conceivable way in which we can describe or measure ourselves. A major concern of human factors is how to design for organisms, no one of which is exactly like any other in physique, ability, mentality, and temperament. The best we can do is to design equipment to fit a *range* of people. In human factors we usually define that range as the difference between the 5th and 95th percentiles. We know that we cannot design to accommodate *everyone*, but if we can accommodate the middle 90 percent of people, then we are doing very well.

Design for the Average? A design approach one sometimes hears is that we should design for the "average person." There are two difficulties with that approach. The first is that, in so doing, the physically smaller 50 percent of users may be unable to reach controls to do their work effectively and the physically larger 50 percent may not have sufficient room to move around or do their work effectively.

The second difficulty is that there may be no such person, that is, a person who is "average" in every conceivable way. In fact, it is difficult to find even a small number of people who are average in more than a few ways. The explanation for the statement lies in the nature of the statistical relationships between variables. Figure 5.5 will help to explain what happens. This is a scatter diagram of measurements on two body dimensions, crotch height and waist circumference, for 6,682 men. If we define "average" liberally as anyone in roughly the middle 25 percent of a distribution, 1,901, or 28.4%, of the men have "average" waist circumferences. In the other dimension, 1,714, or 25.6%, have "average" crotch heights. But only 468, or 7.7%, of the men are "average" in both dimensions. The general rule is that, when the correlation between any two variables is less than perfect (and that is always the case for all correlations between human variables), the numbers of individuals at the intersection of values from the two dimensions is always less, and usually much less, than the numbers of individuals at either value separately.

A result of considering more than two variables was shown in a clever little study by Daniels and Churchill (1952). They started with a sample of 4,063 male Air Force personnel and defined "average" as approximately the middle 30 percent of flyers on any dimension. Table 5.1 shows what happened as they took into account more and more body dimensions. By the time they took 10 dimensions into account, not a single one of the original sample of 4,063 men was left. It may be possible to find individuals who are average in a number of dimensions and characteristics, but, despite what Abraham Lincoln reputedly said, they are not "common."

The results of many studies in the human-factors literature report averages, and many human-factors guidelines and recommendations are based on average data. One should keep the foregoing discussions about individual differences and averages in mind when consulting the literature, or when using current guidelines and recommendations. In addition, one should always be sure to design for the middle 90 percent of users, and not just the "average" ones.

Factors Contributing to Individual Differences A number of variables are responsible for the individual differences seen in the population as a whole. Some of the most important ones are sex, age, national origin, and education.

Sex. Differences in anthropometric and biomechanical characteristics are, of course, prominent between the two sexes. A number of these are illustrated later in the section on anthropometry. Some differences are, however, much more subtle. For example, only about 0.5 percent of women have color-vision defects as compared with about 7 or 8 percent of men—something that should be kept in mind for all kinds of designs in which colors are used. On the whole, however, as more and more women enter the labor force in occupations that had formerly been denied them, it has become clear that they are capable and effective workers, provided that workstations and tasks have been appropriately human-engineered to take account of their structural differences from men.

CROTCH HEIGHT (CM)

WAIST CIRCUMFERENCE (CM)	65.50	67.00	68.50	70.00	71.50	73.00	74.50	76.00	77.50	79.00	80.50	82.00	83.50	85.00	86.50	88.00	89.50	91.00	92.50	94.00	95.50	97.00	98.50	100.00	101.50	TOTALS
127.50						1																				1
125.00																										0
122.50																										0
120.00						1			1																	2
117.50												1		1					1							3
115.00											1	1				1	1									4
112.50										1	1	1	2		2	1	1									9
110.00	1					1		1	1	2			2	4		3	2		1							18
107.50							1	1	4	2		3	1	2	1		1	1								17
105.00			1	1		1		3	3	3	3	4	1	4	1	2	4	1				2				36
102.50							2	4	2	3	5	7	7	11	3	3	2	1	4	2						56
100.00					1		2	4	5	5	4	11	8	7	4	10	4	4	1	2			1			73
97.50						1	2	4	6	8	11	14	7	13	7	8	3	5		4						93
95.00			1		2	3	3	3	7	13	9	21	19	19	19	11	14	5	4	2	1	1				158
92.50				1	3	4	3	11	11	12	12	32	24	23	23	25	5	8	2	5	1	1				206
90.00		1				6	5	4	11	20	29	44	33	33	28	27	15	11	7	5	1	2	1	1		284
87.50		1		2		1	9	12	27	21	37	45	55	40	39	42	27	19	11	7		1	1			397
85.00				2	2	5	9	13	20	39	49	65	63	60	55	41	32	29	14	4	2		1			503
82.50					3	6	16	10	32	37	73	70	97	74	84	49	41	24	16	7	4	1	2		1	647
80.00			2	1	2	5	13	20	38	62	90	119	107	110	94	86	72	36	27	10	8	4		2		908
77.50				2	1	8	4	19	43	60	122	100	119	140	91	87	57	43	23	15	9	1	1	1		937
75.00				1	2	2	13	22	40	62	72	131	139	114	102	89	57	48	23	19	8	4	1			964
72.50			1	2	2	6	5	18	38	50	62	84	96	104	80	54	48	18	12	12	4	3				699
70.00			1		1	6	12	12	36	40	41	56	58	63	15	33	17	19	3	1	2	1		1		448
67.50				1		3	7	9	5	10	18	17	27	15	20	13	6	5	4							160
65.00					2	6	2	3	3	7	5	6	1	4	2	2	2	5	1							46
62.50						1				1		1	3		2											8
60.00											1	1	1													3
TOTAL	1	2	6	13	21	64	110	174	333	468	645	834	872	842	702	587	411	272	154	95	40	21	8	5	1	6682

28.4%

25.7%

7.7%

Figure 5.5 Scatter diagram of two body dimensions on 6,682 men. If we say that men who fall into about the middle 25 percent of each dimension are average, only about 8 percent will be "average" in both dimensions. (Based on data from White & Churchill, 1971)

TABLE 5.1 Results of a Search for an "Average" Man*

	Percent of Original Sample
• Of an original sample of 4,063 men, 1,055 were about average in height.	25.9
• Of the 1,055 men who were about average in height, 302 also had about average chest circumferences.	7.4
• Of the 302 men who were average in height and chest circumference, 143 also had average sleeve lengths.	3.5
• Of the 143 men, 73 also had average crotch heights.	1.8
• Of the 73 men, 28 also had average torso circumferences.	0.69
• Of the 28 men, 12 also had average hip circumferences.	0.29
• Of the 12 men, 6 also had average neck circumferences.	0.14
• Of the 6 men, 3 also had average waist circumferences.	0.07
• Of the 3 men, 2 also had average thigh circumferences.	0.04
• Of the 2 men, neither had an average inseam length.	0.00

*From Daniels & Churchill (1952).

Age. Changes in body structure as people grow and mature are obvious. Equally important, and sometimes not so obvious, are changes in sensory, motor, and intellectual capacities. Figure 5.6, for example, shows average distances at which 18 highway signs could be perceived by three age groups of subjects: young (18–25), middle-aged (40–55), and old (65–79). Older people have decreased sensitivity under dim lighting conditions, especially when there is glare. Similar results have been reported by Freedman, Zador, and Staplin (1993), who found that elderly drivers (ages 75–90) were much less likely than younger drivers to detect objects such as a pedestrian, seated child, or bicyclist through variously tinted rear windows of automobiles.

Another major change in visual capacity that occurs with aging is presbyopia, the inability of the eye to focus clearly on near objects because of decreased flexibility of the lens of the eye. Presbyopia is easily corrected with eyeglasses, but decreased night sensitivity and increased sensitivity to glare are not. All three conditions are relevant for certain kinds of tasks, for example, driving at night.

Figure 5.7 shows hearing losses for females and males as a function of age. You can see that hearing losses are much greater for men than for women, and that hearing losses are much greater at higher frequencies than at lower ones. Much more prominent, however, are the increases in hearing losses that occur with increasing age. It is possible that hearing losses for women will be greater in the future because of their greater exposure to high noise levels in occupations (for example, mining, construction, servicing aircraft) they are now entering. There is also some evidence that hearing losses may be occurring earlier in life, because of

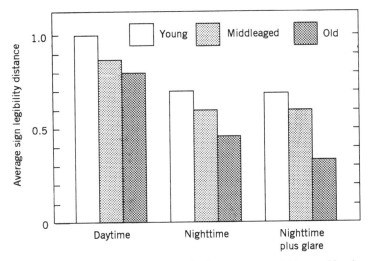

Figure 5.6 Average distances at which 18 highway signs could be read by three groups of subjects under varying conditions of illumination. The greatest distance, for young subjects under daylight conditions, has been arbitrarily set to 1.00, and all other distances have been scaled accordingly. (Constructed from data in Schieber & Kline 1994)

the intense volume levels of music to which many younger people have been exposing themselves. One implication of these data is that, to be detectable by both older and younger persons, alerting and warning devices—for example, smoke alarm sounds (Huey et al 1994)—should use lower frequencies rather than higher ones.

Figure 5.8 shows changes in mental abilities as a function of age. The WAIS test consists of eleven subtests which I have divided here into two groups: the six involving primarily verbal abilities (for example, vocabulary, verbal comprehension, and information) and the five involving performance abilities (for example, digit symbol association, object assembly, and picture completion) for which verbal abilities are relatively unimportant. Scores on the subtests have been normalized so that they are comparable. Notice that performance reaches a maximum at about age 18 years and declines significantly with advancing age. Verbal abilities, on the other hand, reach their peak at about age 30 and, although they also decline with age, the decline is not so rapid.

These data are consistent with findings on expert performance (Ericsson & Charness, 1994). Peak performance for vigorous sports occurs in the twenties, and it is rare for athletes above the age of 30 to reach their best performance, or even to remain competitive with their younger colleagues. On the other hand, in writing—novels, history, philosophy—peaks occur in the 40s and early 50s. For some other types of cognitive tasks, peak performance also occurs in the late 40s and early 50s. In fact, according to Schaie (1994), age decrements in most mental abilities cannot be reliably confirmed before age 60.

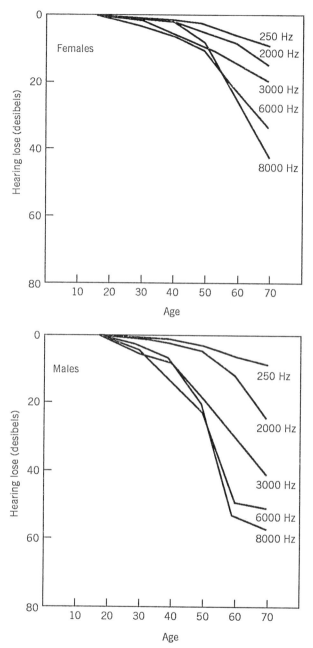

Figure 5.7 Hearing losses at various frequences for males and females as a function of aging. Men suffer greater hearing losses than women, but for both sexes hearing losses increase with age and are greater for higher frequencies. (Constructed from data in Hinchcliffe, 1959).

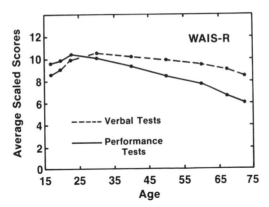

Figure 5.8 Mental abilities as a function of age on two groups of subtests in the Wechsler Adult Intelligence Scale (Revised). Mean abilities decline with age, but peak performance for verbal abilities occurs later and the decline is not as pronounced as for performance tests. (Based on data in Wechsler, 1981).

What conclusions can we draw from the data on aging? They are:

1. There are large individual differences among people at all age groups.
2. Although most abilities decline with advancing age, some people at age 65 or 70 have better functional capacities than some persons who are only 30 years old. [See, for example, Miles (1931); (1933).]
3. We do not age uniformly. Some abilities and capacities change more gradually than others.

These and other effects of age have to be considered in the design of many systems, to take account of the increasing proportion of elderly people in our population. Not only are people living longer but they are also generally healthier than had been the case decades ago. Moreover, laws, such as the Age Discrimination in Employment Act of 1967, say we may not discriminate just because of age. Equipment, workplaces, and tasks must be designed to match the capacities of workers if they are otherwise fit and willing to work. Naturally, living spaces and recreational equipment must also be so designed.

National Origin. Chapter 3 pointed out that differences in national laws and standards must be considered in the design of systems for use in countries other than the United States. When equipment is designed for multinational use, human-factors professionals must take into account a large number of other factors that differ among various ethnic groups: differences among body sizes, in the way people learn, in attitudes toward maintenance and repair, and in the way people perceive graphic representations (Sinaiko 1975; Wyndham 1975). Sometimes even economic and political considerations influence the way equipment is used. Wisner (1985), for example, has described a large number of instances in which advanced technology and modern systems encountered unforeseen difficulties when they were transplanted into developing countries.

Language factors are especially important because the quality of translations of instructions from one language to another is intimately related to work performance. Better translations produce better work, but bad translations can and have resulted in inefficiencies, misuses of equipment, and accidents. Unfortunately, current procedures for getting good translations are complex and time consuming (see, for example, Angiolillo-Bent, 1985). Machine translation helps, but even the best computer programs still cannot produce translations equal to those by skilled human translators.

Education. Differences in educational attainment are among the largest sources of individual differences in skills and abilities. To paraphrase H. G. Wells, the situation these days " . . . becomes more and more a race between education and catastrophe." Computer technology and automation have taken the sweat and tedium out of many jobs, but have made those jobs far more complex and mentally demanding. According to the U.S. Labor Department, 6 percent of jobs in 1984 required workers with the two highest skill levels (see Table 5.2), but for those jobs created between 1984 and 2000 that figure will rise to 13 percent.

Unfortunately, our educational system has not kept pace with those changing requirements. For example, a survey by Coopers and Lybrand in 1987 of more than 300 senior manufacturing executives concluded that "lack of skilled personnel is limiting the implementation of available manufacturing technology such as computer-integrated manufacturing." In commenting about a report titled "What Work Requires of Schools: A SCANS Report for America 2000," Labor Secretary Lynn Martin commented that, "More than half of young people leaving high school won't have the knowledge or foundation . . . to compete in the workplace of the future . . . The world's technological explosion has caught many of our students by surprise."* Finally, a report from the U.S. Department of Education's National Center for Education Statistics, released on September 1, 1993, stated that "Few students in the United States can solve problems that require more than an educated guess . . . Only 16 percent of fourth-graders, 8 percent of eighth-graders and 9 percent of high school seniors could answer math questions requiring problem-solving skills."

Figure 5.9 shows changes in Scholastic Aptitude Test (SAT) scores over a number of years. Many explanations have been advanced to account for these changes, a sophisticated one being that the means have decreased because the tests were administered to a much larger proportion of the population in later years. Since the variances of scores have remained about the same over the years, that explanation is unlikely. For practical purposes, the exact reason for the changes is unimportant. What is important is that the population of potential operators and maintainers that we draw from today is not as well educated as it was in the 1960s. Equipment today must be designed to cope with that reality.

---

*Reported in the *New York Times,* July 3, 1991, p. A17.

TABLE 5.2 Definitions of Skill Levels According to the U.S. Labor Department

Skill Level	Language	Math
6	Reads literature, book and play reviews, scientific and technical journals, financial reports and legal documents. Writes novels, plays, editorials, speeches, critiques.	Advanced calculus, modern algebra, and statistics.
5	Same as level 6, but less advanced.	Knows calculus and statistics, econometrics.
4	Reads novels, poems, newspapers, manuals, thesauri and encyclopedias, prepares business letters, summaries, and reports. Participates in panel discussions and debates. Speaks extemporaneously on a variety of subjects.	Is able to deal with fairly complex algebra and geometry, including linear and quadratic equations, logarithmic functions, and deductive axiomatic geometry.
3	Reads a variety of novels, magazines and encyclopedias, as well as safety rules and equipment instructions. Writes reports and essays with proper format and punctuation. Speaks well before an audience.	Understands basic geometry and algebra. Calculates discount, interest, profit and loss, markup and commissions.
2	Recognizes the meaning of 5,000–6,000 words. Reads at a rate of 190–215 words per minute. Reads adventure stories and comic books, as well as instructions for assembling model cars. Writes compound and complex sentences, with proper end punctuation and using adjectives and adverbs.	Adds, subtracts, multiplies and divides all units of measure. Computes ratio, rate, and percent. Draws and interprets bar graphs.
1	Recognizes the meaning of 2,500 two- or three-syllable words. Reads at a rate of 95–100 words per minute. Writes and speaks simple sentences.	Adds and subtracts two-digit numbers. Does simple calculations with money and basic units of volume, length and weight.

People Change

To add still more to the problems faced by the human-factors professional, people change. Many changes are long-term changes—changes in body dimensions, abilities and skills—as we develop, grow, and age. Other changes are short-term changes and the word *short* here means exactly that, because we change from

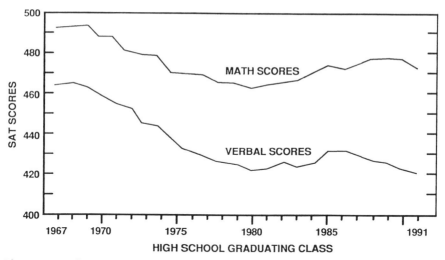

Figure 5.9 Changes in Scholastic Aptitude Test (SAT) scores over the years. (Source: College Entrance Examination Board)

moment to moment. We are different now than we were a moment ago. These short-term changes are due to such things as boredom, fatigue, and lapses of attention. But they are also due to interactions between ourselves and other people, between ourselves and the things we deal with, and between ourselves and our environments. These short-term changes are highly variable. A perplexing response from a computer might intrigue us the first time it appears, but annoy us exceedingly if it continues to reappear. Although you may have successfully obtained money from an automated teller machine a dozen or so times, suddenly one day you can't remember your secret Personal Identification Number. The list could go on and on. What this means is that we must try to generalize about organisms that are annoyingly inconstant.

Everyone Is an "Expert"

Few people would claim to be expert chemists, physicists, or computer programmers if they were not. But all of us think we are experts about human behavior. The reason is not hard to understand. Since we are all human, and have thoughts, feelings, and ways of coping with the world, we tend to generalize from ourselves to other people. We feel "from experience" that we are good judges of other people, and we have "explanations" for why other people behave as they do. We search constantly for working principles that can help us understand why our wives, children, managers, and subordinates behave as they do, why there are so many accidents, why there is so much crime on the streets, and why people don't use equipment the way they are supposed to. In our search we develop strong convictions about the effects of age, sex, race, fatigue, climate, and the environment on human behavior—all largely based on how we perceive and respond to those variables. These convictions are often hasty generalizations, half-truths, prejudices, and even superstitions that are hard to eradicate.

In our meetings we human-factors professionals usually grumble that engineers and designers don't take our advice because they think they are human-factors experts and know it all. However much truth there may be in that complaint, we should acknowledge that we ourselves are also human. For that reason, it behooves us to approach our work with a certain amount of humility and to question our own pronouncements and recommendations to be sure that they are solidly based on fact, and not merely expressions of our own fallible opinions.

ANTHROPOMETRY: THE SCIENCE OF HUMAN DIMENSIONS

The dimensions of the human body—heights, breadths, depths, distances, circumferences, and curvatures, are typically measured with reference to solid identifiable landmarks on bones. These dimensions, and their relationships to one another at skeletal link joints, are useful in the design of workstations and equipment.

Basic Design Options

Equipment, whether it be tools, machinery, workstations, or clothing, can be designed to accommodate body differences in three ways:

1. *Make a single size fit all users.* Workstations having controls within reach of the smallest, or, practically speaking, a 5th-percentile person, will allow most people to reach the controls. A potential difficulty with this design option, however, is that it may create a workstation that cramps operators who are above average in body size. Many architectural features (doors, passageways) are, or should be, designed to accommodate the largest, or, practically speaking, a 95th-percentile person. If you are above average height, however, I'm sure you have found that seating in some commercial aircraft seems to fall short of meeting that criterion.

2. *Make equipment adjustable.* Automobile seats and chairs for computer workstations are common examples of this design option. Ideally, adjustability should accommodate all potential users, but, since the absolute extremes of any body dimension are difficult to define and measure, much less to design to, human factors aims at fitting the middle 90 percent (that is, from the 5th to the 95th percentiles) of the population of potential users, or sometimes the middle 98 percent (that is, from the 1st to the 99th percentiles).

Universal, or nearly universal adjustability is especially important for equipment that may be used interchangeably by different operators. Failure to provide adequately for operator differences contributes to operator dissatisfaction and discomfort and can jeopardize performance, safety, and reliability.

Because so many women are now entering the labor force in occupations from which they had formerly been excluded, such as heavy construction, and police and military jobs, the design of most equipment these days must cover a range of dimensions that extends at least from the 5th-percentile female to the 95th-percentile male. Keep this requirement in mind when you use all tabled data of anthropometric measurements.

3. *Make equipment in several sizes.* This is typically the design solution used for clothing and personal equipment.

Measurement Techniques

Body measurements are usually taken between two end points, such as shoulder to fingertip. The following measurement terms refer to a standing subject:

Height is a straight-line, point-to-point vertical measurement. One height measure, stature, starts at the floor on which the subject stands and extends to the highest point on the skull.

Breadth is a straight-line horizontal side-to-side measurement across a body segment.

Depth is a straight-line, point-to-point horizontal measurement across the body, front to back.

Distance is a straight-line, point-to-point measurement between landmarks on the body.

Curvature is a point-to-point measurement that follows a body contour, for example, around the front of the chest from right to left. This measure is not closed and not usually circular.

Circumference is a closed measurement that follows a body contour, for example, completely around the chest.

Reach is a point-to-point measurement following the long axis of an arm or leg.

One complexity is that stature has been measured four ways: with the subject standing upright, but not stretching; standing freely but stretching to maximum height; leaning against a wall with the back flattened and stretching to maximum height; and lying supine. Differences between the standing measures may be as great as 2 cm. Lying supine yields a taller measure; standing "slumped" reduces stature by several centimeters.

In most cases, subjects are measured in an upright straight posture, with body segments at 0, 90, or 180 degrees to each other. The head is normally positioned so that the pupils, right tragion (the "ear hole"), and the lowest point of the right orbit (eye socket) are all on horizontal planes. When measures are made with the subject seated, the thighs are horizontal, the lower legs vertical, and the feet flat on the floor. Subjects are typically measured nude, or nearly so, and unshod (see Figure 5.10).

Figure 5.11 shows reference planes and descriptive terms often used in anthropometry. Figure 5.12 shows important landmarks of the human body in the sagittal (side) view, and Figure 5.13 shows landmarks in the frontal view.

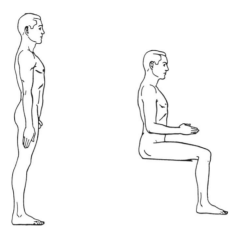

Figure 5.10 Postures typically assumed by subjects for anthropometric measurements.

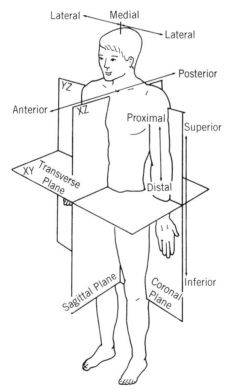

Figure 5.11 Anatomical planes and orientations used in anthropometry. (From NASA-STD-3000)

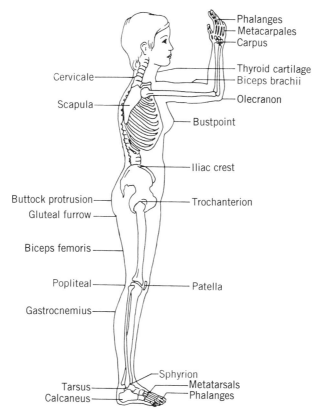

Figure 5.12 Some anatomical and anthropometric landmarks. (From NASA-STD-3000)

Anthropometric Data

Using and Interpreting Anthropometric Data Hundreds of thousands of anthropometric measurements have been made by many different investigators and almost every human-factors textbook, standard, and set of guidelines contains representative dimensions and recommendations for various design purposes. Engineers and designers, and even human-factors professionals, are sometimes dismayed to find that the measurements given in various sources do not necessarily agree with one another exactly. In fact, I have sometimes heard impassioned arguments among professionals over these discrepancies.

Anthropometric measurements, even those in computer data banks or in anthropometric models, do not have the precision of physical constants. Measures differ, and can differ greatly, depending on the size and composition of the group on which measurements were made, the kinds of tools used to make measurements, and the skill of the measurers in locating body landmarks precisely. As I shall show in the

162   5: HUMAN PHYSICAL CHARACTERISTICS

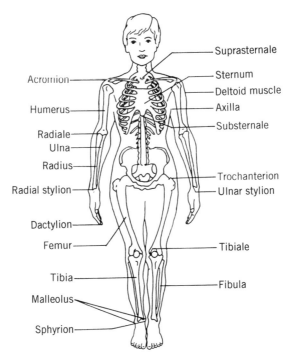

Figure 5.13   Some anatomical and anthropometric landmarks. (From NASA-STD-3000)

text that follows, measures differ between the sexes, among various ethnic and national groups, and among persons of different ages. Not only that, but they change over time. Since there is no standard or reference population for measurement purposes, all anthropometric measures (means, medians, percentiles, standard deviations) are only statistical approximations to some "true" or "correct" value, and, like all measures of human variables, should properly be reported with their appropriate standard errors of measurement. Even though these standard errors are seldom computed or even alluded to, one should understand that they belong there.

To use and interpret anthropometric data sensibly, find data that were obtained on a sample matching as closely as possible the population of users for which you are designing, and remember that, at best, the data are only approximate. When (1) it is important to have as accurate data as possible, (2) you do not have data on the exact population you are designing for, and (3) you cannot measure your target population, it is possible to use one or more sophisticated forecasting methods to adjust available data and predict what measurements should be in that other population. Forecasting methods depend on complex statistical manipulations that are described in detail by Roebuck (1995).

Some Representative Data There are abundant anthropometric data on military populations, but those on civilian populations are sparse. Table 5.3 gives some of the best data available for American civilians.

From what I said earlier about the correlations between dimensions (refer to Figure 5.5 and my discussion of it), you should realize that body measurements on a 5th-percentile person are not all at the 5th percentile. For example, 5th-percentile leg length plus 5th-percentile torso length plus 5th-percentile head height do not add up to 5th-percentile stature. Percentiles can be added only if they have been predicted from regression equations between measures of interest. For more detailed data and applications, consult the comprehensive three-volume NASA sourcebook on anthropometry (Staff of Anthropology Research Project, 1978), which contains hundreds of measurements on diverse populations, information on how they were obtained, and instructions on how they are to be used.

Populations Differences The population of the United States is a composite of persons of many different races and ethnic origins. Although the 1990 census contains such data, they are neither very detailed nor reliable, since they are based on self-reports. A better source of information comes from a survey of U.S. Army personnel (Gordon et al., 1989). Subjects in this sample were classified in nine major ethnic groups: Arabian, Asian, Caribbean Islanders, Central and South Americans, Europeans, Mixed/Other, Native American, North American but not Native American, and Pacific Islanders. They were then further subdivided into 100 categories, ranging from Acoman (a type of American Indian), Antiguan, Apache, Arab, Argentinian, and Austrian, to Vietnamese, Virgin Islander, Welsh, West Indian, and Yugoslavian. Keep in mind, though, that these are Army, not civilian, data.

This diversity is the source of a number of human-factors problems. One is the difficulty of defining exactly what the American civilian population consists of. Selective immigration and disparate birth rates among various subgroups in this country are continually changing its composition. Another difficulty arises from the significant anthropometric differences among various ethnic groups (see Figure 5.14). Not only do various ethnic groups vary in height, but they also vary in body proportions. For example, Germans are relatively more long-legged, as measured by the ratio of leg length to stature, than most other ethnic groups. Japanese, on the other hand, are relatively short-legged. This means that, if one were to take an automobile designed for a German population and scale the interior down to the average height of the Japanese, many Japanese drivers would not be able to reach the foot pedals.

Such differences have to be taken into account in the design of products manufactured for worldwide distribution. Computer workstations, automobiles, and aircraft, designed exclusively for use in Italy, Japan, or other Asian countries, have to be designed to different proportions than those designed for use in this country. In the case of products designed exclusively for use in the U.S., population differences are generally already factored into anthropometric data.

TABLE 5.3 Body Dimensions (in cm) of U.S. Female and Male Adults*

	Percentiles						Standard Deviation S	
	5th		50th		95th			
	Female	Male	Female	Male	Female	Male	Female	Male
Heights								
Standing								
Stature[f]	152.78	164.69	162.94	175.58	173.73	186.65	6.36	6.68
Eye height[f]	141.52	152.82	151.61	163.39	162.13	174.29	6.25	6.57
Shoulder (acromial) height[f]	124.09	134.16	133.36	144.25	143.20	154.56	5.79	6.20
Elbow height[f]	92.63	99.52	99.79	107.25	107.40	115.28	4.48	4.81
Wrist height[f]	72.79	77.79	79.03	84.65	85.51	91.52	3.86	4.15
Crotch height[f]	70.02	78.44	77.14	83.72	84.58	91.64	4.41	4.62
Sitting								
Height[s]	79.53	85.45	85.20	91.39	91.02	97.19	3.49	3.56
Eye height[s]	68.46	73.50	73.87	79.20	79.43	84.80	3.32	3.42
Shoulder (acromial) height[s]	50.91	54.85	55.55	59.78	60.36	64.63	2.86	2.96
Elbow height[s]	17.57	18.41	22.05	23.06	26.44	27.37	2.68	2.72
Thigh height[s]	14.04	14.86	15.89	16.82	18.02	18.99	1.21	1.26
Knee height[f]	47.40	51.44	51.54	55.88	56.02	60.57	2.63	2.79
Popliteal height[f]	35.13	39.46	38.94	43.41	42.94	47.63	2.37	2.49
Depths								
Forward (thumbtip) reach	67.67	73.92	73.46	80.08	79.67	86.70	3.64	3.92
Buttock–knee distance (sitting)	54.21	56.90	58.89	61.64	63.98	86.74	2.96	2.99
Buttock–popliteal distance (sitting)	44.00	45.81	48.17	50.04	52.77	54.55	2.66	2.66

Elbow–fingertip distance	40.62	44.79	44.29	48.40	48.25	52.42	2.34	2.33
Chest depth	20.86	20.96	23.94	24.32	27.78	28.04	2.11	2.15
Breadths								
Forearm–forearm breadth	41.47	47.74	46.85	54.61	52.84	62.06	3.47	4.36
Hip breadth (sitting)	34.25	32.87	38.45	36.68	43.22	41.16	2.72	2.52
Head dimensions								
Head circumference	52.25	54.27	54.62	56.77	57.05	59.35	1.46	1.54
Head breadth	13.66	14.31	14.44	15.17	15.27	16.08	0.49	0.54
Interpupillary breadth	5.66	5.88	6.23	6.47	6.85	7.10	0.36	0.37
Foot dimensions								
Foot length	22.44	24.88	24.44	26.97	26.46	29.20	1.22	1.31
Foot breadth	8.16	9.23	8.97	10.06	9.78	10.95	0.49	0.53
Lateral malleolus height[f]	5.23	5.84	6.06	6.71	6.97	7.64	0.53	0.55
Hand Dimensions								
Circumference, metacarpale	17.25	19.85	18.62	21.38	20.03	23.03	0.85	0.97
Hand length	16.50	17.87	18.05	19.38	19.69	21.06	0.97	0.98
Hand breadth, metacarpale	7.34	8.36	7.94	9.04	8.56	9.76	0.38	0.42
Thumb breadth, interphalangeal	1.86	2.19	2.07	2.41	2.29	2.65	0.13	0.14
WEIGHT (in kg)	39.2[a]	57.7[a]	62.01	78.49	84.8[a]	99.3[a]	13.8[a]	12.6[a]

*From Kroemer, Kroemer, & Kroemer-Elbert (1994). *Ergonomics: How to Design for Ease and Efficiency* (p. 30). Englewood Cliffs, NJ: Prentice-Hall. Used with permission.

[a]Estimated (from Kroemer, 1981). [f]Above floor. [s]Above seat.

Note. In this table, the entries in the 50th percentile column are actually "mean" (average) values. The 5th and 95th percentile values are from measured data, not calculated (except for weight). Thus, the values given may be slightly different from those obtained by subtracting 1.65 S from the mean (50th percentile), or by adding 1.65 S to it.

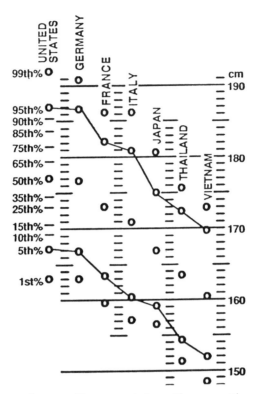

Figure 5.14 Statures of seven military populations. The percentile value on the left are for U.S. Air Force flight personnel. (From Kennedy, 1975)

Age Differences Table 5.4, showing heights and weights of white male and female Americans of various ages, was compiled from data in several sources. During childhood and the pre-adult years, stature, weight, and other body dimensions increase rapidly, then stabilize in the early twenties, and decrease slightly with increasing age. The age at which growth stabilizes varies from individual to individual, and is different for different body dimensions.

Sex Differences As is evident from Tables 5.3 and 5.4, men are larger than women in most body dimensions, but the differences vary considerably. Women are larger only in hip breadth, hip circumference, and thigh circumference. Men's legs, on the other hand, are not only longer than women's, but are also longer relative to their standing and sitting heights.

Changes over Time The body dimensions of various civilian populations are not stable over long periods of time. The average increase for nine European and Japanese populations over roughly the last hundred years has been about 1 cm per decade (Roebuck, Smith, & Raggio, 1988). Similar changes have been happening

TABLE 5.4 Heights and Weights of White Male and Female Americans at Different Ages*

Age (yrs.)	Male				Female			
	Height (in.)		Weight (lb.)		Height (in.)		Weight (lb.)	
	Mean	SD	Mean	SD	Mean	SD	Mean	SD
1	29.7	1.1	23	3	29.3	1.0	21	3
2	34.5	1.2	28	3	34.1	1.2	27	3
3	37.8	1.3	32	3	37.5	1.4	31	4
4	40.8	1.9	37	5	40.6	1.6	36	5
5	43.7	2.0	42	5	43.8	1.7	41	5
6	46.1	2.1	47	6	45.7	1.9	45	5
7	48.2	2.2	54	7	47.9	2.0	50	7
8	50.4	2.3	60	8	50.3	2.2	58	11
9	52.8	2.4	66	8	52.1	2.3	64	11
10	54.5	2.5	73	10	54.6	2.5	72	14
11	56.8	2.6	82	11	57.1	2.6	82	18
12	58.3	2.9	87	12	59.6	2.7	93	18
13	60.7	3.2	99	13	61.4	2.6	102	18
14	63.6	3.2	113	15	62.8	2.5	112	19
15	66.3	3.1	128	16	63.4	2.4	117	20
16	67.7	2.8	137	16	63.9	2.2	120	21
17	68.3	2.6	143	19	64.1	2.2	122	19
18	68.5	2.6	149	20	64.1	2.3	123	17
19	68.6	2.6	153	21	64.1	2.3	124	17
20–24	68.7	2.6	158	23	64.0	2.4	125	19
25–29	68.7	2.6	163	24	63.7	2.5	127	21
30–34	68.5	2.6	165	25	63.6	2.4	130	24
35–39	68.4	2.6	166	25	63.4	2.4	136	25
40–49	68.0	2.6	167	25	63.2	2.4	142	27
50–59	67.3	2.6	165	25	62.8	2.4	148	28
60–69	66.8	2.4	162	24	62.2	2.4	146	28
70–79	66.5	2.2	157	24	61.8	2.2	144	27
80–89	66.1	2.2	151	24				

*From Stoudt, Damon, & McFarland (1960).

to Americans. The data in Figure 5.15, although for a selected population, are typical. Since the majority of products are designed with the expectation that they will have relatively short lives, these changes over time are of little importance for most engineers, However, some systems, for example, aircraft, power plants, and large vessels, have lives extending over several decades. For such systems, these temporal changes should probably not be ignored.

Effects of Clothing As I said earlier, most anthropometric measurements are made on nude, or nearly nude, persons, a state in which operators of machinery or

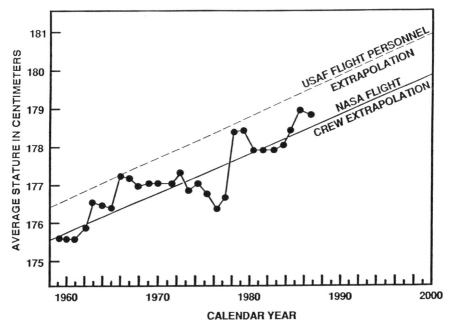

Figure 5.15 Average statures of USAF flying personnel and NASA flight crews over a number of years. (Reprinted with permission from Proceedings of the Human Factors Society 32nd Annual Meeting, 1988. Copyright 1988 by the Human Factors and Ergonomics Society. All rights reserved.)

equipment are seldom found. Light indoor clothing adds little to body dimensions, but work clothes and clothing worn by operators in certain extreme environments, such as in the arctic regions or in space, may increase some body dimensions by several inches. In addition, some operators may be encumbered with equipment, such as anti-g suits, parachutes, or oxygen equipment, that increase their spatial dimension even further. These considerations obviously have to be factored into all system designs.

Functional Anthropometry The data presented so far are structural, or static, dimensions of the human body in standard postures. They do not describe functional dimensions, that is, the space needed by the body when it is moving and doing work. Functional dimensions are particularly important for:

1. Working positions, especially in restricted areas, such as may have to be assumed by maintainers, plumbers, and pipefitters.
2. The placement of controls within easy reach and grasp.
3. Aperture sizes and depths of reach through apertures for maintenance.

Most data on reach envelopes are taken from the Seat Reference Point (SRP), the point where lines down the seatback and along the seat pan meet (Figure 5.16). You can find numerous other reach envelopes in NASA-STD-3000.

ANTHROPOMETRY: THE SCIENCE OF HUMAN DIMENSIONS   169

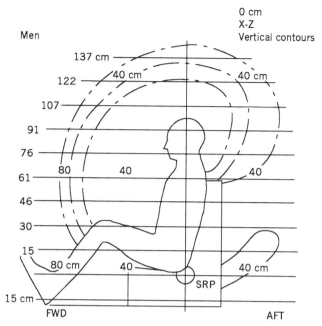

Figure 5.16 A set of reach envelopes for 5th, 50th, and 95th percentile Air Force males. SRP is the seat reference point. (From NASA-STD-3000)

Space needs and workplace dimensions for the body in common working postures are harder to specify because those postures vary so much from task to task. Figure 5.17 and Table 5.5 provide typical data on dimensions for some working positions. For other data of this type see NASA-STD-3000, and Hertzberg (1972).

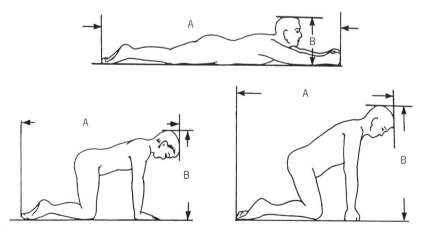

Figure 5.17 Some typical functional working positions: prone (*upper figure*), crawling (*lower left*), and kneeling crouched (*lower right*). In each case A is a length dimension and B a height. Corresponding dimensions are given in Table 5.5. (From Hertzberg, 1972)

TABLE 5.5 Working-Position Dimensions for U.S. Air Force Male Personnel*

Dimensions	Percentiles (in.)			Standard Deviation
	5th	50th	95th	
Prone:				
Height	12.3	14.5	16.4	1.28
Length	84.7	90.1	95.8	3.41
Crawling:				
Height	26.2	28.4	30.5	1.30
Length	49.3	53.2	58.2	2.61
Kneeling Crouched:				
Height	29.7	32.0	34.5	1.57
Length	37.6	43.0	48.1	3.26

*Data of Hertzberg (1956), from Van Cott & Kinkade (1972).

THE SKELETAL SYSTEM

The skeletal system of the human body consists of approximately 200 bones, their articulations, and connective tissues. Its main function is to provide the internal framework for the body. Without it the entire body would collapse into a heap of soft tissue. The long bones provide links between the joints and are the lever arms to which tendons attach. Ligaments connect bones at their articulations, the points where they come together.

Although bones are firm and hard, and so can resist high strains, they have some elastic properties, particularly in childhood, when mineralization is still relatively low. Growth takes place until about 30 years of age, after which a decrease in the mass of bones and other subtle changes tend to make them hollower and more brittle.

Mobility Mobility, also referred to as flexibility, is the range through which various connected body segments can move. Figure 5.18 illustrates some possible motions. Tabled data on mobility, or data from some models and manikins, are not as straightforward as they appear. The actual point of rotation of a body member moves with the motion. The actual arms of rotation are not well defined, and the range of motion varies greatly from person to person.

More often than not, human motion involves the interaction of two or more joints and muscles. Thus the movement range of a single joint may be dramatically changed by the movement of an adjacent one. In other words, joint movement ranges are neither independent nor additive. For example, if you were using a scaled manikin to make an engineering layout, you might conclude that a foot control would be reached with a hip flexion of 50 degrees and with the knee extended (0 degrees flexion). Both of these are individually correct. Taken together, however, they are not. Hip flexion is reduced by over 30 degrees when the knee is extended.

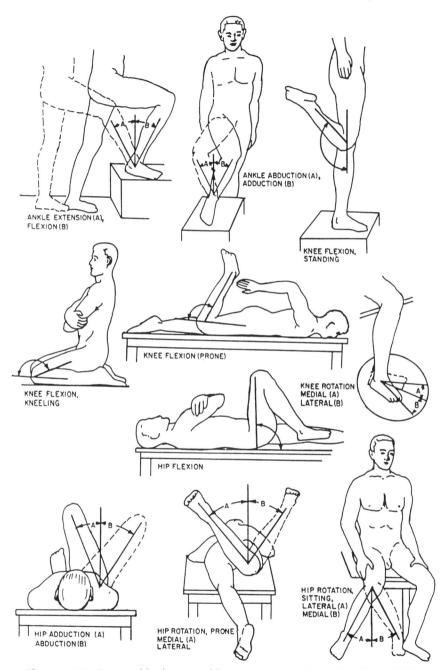

Figure 5.18 Some ankle, knee, and hip movements. (From Hertzberg, 1972)

As a result, an operator would not be able to reach the control as anticipated. NASA-STD-3000 has data on the changes in range of movement of certain joints with movements of adjacent ones.

Joint mobility decreases only slightly in *healthy* persons between the ages of 20 and 60. However, the incidence of arthritis increases so markedly beyond age 45 that the *average* joint mobility for any older population will be considerably decreased. This should be kept in mind when designing for a general population. In general, women have greater mobility than men at all joints except the knee.

Slender men and women have the widest range of joint movement. Persons with average and muscular body builds have lesser ranges of movement, and the differences between thin and fat persons may result in variations greater than 10 degrees. Physical exercise can increase joint mobility, but excessive exercise can result in a person becoming "muscle bound," with a consequent reduced range of motion.

Design Considerations In general, designs should be based on 5th-percentile limits when personnel must position their bodies to operate or maintain equipment, and on the 95th percentile to accommodate a full range of unrestricted movement. As usual, you have to consider these limits for the combined male and female population.

The Spinal Column The spinal column is a bony system of particular concern to designers, because it is the site of many work-related injuries—a very costly industrial health problem these days. Many of these injuries seem to arise from acute or cumulative high and/or asymmetric external loads that affect the muscles or cartilage stabilizing the bony stack of vertebrae. Some, but not all such work-related injuries can be avoided by the proper design of work stations, equipment, and tasks (Kroemer, 1985). Work stations and procedures can also be engineered to allow persons with persistent back pain to perform satisfactory physical work (Rodgers, 1985).

SKELETAL MUSCLES

Skeletal muscles connect body segments across their joints. They are of engineering interest because they move segments of the body under voluntary control and allow us to exert force on objects outside of the body. The human body contains several hundred skeletal muscles known by their Latin names. Cardiac and smooth muscles, such as those that control the behavior of our heart and internal organs, are almost entirely beyond voluntary control and so are of little concern for design purposes. Muscles act only by contraction; elongation is produced by external forces. Skeletal muscles are usually arranged in pairs, so that a contracting muscle is counteracted by its opponent. The muscle characteristics of greatest relevance to design are muscle strength and working capacity.

Muscle Strength and Working Capacity

Strength is the greatest amount of force that can be exerted by a muscle instantaneously. As a practical matter, this means the force that can be exerted for a relatively few seconds. It is typically measured isometrically, that is, by the greatest amount of force a muscle can exert without shortening its length, and so is referred to as "static" force. Units of isometric force are ounces and pounds in the English system, ponds and kiloponds in the metric system.

Working capacity is the amount of external physical work an individual can perform. It is measured in units of energy such as ft.-lb., kp.-m., or joules. Power is the amount of work done per unit time and is expressed in horsepower or watts. The working capacity of an operator is normally limited by his or her endurance (the efficiency of the circulatory system), and not by muscle strength per se. Physical work is often described as "dynamic" work, since muscles contract and relax alternately, that is, they both tense and change length.

Since there are definite differences between an operator's static strength and the ability to do prolonged work, we must consider three different kinds of data:

1. Maximal static forces.
2. Weight-lifting capacity.
3. Dynamic working capacity.

Maximal Static Force Static (isometric) forces are measured when force is applied to a control, but the control does not move, or moves negligibly, in relation to the operator's body. Table 5.6 is an example of the kind of static force data you can find in several sources.

Weight-lifting Capacity Figure 5.19 shows lifting capacities for 5th-percentile males and females. As you can see, our ability to lift weights depends on where the weight is located with respect to the body. Other data of this kind are available in Bailey (1982), NASA-STD-3000, and Ciriello and colleagues (1993).

Dynamic Working Capacity Here are some estimates of usable external power outputs:

1. Up to 6 hp. for single movements lasting less than 1 second.
2. Between 0.5 and 2 hp. during brief periods of exercise lasting between 0.1 and 5 minutes.
3. Between 0.4 and 0.5 hp. during steady-state work of 5 to 150 minutes.
4. Up to 0.2 hp. for long-term work lasting all day.

Since these values were measured on champion athletes under ideal conditions, they should be reduced by about 20 to 30 percent for ordinary healthy men, and

TABLE 5.6 Maximal Static Grip Forces by Males*

	Percentiles (lb)			
Population	5th	50th or Mean	95th	SD
Air Force personnel, general:				
Right hand	(59)**	104	(148)**	27.3
Left hand	(56)**	94	(134)**	23.7
Air Force personnel, aircrewmen:				
Right hand	105	134	164	18.0
Left hand	96	124	154	16.0
Air Force rated officers:				
Preferred hand	98	124	154	16.8
Army personnel:				
Right hand	106	137	172	—
Left hand	99	132	168	—
Navy personnel:				
Mean of both hands	95	119	143	14.4
Industrial workers:				
Preferred hand	92	117	143	15.4
Truck and bus drivers:				
Right hand	91	121	151	18.1
Left hand	86	113	140	16.4
Rubber industry workers:				
Right hand	(89)**	124	(159)**	21.2
Left hand	(86)**	122	(159)**	22.2
University men:				
Right hand	(74)**	108	(142)**	21.0
Left hand	(65)**	95	(124)**	18.0
Same subjects, force exerted over one minute:				
Right hand	(42)**	62	(82)**	12
Left hand	(39)**	55	(71)**	10

*From Hertzberg (1972).
**Percentiles in parentheses were computed from the 50th percentile using the standard deviation.

about 40 to 60 percent for women. Rotating a crank by legs or arms is the most effective and least fatiguing way of exerting muscular power repeatedly over a long period of time. Reciprocal motions of a lever are less satisfactory.

Modifying Factors A large number of factors affect muscle strength and working capacity.

Biological Factors. Age and sex are two biological variables of importance. At about age 10, boys and girls are roughly equal in strength. Strength increases rapidly in the teens, but much more so for boys than for girls. For men, strength

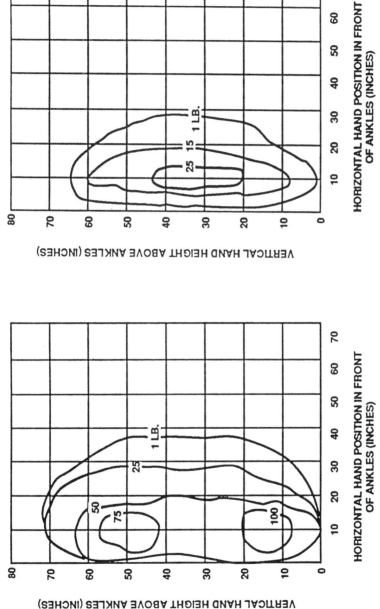

Figure 5.19 Lifting capabilities for 5th percentile males (*left*) and 5th percentile females (*right*). (From Bailey, 1982)

175

reaches a maximum in the middle to late twenties, remains at this level for five to ten years, and then drops slowly but increasingly. Men have about 95 percent of their earlier strength still available at age 40, and about 80 percent at age 50 to 60. Women also reach their maximal strength in the early twenties and remain at this level for about 10 years. The decline in subsequent years is much more rapid for women than for men. At about age 30 women can exert about two-thirds of men's forces, but at age 50 only about half.

Decrements in muscle strength do not proceed at the same rate in all parts of the body. The muscles of the hands and arms are less affected by age than are those of the trunk and legs.

There are significant but low correlations between body build and strength. For endurance and sustained dynamic work, normal, or even slender body build seems to be advantageous. Handedness has such a relatively small effect on strength that differences in strength between the two sides of the body can be safely neglected. Exercise (training) can improve an operator's strength and working capacity within an individual's innate physical potential.

Environmental Factors. Altitude affects physical capabilities, especially those for sustained energy expenditure. Endurance time for moderate muscular work may decrease at altitudes as low as 6,500 ft. and, at 10,000 to 12,000 ft., work capacity is about 10 to 15 percent less than at sea level. Acclimatization to high altitudes can restore sea-level capacities for work, but this acclimatization may take weeks or months. Even fully acclimatized men have only about half as much muscular endurance at 20,000 ft. as at sea level.

High ambient temperatures, above about 85° F, especially when combined with high humidity, reduce muscular endurance, but do not have much effect on short bursts of energy. Unacclimatized men tolerate heat better than unacclimatized women. Low ambient temperatures per se do not greatly reduce the capacity for work, but the bulkiness of clothing can.

Occupational Factors. Exertable forces and working efficiency can be markedly affected by the body positions operators must assume because of their working environments or the design of equipment they use. For example, forces that operators can exert are greater if the operator is braced or restrained than if he or she is standing upright and unbraced. Similarly, working efficiency is better if operators have room to move freely than if they are in narrow, confined spaces.

Fatigue Despite centuries of study and research " . . . no one has yet come up with either a precise definition of fatigue and/or reliable ways to measure it. The very broad and general definition of fatigue—'fatigue denotes a state represented by a loss in efficiency and a general disinclination to work'—is perhaps as valid today as it was 20 or 30 years ago." (Mital & Kumar, 1994). Most scientists agree, however, that there appear to be two different kinds of fatigue: central and peripheral. Although the two often occur together, central fatigue is the result of prolonged mental or perceptual effort; peripheral fatigue the result of muscular effort.

Three general methods have been used to measure fatigue: physiological (for example, depletion of nutritive material to the muscles, formation of carbon dioxide, heart rate), performance (for example, change in the rate, quality, or quantity of work), and subjective (feelings of tiredness or heightened exertion). Oddly, the three measures do not necessarily agree. A worker whose output may decline and who feels exhausted at the end of a work day may suddenly revive to go jogging or bowling.

Some observable indices of fatigue are:

- Shaking a limb, stretching, and yawning;
- Frequent breaks from work, for example, to get a drink of water;
- Declining productivity or performance, or increasing errors and accidents toward the end of a work period;
- Absenteeism or personnel labor turnover.

Muscular Fatigue. Muscular, or peripheral fatigue is most directly related to the adequacy of the blood supply to the muscles involved. A copious blood flow is needed to supply the muscles with oxygen and high-energy phosphates, and to carry away the products of muscle metabolism. Since muscle contractions reduce blood flow, an important factor in preventing muscle fatigue is designing jobs to reduce requirements for static muscle loading, such as gripping, extended reaches, and awkward postures. Dynamic or rhythmic movements allow blood to flow between contractions, and so are generally less fatiguing than static contractions. Even so, dynamic movements that require a great amount of force or that occur at a high rate may still induce rapid fatigue.

There is a nonlinear, inverse relationship between the force a person can exert and the time during which it is exerted. Maximal force (100% of strength) can be exerted for only a few seconds, 50 percent of strength is available for about a minute, and only about 15 percent to 20 percent of maximal strength can be maintained over many hours without fatigue. Contracted muscles compress their own blood vessels, shutting down circulation which, in turn, cuts off supplies of needed energy and removal of metabolic by-products. Such fatigue sometimes occurs even when a muscle is not working very hard. You notice it when you work overhead with your arms raised, for example, in replacing a light bulb in an overhead light socket. Fatigue in the shoulder muscles makes it impossible to keep your arms raised for more than a minute or two, even though the muscles are still receiving nerve impulses.

Central Fatigue. Central fatigue may be due to changes in the nervous system but its measurement is complicated by motivational factors and the inherent interest of work. Alternating work and rest periods helps to relieve fatigue from any source.

Applications to Design

Data on voluntary muscle strength help to determine maximum and optimum control resistances, forces required for a variety of manual tasks, and weights that can be lifted or carried safely and efficiently. The maximum resistance of a control should be low enough so that it can be overcome by the weakest (usually 5th-percentile) operator. Operational resistances should not require the application of maximum power by any operator, and should be low enough to prevent fatigue or discomfort, but high enough to prevent inadvertent activation and to provide kinesthetic feedback cues about control movements.

Specifications for other physical tasks, such as lifting and carrying, should also be made with regard to the capability of the weakest potential (usually 5th-percentile) operator. To avoid fatigue, rapid decrements in performance, and possible injury (muscle strain or rupture, strained or torn ligaments), no task should require performance at or near the limits of physical abilities. Recommendations about the design of devices for sustained work can be found in Van Cott and Kinkade (1972).

Cumulative Trauma Disorders

Cumulative trauma disorders (CTDs) are disorders of the muscles, tendons, nerves, and blood vessels caused, precipitated, or aggravated by repeated exertions or movements of the body. The most common types of CTDs are tendon and nerve disorders of the upper extremities: tendinitis, tenosynovitis, De Quervain's disease, trigger finger, epicondylitis, carpal tunnel syndrome, and cubital tunnel syndrome. According to statistics from the Bureau of Labor Statistics, there were 302,400 "repeated trauma injuries" in 1993, accounting for nearly two-thirds of total illnesses and almost five percent of all injuries and illnesses. Between the ages of 18 and 64, more people are disabled from musculoskeletal problems than from any other category of disorder.

The incidence of CTDs in industry has increased dramatically in recent years due to higher production rates; increased awareness among workers, practitioners, and the press; better statistics and coding; expanded workers' compensation laws; and less worker tolerance. CTDs develop with the use of diverse products. Repeated and constant use of computer terminals, or repeatedly sweeping grocery items across a scanner, often result in injuries to the nerves of the wrist—known as carpal tunnel syndrome, a compression of the median nerve that causes pain, tingling, or numbness in the hand and is sometimes severely debilitating. Knives used in the meat-packing industry, scissors in the garment industry, and power tools of many types often produce CTDs. Even recreational activities can produce such injuries—witness the tennis and golfer's elbow.

Design solutions for these problems generally fall into three categories: (1) reducing the frequency with which movements must be made through automation, mechanization, job enlargement, or job rotation; (2) changing the workstation, tool, or object used, to produce more favorable postures and positions of body members; and (3) reducing the amount of force that must be exerted on controls and tools, and the duration of their usage.

OTHER MACHINERY OF THE BODY

The nervous, respiratory, circulatory, and metabolic systems are some of the other major parts of the machinery of the body. They will be given superficial coverage here, not because they are unimportant, but only because they do not figure heavily in system design.

The Nervous System

The nervous systems consists of three main anatomical parts: the *central* nervous system, which includes the brain and spinal cord; the *peripheral* nervous system, which includes the cranial and spinal nerves; and the *autonomtic* nervous system, which, in turn, is made up of the *sympathetic* and *parasympathetic* systems.

The peripheral system receives information from our various senses, transmits it to the central nervous system for control and decision-making, and sends nerve impulses back to the muscles. The autonomic system regulates involuntary functions of cardiac muscles, blood vessels, digestion, and glucose release by the liver. It is the system that is responsible for the "flight, fright, or fight" reactions that we experience in emergency or dangerous situations.

The Respiratory and Circulatory Systems

The respiratory system provides oxygen for energy metabolism and dissipates the by-products (carbon dioxide, water, and heat) of metabolic action. The circulatory system provides the means by which these products are transported between the lungs, heart, muscles, and other body cells. Both systems increase their activity under conditions of stress or work. In performing heavy work, for example, the blood supply to the muscles may increase by a factor of five or more, while considerably less blood is delivered to the digestive tract, liver, and kidneys.

The Metabolic System

The metabolic system of the body transforms energy into several outputs according to the following equation:

$$I = M + W + S$$

where I is energy supplied to the body by food or drink, M is metabolic energy, W is work performed, and S is energy stored in the body (this may be positive or

TABLE 5.7 Energy Expenditures and Heart Rates for Various Categories of Work Performed over a Whole Work Shift*

	Total Energy Expenditure (kJ per minute)	Heart Rate (beats per minute)
Light work	10	90 or less
Medium work	20	100
Heavy work	30	120
Very heavy work	40	140
Extremely heavy work	50	160 or more

*From Kroemer, Kroemer, & Kroemer-Elbert, ERGONOMICS: How to Design for Ease & Efficiency, ©1994, p. 30. Reprinted by permission of Prentice Hall, Englewood Cliffs, New Jersey.

negative). Human work efficiency (e) is the ratio of work performed to energy input, that is:

$$e = 100 \ (W/I)$$

Energy is measured in Joules (J) or calories (cal) and the following equation shows the relationship between them and other physical measures:

$$1J = 0.239 \text{ cal} = 10^7 \text{ ergs} = 0.948 \text{ BTU} = 0.738 \text{ ft.-lb.}$$

Only about 5 percent of energy input is converted into work in everyday activities. Under favorable circumstances, however, highly trained athletes may achieve efficiency ratios as high as 25 percent. Table 5.7 shows approximate energy expenditures and heart rates for various categories of work performed over a normal work shift. Operators who are capable and trained can maintain a balance between body needs and output in doing light, medium, and even heavy work throughout a full work day. That is not the case with very heavy work. The lungs and heart are no longer able to supply the body with all the oxygen it needs, resulting in what is called an oxygen deficit. The operator must now take intermittent rest pauses. With extremely heavy work, the accumulation of waste products from the muscles, chiefly lactic acid, and the oxygen deficit become so great that frequent rest pauses are necessary, and even highly trained persons may not be able to continue at this rate throughout a full work shift.

INTERACTIONS WITH THE ENVIRONMENT

The human body functions best within fairly narrow ranges of environmental conditions. Yet many systems require workers or operators to perform in environments that exceed those limits and that, in some cases, are dangerous and hostile. A deep mine is a dark, dirty, noisy, hot, and often contaminated place in which to work. The almost constant threat of cave-ins and explosions add to those stresses. At another extreme, activities in space take operators to regions where there is no air at all, and where reduced or zero gravity creates a unique set of problems that affects locomotion and all forms of muscular activity.

The environments in which we live and work can affect performance in several ways. Some environments, for example, those with very high or very low light levels, high noise levels, some noxious gases, and reduced oxygen concentrations, may seriously interfere with our ability to sense information. Other environmental conditions can affect our thinking and decision-making abilities, and still others our ability to make movements effectively. Extreme deviations from tolerable environmental conditions may cause physical injury or endanger life. Perhaps the most insidious aspect of some environments is that they contain hidden hazards. There is no way, for example, that a person can see, hear, smell, or feel lethal radiation.

Tolerance limits to environmental stresses are inversely related to the duration of the stress. For very short periods of time, people can withstand environments that greatly exceed generally accepted limits. For example, a properly clothed person can tolerate temperatures ranging from $-60°$ F to $+140°$ F without injury for about an hour. For shorter periods, this range can be extended from $-100°$ to over $+200°$ F. As another example, a man lying supine can survive transverse accelerations of over 20 g for about 30 seconds, and over 50 g for 1 second or less. For the most part, however, our concerns are with much narrower environmental ranges, because well-designed systems should not expose operators to environments that exceed safe tolerance limits.

The Thermal Environment

Variations in the thermal environment do not interfere directly with our sensory or motor performance, but, as you know from first-hand experience, they can dramatically influence your comfort and psychological functioning.

Measures of the Thermal Environment Four factors determine the physical conditions of the climatic environment and our perceptions of it: the air (or water) temperature, humidity, air (or water) movement, and temperature of surrounding surfaces.

Air temperature is measured with thermometers, thermistors, or thermocouples. For measurements of dry-bulb temperature, the sensor is kept dry and shielded from radiation.

Humidity is measured with a psychrometer, hygrometer, or appropriate electronic device. These instruments make use of the principle that the cooling effect of evaporation is directly proportional to the humidity in the air, that is, evaporative cooling decreases as the water vapor pressure in the air increases. The highest absolute content of vapor in the air is reached when any further increase would result in the formation of water droplets falling out of the air. That, in turn, depends on the temperature of the air: higher temperatures allow more water vapor to be retained. The usual measure is "relative humidity," the actual vapor content as a percentage of the maximum possible at any given temperature and pressure.

Air movement is measured with various types of anemometers. Surface temperature is measured with thermometers.

Heat Exchanges with the Environment Body heat is exchanged with the environment through radiation (R), convection (C), conduction (K), and evaporation (E). Radiation depends primarily on the difference in temperature between two surfaces, for example, a person's skin and a wall. Heat is always radiated from the warmer to the cooler surface, and does not depend on the temperature of the air between the two surfaces.

Heat exchanges by convection and conduction are proportional to the area of the skin involved and the temperature difference between the skin and the layer of the medium with which it is in contact. Convection occurs when the skin is in contact with air or fluids, conduction when it is in contact with a solid body.

Evaporation results only in heat losses from the body and is a function of the body surface involved in the evaporation process, the humidity and vapor pressures of the skin, and surrounding air. Convection (movement of the air layer at the skin) increases heat loss by evaporation if humid air is replaced by dryer air.

Acclimatization Continuous or repeated exposures to hot conditions results in a gradual adjustment of body functions so that an individual is better able to tolerate and work effectively in heat. Significant acclimatization occurs within a week and is complete within about two weeks. Interruption of exposure for even a few days results in a substantial loss of acclimatization and the effects are completely lost after about two weeks.

Healthy and well-trained individuals acclimatize more easily than people in poor physical condition, but training cannot replace acclimatization. Heat acclimatization takes place in both hot and dry, and hot and humid, environments, and seems to be unaffected by the type of work performed. Acclimatization to cold is so much less pronounced than acclimatization to hot that there is even some question whether it even takes place when appropriate clothing is worn. There are only slight differences between men and women in their ability to adapt to hot and cold conditions.

Thermal Comfort Comfort is determined more by the conditions immediately surrounding an individual—the microclimate, and less by the climate in general—the macroclimate. Comfortable microclimates are highly variable and depend only in part on the macroclimate. The gender and age of the individual, the clothing worn, and the type and amount of activity being performed, are significant variables.

Various combinations of climatic factors can subjectively feel the same, and there have been a number of formulas proposed to show how these factors combine. One generally accepted nomograph combines dry air temperature, humidity, and air movement into a measure called "effective temperature" (Figure 5.20). A given effective temperature produces sensation equal to what a person wearing indoor clothing, doing sedentary or light work, would feel if there were no air movement at that temperature.

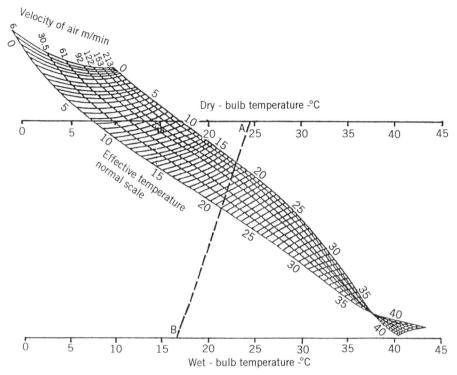

Figure 5.20 Effective temperature nomograph for persons wearing customary indoor clothing, engaged in sedentary or light muscular work. Example: For a dry-bulb temperature of 24.5° (A), wet-bulb temperature of 16.5° (B), and air velocity of 30.5 m/min, the effective temperature is 20.5°. (From MIL-STD-1472D)

Figure 5.21 shows comfort and discomfort zones, expressed in degrees Celsius of effective temperature (ET). Effective temperatures from about 21 to 27° ET feel comfortable in the summer for persons appropriately clothed and engaged in light work. Comparable values are 18° to 24° ET in a cool climate or during winter. Preferred ranges of relative humidity are between 30 and 70 percent.

Design Considerations A designer has many options available to produce a thermal environment that will be comfortable for operators. Table 5.8 shows ways in which the thermal environment can be designed to increase or decrease body heat.

Generally speaking, design should aim at producing environments within the comfort zones in Figure 5.21. Air temperatures at floor and head levels should not differ by more than about 6° C. Differences between the temperatures of body surfaces and side walls should not exceed 10° C, and air velocities should be kept below 0.5 m/s. For more detailed information, consult ANSI-ASHRAE Standard 55.

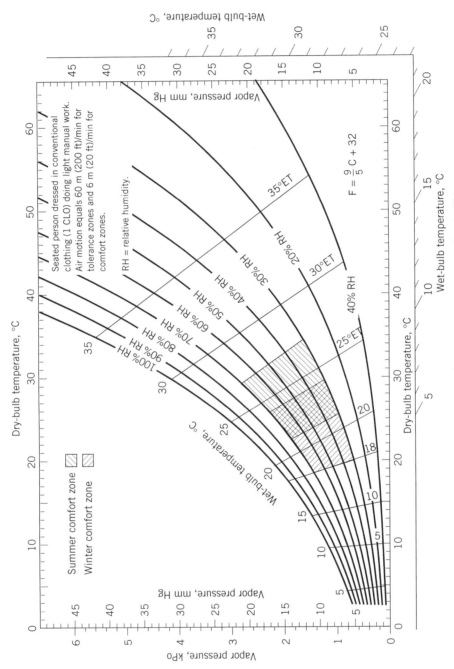

Figure 5.21 Summer and winter comfort zones. (From MIL-STD-1472D)

TABLE 5.8 Designing the Thermal Environment to Increase (+) or Decrease (−) Body Heat*

Transfer Mechanism	Humidity		Air Movement		Temperature (as compared with skin temperature) of					
					Air or Water		Solids		Opposing Surface	
	Dry	Moist	Fast	Calm	Hotter	Colder	Hotter	Colder	Hotter	Colder
Radiative	no direct effect		no direct effect		not applicable		not applicable		+	−
Convective	no direct effect		−	(−)	+	−	not applicable		not applicable	
Conductive	not applicable		not applicable		not applicable		+	−	not applicable	
Evaporative	−	(−)	−	(−)	−	(−)	not applicable		not applicable	

*From Kroemer, Kroemer, & Kroemer-Elbert, ERGONOMICS: How to Design for Ease & Efficiency, ©1994, p. 30. Reprinted by permission of Prentice Hall, Englewood Cliffs, New Jersey.

(−) indicates a heat loss not as pronounced as in the corresponding condition.

Composition of the Air

We are biologically adapted to breathing air that contains 21 percent oxygen by volume at sea level, which figures out to an oxygen partial pressure of 152 mm Hg (3 psia). Significant deviations from that norm occur from ascent to high altitudes, descent below sea level, or from dilution of the air with other gases. Too little oxygen (hypoxia) induces sleepiness, headache, the inability to perform simple tasks and, in the extreme, loss of consciousness and death. Too much oxygen (hyperoxia) can also be harmful. Prolonged breathing of pure oxygen at sea level or breathing air in deep-sea diving can eventually produce inflammation of the lungs, respiratory disturbances, various heart symptoms, blindness, and loss of consciousness. Figure 5.22 shows ranges of acceptable oxygen levels and conditions that produce an excess (hyperoxia) or a deficiency (hypoxia) of oxygen.

The hypoxia resulting from ascent to high altitudes can be counteracted by breathing pure oxygen at atmospheric pressures, by breathing oxygen at greater than atmospheric pressures (pressure breathing), or by pressurizing living and working areas (as is done with most commercial aircraft). The hyperoxia resulting from deep-sea diving operations can be counteracted by diluting the air with physiologically inert gases: nitrogen, helium, neon, krypton, xenon, or hydrogen. In excess, even these gases can be toxic so that their use must be regulated carefully (see, for example, NASA-STD-3000).

Our industries and systems produce a host of pollutants that contaminate the air and have serious effects on health and performance. Carbon monoxide, for example, one of the gases emitted by automobiles and other sources of combustion, is

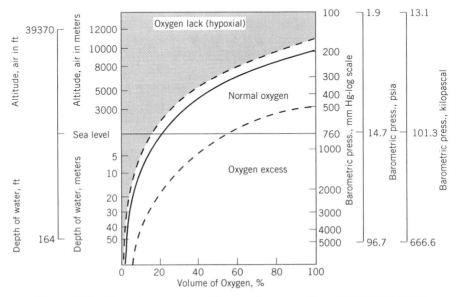

Figure 5.22 Hypoxia, normal and hyperoxic zones. (From NASA-STD-3000)

particularly dangerous because it is colorless and odorless, and the onset of symptoms is insidious. Once it combines with the hemoglobin of the blood to produce carboxyhemoglobin, it is difficult to get rid of.

In recent years there has been a great deal of concern about gaseous contaminants from computers, but the scientific evidence seems to show that these concerns are unwarranted.

Noise

Our civilization is a noisy one. We are constantly being assailed by the sounds of machines and systems that we use for constructive purposes, and by music and speech issuing from amplifying systems that often subject us involuntarily to sounds at disagreeable intensities. These sounds can produce hearing losses, disrupt speech communication, affect psychomotor performance, and cause annoyance and stress (see Kryter 1985 and Harris, 1991 for comprehensive reviews of these effects). For our purposes, we may define noise simply as unwanted sounds.

Acoustic sounds or noises are produced by the impact of pressure fluctuations in the atmosphere on our ears. These pressure fluctuations are measured in decibels (dB) of sound pressure level. The threshold of hearing in the frequency range of 1,000 to 5,000 Hz is about 20 μPascals (2.9×10^{-9} psi). Sound pressures of about 200 Pascals (0.029 psi) are painful and may cause permanent hearing damage. Sound pressure level in dB is a ratio between any two sound pressures, one of which is a reference sound pressure, usually the threshold of hearing (20 μPascals):

$$\text{SPL} = 20 \log_{10} (P/P_0) \text{ dB}$$

where P is the root-mean-square sound pressure in Pascals for the sound in question, and P_0 is the reference sound pressure level, usually 20×10^{-6} Pascals. Figure 5.23 shows the relationship of sound pressure level in decibels to sound pressure level in Pascals and some approximate equivalents in familiar noises.

Since the ear is not equally sensitive to all frequencies, two noises having the same sound pressure levels may or may not sound equally loud, depending on their frequency compositions. A frequency-weighting network, the A-weighting function in some sound level meters adjusts measures at various frequencies so that resultant integrated levels bear a closer relationship to perceived loudness. In other words, noises having the same A-weighted levels, expressed in dB_A, should sound about equally loud to the average listener.

Physiological Effects of Noise Intense sounds may result in temporary or permanent hearing loss, the severity being dependent on the duration of the exposure, the physical characteristics of the sound (its intensity, frequency, and whether pure-tone or wide-band), and the nature of the exposure (continuous or intermittent). For any given exposure time and sound pressure level, continuous noise causes greater hearing losses than does impulse noise. Intense noises also cause a number of other physiological effects, among them, changes in the concentrations of corticosteroids

188 5: HUMAN PHYSICAL CHARACTERISTICS

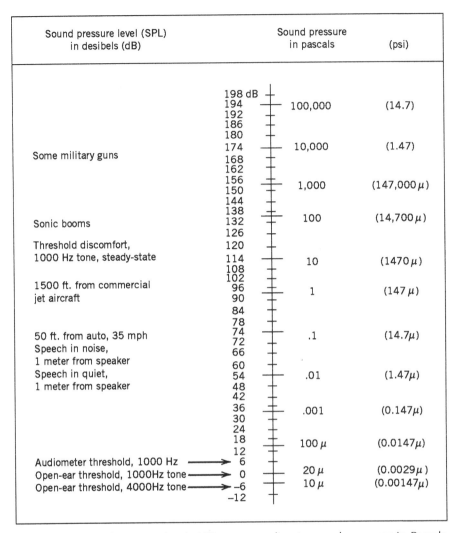

Figure 5.23 Sound pressure levels (dB) corresponding to sound pressures in Pascals and psi. (From NASA-STD-3000)

in the blood and brain, changes in the liver and kidneys, fluctuations in blood pressure, and abnormal heart rhythms. At extremely high intensities (160 dB), noise is lethal.

Effects of Noise on Performance One of the principal undesirable effects of noise on performance is interference with speech and auditory communication. These effects begin at speech-interference levels of about 40 dB and become severe at about 60 dB. At about 80 to 85 dB, face-to-face communication is difficult, even when communicators are separated by only one or two feet and shout at each other.

Other effects of noise at intensities of about 90 dB or more are vigilance decrement, altered thought processes, interference with mental work, and decreased visual capacities.

Annoyance Effects of Noise Sounds become annoying when they are unwanted, objectionable, or unacceptable. These annoyance effects can occur at rather low intensities, can interfere with rest and sleep, and stimulate emotions such as anger and frustration.

Design Considerations Acoustics and acoustical engineering are highly developed disciplines with a wealth of data and design principles. One important set of principles are hearing-conservation criteria. These are comprehensive statements of the relations between various measures of noise exposure, such as sound pressure level and exposure time, and the probability of temporary or permanent hearing loss or other undesirable effects. Figure 5.24, for example, shows indoor noise criteria curves for continuous wide-band noises. The sum of all individual sound pressure levels from all operating systems and subsystems should not exceed the NC 50 contour for work periods and the NC 40 curve for sleep.

There are three basic methods of controlling noises:

1. Reduce noise at its source.
2. Interrupt or absorb noise along its path of transmission.
3. Provide personal hearing protection.

Sources of noise from equipment are vibration, impact, friction, and fluid-flow turbulence. Since their control is primarily an engineering matter, involving no human-factors principles, they are not discussed here. Methods for correcting noises at their sources are given in a number of sources (for example, NASA-STD-3000, OSHA 1980, Harris, 1991).

Sources of noise can radiate sound directly into the air or induce vibrations in structures that, in turn, radiate into the air. Airborne noise can be reduced by (a) enclosing the noise source in a barrier, (b) sound absorbing linings, and (c) sealing openings in walls. Structurally transmitted vibrations and noises can be reduced by (a) isolating machinery supports and panels (b) damping panels, (c) decoupling pipes from pumps with a section of hose, and (d) detuning vibration frequencies by stiffening panels.

If none of the measures above is effective in reducing noise to acceptable levels, as a last resort, one can use personal hearing protective devices. These are earplugs, ear muffs, noise-attentuating helmets, or some combination of them. Nixon and Berger (1991) provide information about acceptable hearing-protective devices.

Vibration

Vibration is the alternating motion of a surface or body with respect to a reference point. The range of interest is generally between 0.1 and 80 Hz. Most vehicles

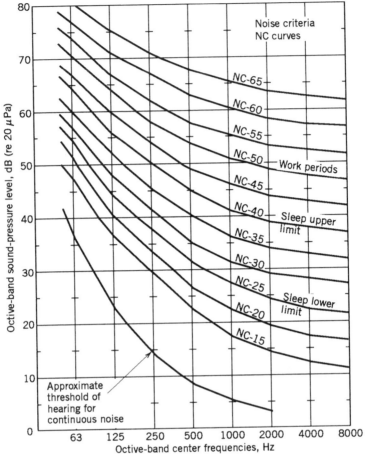

Figure 5.24 Indoor noise criteria (NC) curves. (From NASA-STD-3000)

(automobiles, railway trains, aircraft, motorcycles), machines (printing presses, lathes, threshing machines), tools, and appliances (rotary drills, jackhammers, mixers, washing machines), and even some buildings (electrical power-generating plants) vibrate. Vibration may be transmitted to operators by direct mechanical paths or through several flanking paths. For example, intense airborne noises may impinge on a structure and cause structural vibration that is then transmitted to operators. Vibrating objects may cause the whole body, or only parts of it, to vibrate.

Human Responses to Vibration The physical responses of the body to vibration are the results of complex interactions between body masses, elasticities, damping, and couplings in the low-frequency range, generally up to 50 Hz. Many organs of the body—head, eyeballs, shoulder, hand, abdominal viscera—have resonant frequencies that amplify the motions transmitted to them. These can be modified by cushioned seats, foot restraints, handholds, and mountings. A number of tools used in the manufacturing and construction industries, such as chain saws, pneumatic compactors, and electric hand tools, transmit vibrations to the hands and arms and are often responsible for the cumulative trauma disorders described earlier.

Vibration may affect performance either by modifying perception, blurring visual images of dials and indicators, or affecting control movements. It may also produce physiological and biodynamic, as well as subjective or annoyance effects. Excessive vibration can be unpleasant, painful, and even hazardous. Table 5.9 shows some effects of vibration.

Vibration Exposure Limits. Where proficiency is required for operational or maintenance tasks, whole-body vibration should not exceed the acceleration values in Figure 5.25. If comfort is the objective, those acceleration values should be divided by 3.15. Avoid equipment vibrations in the shaded area of Figure 5.25 whenever controls must be manipulated or numerals or letters must be read. These curves do not necessarily represent limits of safe exposure for all individuals and working conditions. The direction of the vibration, method of working, proficiency of the operators, and climatic conditions all modify the biological effects of vibration. Adhering to these limits should, however, reduce the risks of impairment or injuries to acceptable or negligible levels.

Building Vibration. Vibration in buildings may be caused by airborne sounds—blasts, sonic booms, low-frequency noises from aircraft—or vibration transmitted through the structure of the building from outside sources—movements of heavy vehicles. Methods for assessing the human acceptability of such vibrations are given in ISO Standard 2631 and later addenda. Table 5.10 shows maximum acceleration values for acceptable vibration environments. Figure 5.26 shows the probability of complaints from residents to impulsive shock excitations.

TABLE 5.9 Some Activities Affected by Various Vibration Frequencies*

Activity	Frequency Range (Hz)
Equilibrium	30–300
Tactile sensing	30–300
Speech	1–20
Head movements	6–8
Reading text	1–50
Tracking	1–30
Reading instruments	6–11
Depth perception	25–60
Grasping handles	200–240

*Adapted from NASA-STD-3000.

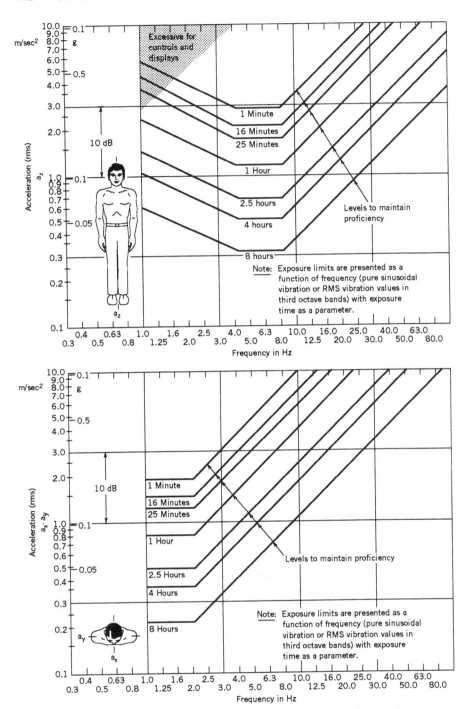

Figure 5.25 Vibration exposure criteria for longitudinal (*upper box*) and transverse (*lower box*) directions with respect to the body axis. (From MIL-STD-1472D)

INTERACTIONS WITH THE ENVIRONMENT   193

TABLE 5.10 Maximum Acceleration Values in m/sec² for Acceptable Environmental Vibration within the Range of 1 to 80 Hz

Environment	Time	Continuous or Intermittent rms Acceleration Amplitude	Impulsive Shock Excitation Peak Acceleration Amplitude
Critical Areas, Such as Hospital Operating Rooms	anytime	0.0036	0.005
Residences	daytime	$\dfrac{0.072}{\sqrt{t}}$	$\dfrac{0.1}{\sqrt{n}}$
	nighttime	0.005	0.01
Offices	anytime	$\dfrac{0.14}{\sqrt{t}}$	$\dfrac{0.2}{\sqrt{n}}$
Factories or Workshops	anytime	$\dfrac{0.28}{\sqrt{t}}$	$\dfrac{0.4}{\sqrt{n}}$

*From von Gierke & Ward *Handbook of Noise Control*, 2nd ed. (1991) with permission from McGraw-Hill, Inc. Reproduced by permission.

Daytime extends from 7:00 A.M. to 10:00 P.M.; nighttime from 10:00 P.M. to 7:00 A.M.

t = time in seconds up to 100 sec.; for times longer than 100 sec., $t = 100$.

n = number of discrete shock excitations 1 second or less in duration; for numbers of excitations greater than 100, $n = 100$.

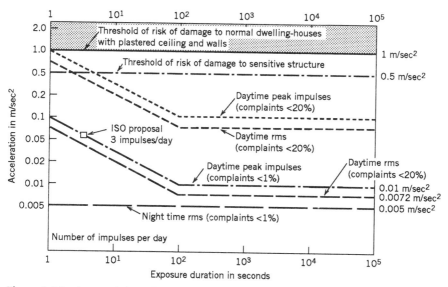

Figure 5.26 Acceptability of acceleration in residential buildings within the frequency range of 1 to 80 Hz. (From von Gierke & Ward, *Handbook of Noise Control*, 2nd ed. 1991, McGraw-Hill, Inc. Reproduced by permission.)

5: HUMAN PHYSICAL CHARACTERISTICS

Motion Sickness. Low-frequency vertical oscillations, especially in the range of 0.1 to 0.3 Hz, often result in motion sickness. At these frequencies, 1 m/sec² rms induces vomiting in about 10 percent of persons. Some approximate comfort reactions to whole-body vibration are:

Vibrational loads	Subjective reactions
Less than 0.153 m/sec²	Not uncomfortable
0.315 to 0.63 m/sec²	A little uncomfortable
0.5 to 1.0 m/sec²	Fairly uncomfortable
0.8 to 1.6 m/sec²	Uncomfortable
1.25 to 2.5 m/sec²	Very uncomfortable
Greater than 2.0 m/sec²	Extremely uncomfortable

For other extensive data on the effects of vibration, consult NASA-STD-3000.

Design Considerations   Vibration can be controlled by (a) reducing vibration at its source, (b) interrupting or absorbing vibration along a transmission path, and (c) protecting the operator. The first two are engineering techniques widely used in industrial and building vibration control, and are the preferred solutions to vibration problems. They do not involve human-factors principles. Residual over-limit vibrations can be controlled, to some extent, by protecting the operator, either through body posture or support. The semi-supine position is generally best for severe vibration in the *x, y,* or *z*-axis. The seated position is worse for *z*-axis vibration; the standing position is worse for *x* and *y*-axis vibration.

Operator supports should have resonant frequencies one-half the lowest vibration frequency of significance. Supports that can be used for this purpose include:

- Contoured seats;
- Contoured and adjustable couches;
- Elastic seat cushions;
- Suspension seats;
- Body restraints;
- Rigid or semi-rigid body enclosures;
- Head restraints; and
- Vibration absorbent hand or foot pads.

Movement and Acceleration

The principal effects of movement are on perception and reaction times. As speeds of human movement increase, less time is available for the perception of stationary objects, for example, highway signs. The solid line in Figure 5.27 shows stopping distances for vehicles traveling at various speeds on wet pavements. This curve takes into account the distance a vehicle will have traveled before a person can perceive a dangerous situation and the need to respond—the driver's reaction time,

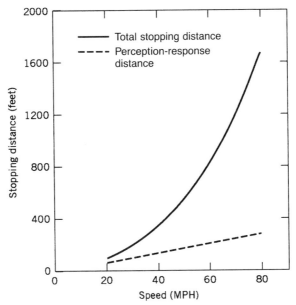

Figure 5.27 The dashed line shows the distance a vehicle will travel before a normal driver can react to an emergency. The solid line shows for design purposes the stopping distances of vehicles traveling at various speeds. (Constructed from data in Olson et al., 1984)

and the time needed for the brakes to decelerate the vehicle and bring it to a stop. These are recommended distances for the design of highways, placement of warning and directional signs, and formulation of driving regulations. Note, however, that improvements or changes in highway surfaces, tires, and vehicles may eventually require modification of these data.

The most severe effects of movement on performance are caused by linear or angular changes in movement, that is, linear or rotational accelerations. Linear accelerations are measured in g's, multiples or fractions of the accelerative force exerted by gravity; rotational acceleration in deg/sec^2. Table 5.11 defines the various directions of acceleration and their resulting effects on the body.

A special class of accelerations, impact accelerations, are involved in automobile and other vehicular accidents. These have an abrupt onset, are of short duration, and of high magnitude.

Human Responses to Linear Acceleration Tolerance to linear acceleration depends on many factors, among them, the magnitude and direction of the force, rate of onset and decline of the applied force, direction of the g vector, type of g-protection device, and body restraint worn by the operator, body position, age and physical condition of the operator, emotional and motivational factors, and training of the operator in techniques of breathing, straining, and muscular control.

Even small g-forces, as low as 1.2 or 1.3 g's, in any direction, have noticeable

TABLE 5.11 The Acceleration Coordinate System Used in NASA-STD-3000 and Their Resultants on the Human Body

	Direction of Acceleration		Inertial Resultant of Body Acceleration	
Linear Motion	Acting Force	Acceleration Description	Reaction Force	Verticular Description
Forward	$+a_X$	Forward	$+G_X$	Eyeballs In
Backward	$-a_X$	Backward	$-G_X$	Eyeballs Out
Upward	$-a_Z$	Headward	$+G_Z$	Eyeballs Down
Downward	$+a_Z$	Footward	$-G_Z$	Eyeballs Up
To Right	$+a_Y$	Right Lateral	$+G_Y$	Eyeballs Left
To Left	$-a_Y$	Left Lateral	$-G_Y$	Eyeballs Right
Angular Motion				
Roll Right	$+\dot{p}$*		$-\dot{R}_X$*	Cartwheel
Roll Left	$-\dot{p}$		$+\dot{R}_X$	
Pitch Up	$+\dot{q}$		$-\dot{R}_Y$	Somersault
Pitch Down	$-\dot{q}$		$+\dot{R}_Y$	
Yaw Right	$+\dot{r}$		$+\dot{R}_Z$	Pirouette
Yaw Left	$-\dot{r}$		$-\dot{R}_Z$	

*Overdots indicate rotational rather than linear forces.

effects on human movement and comfort, and forces of 2 or more g's have severe effects. Upward acceleration forces (+Gz) are tolerated somewhat better than downward (−Gz) forces. Under +3 to +4 g's it is impossible to raise oneself and difficult to raise arms and legs. Progressive dimming and tunneling of vision occurs after 3 or 4 seconds. Under higher g forces, from $4\frac{1}{2}$ to 6 g's one can expect progressive dimming of vision, followed by blackout and unconsciousness. With negative g forces, severe effects, for example, hemorrhaging, may occur under forces of −2 to −3 g's, and −5 g's for five seconds is the limit of tolerance rarely reached by most subjects.

Little information is available on tolerances for lateral acceleration, but they seem to be of about the same order of magnitude as for upward acceleration. In contrast to the values just given, subjects can tolerate, albeit with difficulty, up to 15 g's of forward or backward acceleration.

Human Tolerance to Rotational Acceleration Tolerance to rotational accelerations depends on the center of rotation with respect to the body, the axis of rotation, and the rotation rate. Most subjects, without prior experience, can tolerate rotation rates up to 6 rpm in any axis or combination of axes. Most subjects rapidly become sick and disoriented at rotation rates above 6 rpm unless they have been carefully prepared in a program of graduated exposure.

Human Tolerance to Impact Accelerations Greatest concern with impact accelerations, or decelerations, is not with performance or comfort, but with survival.

Tolerance to impact is determined by skeletal fracture levels: Damage to the vertebrae, often of the cervical vertebrae in the neck, is the most common kind of injury, but, at higher levels of impact, injury to the head is the most frequent and has the most severe consequences. Seat belts and air bags are, of course, the countermeasures now routinely used in automobiles. With proper design, people have survived linear decelerations of up to nearly 200 g's, a figure that is not, however, recommended.

Illumination

Light is necessary in order to see, and it can be used artistically to create aesthetically pleasing effects in homes, restaurants, lobbies, offices, museums, and other spaces. The artistic value of light was recognized even by the ancient Greeks and Romans, who oriented and designed their buildings to make most effective use of natural light. Used improperly, light can annoy, distract, affect our performance, and even blind us.

For purposes of systems design, our interest in light is primarily in its utilitarian aspects. These are best summarized in two comprehensive volumes, published by the Illuminating Engineering Society of North America (Kaufman & Haynes,

TABLE 5.12 Recommended Illuminances for Generic Types of Activities in Interiors. (Adapted with Permission from Kaufman and Haynes, 1981b, Published by the Illuminating Engineering Society of North America, 120 Wall Street, New York, NY 10005)

Type of Activity	Ranges of Illuminance	
	Lux	Footcandles
Moving in spaces with dark surroundings	20–50	2–5
Orienting oneself in spaces during short temporary visits	50–100	5–10
Working in spaces where visual tasks are only occasionally performed	100–200	10–20
Performing visual tasks of high contrast or large size	200–500	20–50
Performing visual tasks of medium contrast or small size	500–1000	50–100
Performing visual tasks of low contrast or very small size	1000–2000	100–200
Performing visual tasks of low contrast and very small size over prolonged periods	2000–5000	200–500
Performing very prolonged and exacting visual tasks	5000–10000	500–1000
Performing visual tasks of extremely low contrast and small size	10000–20000	1000–2000

1981a, 1981b). Volume 1 of the *IES Lighting Handbook* is a reference volume, containing basic information on the physics of light, vision, measurement of light, color, daylighting, light sources, and lighting calculations. Volume 2, the application volume, contains extensive data for indoor and outdoor lighting in offices, educational facilities, institutions, public buildings, industries, residences, sports and recreational areas, roadways, aviation, transportation, advertising, and underwater. Recommendations are given in great detail. Table 5.12 gives recommendations for some general types of activities in interiors.

WORK, REST, AND WORK–REST CYCLES

People, unlike machine systems, cannot continue to function at a sustained level for long periods of time. They have to rest to recover from the effects of continued work. Jobs cover the full range—from those that are physically demanding, and must be continued uninterrupted for relatively long periods of time, such as work on an automobile assembly line, to those that are much less physically demanding, such as composing a book on a word processor, in which the worker has almost complete freedom to engage in a number of different kinds of tasks requiring different levels of effort.

Jobs also differ along a physical–mental continuum. For miners, construction workers, and many other kinds of workers, the stresses are primarily physical. Air traffic controllers, on the other hand, have relatively sedentary jobs, but are subjected to mental stresses that are just as real and intense. Efficient work–rest cycles are different for different kinds of jobs, and it is difficult to come up with prescriptions that will fit them all. In fact, the only real commonality is that they *all* require rest periods, interspersed with bouts of work.

Rest or recovery doesn't mean that a person has to stop all activity or take a scheduled work break. Periods of light activity, such as record keeping, interspersed between periods of heavy activity, can be restful. So, too, can the use of muscle groups that are different from those in the main job.

Biological Rhythms A complicating factor in all decisions about work and rest cycles are cyclic changes that we all go through in our physiological and psychological states. Examples of such cyclic changes are shown in Figure 5.28. Two of the curves, heart rate and skin resistance, are physiological measures; two, arithmetic computation and probability monitoring, are performance measures. These data are averages for two teams of men (11 in all) who were isolated for 15 days in a special chamber, simulating a kind of space vehicle. All curves show regular fluctuations corresponding to earthly day and night cycles, yet the most interesting feature of the data is that the men were not working on a normal Earth schedule. Throughout the entire period, the men alternated four hours of work with two hours of rest or sleep, a schedule that might be used in certain types of missions where equipment has to be monitored around the clock.

The cyclical changes in Figure 5.28 are called *circadian rhythms*. Other

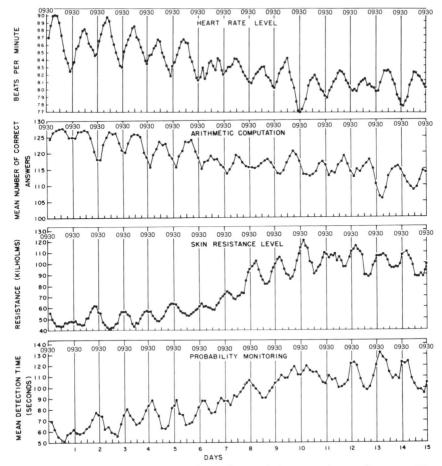

Figure 5.28 Average performance of 11 subjects during a 15-day confinement. (Data of Adams & Chiles, 1961, in Chapanis, 1971. Reprinted with permission of author and publisher)

rhythms, such as hunger cycles, may occur more frequently than daily, and still others, such as menstrual cycles in women, over a period of days. These rhythms are determined primarily by internal biological determinants that are yet only vaguely understood, and by external factors, such as light–dark cycles, seasonal changes, level of activity, times of sleeping, and activity levels. Because they are so strong and difficult to alter, these cycles affect two kinds of situations frequently experienced by system operators.

Time Zone Changes. The speeds of modern aircraft make it possible for air crews and passengers to cross several time zones within the space of a few hours, and anyone who has flown across these zones has experienced, first-hand, the effects of that translation in time. Flying west to east is generally worse than flying in the reverse direction. An early morning trip from New York to London, for example, results in a 5- or 6-hour time shift toward evening. When most local people are ready for bed, the traveler's temperature cycle is near its peak. The next morning, when a business traveler should be alert to make decisions, temperature cycles are near their minimum, resulting in impaired performance. This problem is so important that some business concerns have a policy that no decisions should be made for at least 24 hours after arrival, but even that may not be enough for some travel distances. For air crews, the Civil Aviation Organization has developed a travel-time formula that combines travel time, the number of time zones crossed, departure time, and arrival time, to specify lengths of rest periods needed to allow at least some adjustment of circadian rhythms.

Shift Work. The Bureau of Labor Statistics defines shift work as any work period that starts at a time other than between 7:00 and 9:00 A.M. By that definition, somewhat over a quarter of the American labor force is engaged in some kind of shift work. Some shift work is required to provide 24-hour service, or to provide operators for continuous processes. Examples of the former are police, fire, and hospital services and of the latter air traffic control and electrical power generation. In some industries, shift work is required because of delivery schedules. For example, many newspapers are printed at night or in the early morning, so that they can be delivered to customers in time to be read with the morning coffee. Shift work may also be determined by economic considerations. For example, computer programming and data analysis are often done at night, because computer time is less expensive and more available then.

Ideally, shift schedules should be designed on the basis of job-performance capabilities of operations at different times, the physiological effects of night work on sleep and diet, and psychosocial effects of evening and night work on families and friends. Of these, the acceptability of shift work seems to be determined more by psychological and social factors rather than physiological ones. Figure 5.29 shows three common schedules of two eight-hour shifts. Since day work is generally considered to be more preferable than evening or night work, it is usually assigned to more senior workers as a reward. The principal advantage of these schedules is their predictability. The principal disadvantages of the alternating schedules are the difficulty workers have in adjusting physiologically to the weekly changes in hours of work, and to the disruption of normal family activities. A book by the Ergonomics Group of the Eastman Kodak Company (1986) offers an extensive discussion of many other shift schedules, with their advantages and disadvantages.

In the final analysis, however, there is no ideal shift work schedule. All involve some sort of compromise. The single most important requirement is predictability. In that respect, there are a few cases in which even that consideration is violated, often with unfortunate consequences. One such case (Chapanis, 1980) involved piloting operations on the Chesapeake Bay. The job of piloting ships between Baltimore and Cape Henry (at the mouth of Chesapeake Bay) is a demanding task, requiring the exercise of specialized knowledge and skills, constant alertness and vigilance, and decision-making of the most complex sort. At the time of the study, the middle 50 percent of trips took between 9.5 to 12 hours.

WORK, REST, AND WORK–REST CYCLES   201

Schedule	Crew	Week 1 MTWTF SSu	Week 2 MTWTF SSu	Week 3 MTWTF SSu	Week 4 MTWTF SSu
Fixed shifts, A	1	AAAAA RR	AAAAA RR	AAAAA RR	AAAAA RR
and B	2	BBBBB RR	BBBBB RR	BBBBB RR	BBBBB RR
Weekly alterna-	1	AAAAA RR	BBBBB RR	AAAAA RR	BBBBB RR
tion, A and B (dou- ble-days)	2	BBBBB RR	AAAAA RR	BBBBB RR	AAAAA RR
Weekly, A and C	1	AAAAA RR	CCCCC RR	AAAAA RR	CCCCC RR
	2	CCCCC RR	AAAAA RR	CCCCC RR	AAAAA RR

Figure 5.29 Three common schedules of two eight-hour shifts. Starting times are 8 A.M. for the A shift, 4 P.M. for the B shift, and midnight for the C shift. R indicates rest days. (Reprinted courtesy of Eastman Kodak Company, 1986)

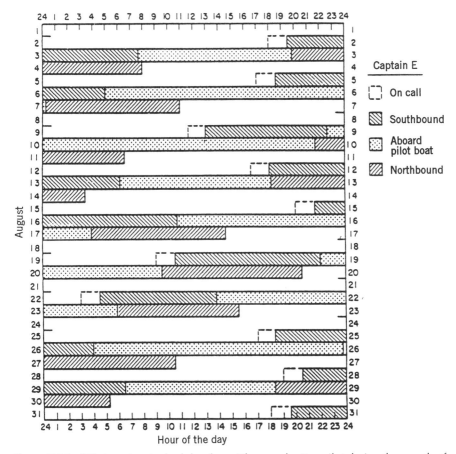

Figure 5.30 Work and rest schedule of one Chesapeake Bay pilot during the month of August 1980. (From Chapanis, 1980)

The typical routine was that a pilot would be "on call" at home. When a ship was ready to depart Baltimore, the captain would receive a call to report for work, pilot the ship southward to Cape Henry, disembark, and rest aboard a pilot (dormitory) boat until an incoming ship appeared, and then pilot that one to Baltimore. Although most southbound trips began between 1600 and 2100 hours, northbound ships tended to arrive at Cape Henry at all hours of the day or night, in a random pattern. The work schedule of one pilot during the month of August 1980 is shown in Figure 5.30. Small wonder the pilots were complaining of fatigue!

SUMMARY

This chapter has surveyed briefly some of the important characteristics of human operators—the raw materials with which human-factors professionals have to work. The emphasis in the chapter has been on some general human characteristics, on the body as a structural and biological entity, and on the multiplicity of environmental factors that affect human functioning. One major theme of the chapter has been diversity, or human variability in every way in which people can be measured. This diversity creates some of the greatest problems for system design and for the human-factors professional. Human variability is not infinite, however. We are able to provide reasonable estimates of practical working limits to human variability. Although these estimates are not as precise as most physical measures, they are good enough to specify design limits that will accommodate a majority of potential users. Equipment designed to such specifications will enable most users to work safely, comfortably, and effectively, and will avoid some of the glaring examples of user–equipment mismatches that have sometimes occurred in the past.

REFERENCES

Adams, O. S., & Chiles, W. D. (1961). *Human Performance as a Function of the Work–Rest Ratio During Prolonged Confinement.* ASD Technical Report 61-720. Wright-Patterson Air Force Base, Ohio: Aerospace Medical Laboratory.

American Standards Association (1954). *The Relations of Hearing Loss to Noise Exposure.* New York: American Standards Association.

Angiolillo-Bent, J. S. (1985). The role of human factors in the translation of operating instructions. In *Proceedings of the Human Factors Society 29th Annual Meeting* (pp. 91–94). Santa Monica, CA: Human Factors Society.

Bailey, R. W. (1982). *Human Performance Engineering: A Guide for System Designers.* Englewood Cliffs, NJ: Prentice-Hall.

Chapanis, A. (1971). The search for relevance in applied research. In W. T. Singleton, J. G. Fox, & D. Whitfield (Eds.), *Measurement of Man at Work* (pp. 1–14). London: Taylor and Francis.

Chapanis, A. (Ed.) (1975). *Ethnic Variables in Human Factors Engineering.* Baltimore: The Johns Hopkins University Press.

Chapanis, A. (1980). *Hours of Work, Work Schedules, and Fatigue in Certain Chesapeake Bay Piloting Operations.* Baltimore: Unpublished manuscript.

Chapanis, A., Garner, W. R., & Morgan, C. T. (1949). *Applied Experimental Psychology: Human Factors in Engineering Design.* New York: John Wiley & Sons.

Chapanis, A., & Moulden, J. V. (1990). Short-term memory for numbers. *Human Factors, 32,* 123–137.

Ciriello, V. M., Snook, S. H., & Hughes, G. J. (1993). Further studies of psychophysically determined maximum acceptable weights and forces. *Human Factors, 35,* 175–186.

Daniels, G. S., & Churchill, E. (1952). *The "Average" Man?* Report Number WCRD-TN-53-7. Wright Patterson Air Force Base, Ohio: Aero Medical Laboratory.

Ericsson, K. A., & Charness, N. (1994). Expert performance: its structure and acquisition. *American Psychologist, 49,* 725–747.

Freedman, M., Zador, P., & Staplin, L. (1993). Effects of reduced transmittance film on automobile rear window visibility. *Human Factors, 35,* 535–550.

Gordon, C. C., Churchill, T., Clauser, C. E., Bradtmiller, B., McConville, J. T., Tebbetts, I., & Walker, R. A. (September 1989). *1988 Anthropometric Survey of U.S. Army Personnel: Methods and Summary Statistics.* Technical Report NATICK/TR-89/044. Natick, MA: United States Army Natick Research, Development and Engineering Center, Science and Advanced Technology Directorate.

Harris, C. M. (Ed.) (1991). *Handbook of Acoustical Measurements and Noise Control.* (3rd Edition). New York: McGraw-Hill.

Hertzberg, H. T. E. (1972). Engineering anthropology. In H. P. Van Cott & R. G. Kinkade (Eds.), *Human Engineering Guide to Equipment Design* (pp. 467–584). (Revised Edition). Washington, DC: Government Printing Office.

Hinchcliffe, R. (1959). The threshold of hearing as a function of age. *Acustica, 9,* 303–308.

Huey, R. W., Buckley, D. S., & Lerner, N. D. (1994). Audible performance of smoke alarm sounds. In *Proceedings of the Human Factors and Ergonomics Society 38th Annual Meeting* (pp. 147–151). Santa Monica, CA: Human Factors and Ergonomics Society.

Kaufman, J. E., & Haynes, H. (Eds.) (1981a). *IES Lighting Handbook: Reference Volume.* New York: Illuminating Engineering Society of North America.

Kaufman, J. E., & Haynes, H. (Eds.) (1981b). *IES Lighting Handbook: Application Volume.* New York: Illuminating Engineering Society of North America.

Kennedy, K. W. (1975). International anthropometric variability and its effects on aircraft cockpit design. In A. Chapanis (Ed.) *Ethnic Variables in Human Factors Engineering* (pp. 47–66). Baltimore: The Johns Hopkins University Press.

Kroemer, K. H. E. (1985). *Ergonomics Manual for Manual Material Handling.* (3rd rev. ed.). Blacksburg, VA: Ergonomics Laboratory, IEOR Department, Virginia Polytechnic Institute and State University.

Kroemer, K. H. E., Kroemer, H. J., & Kroemer-Elbert, K. E. (1986). *Engineering Physiology: Physiologic Bases of Human Factors/Ergonomics.* New York: Elsevier.

Kroemer, K., Kroemer, H., & Kroemer-Elbert, K. (1994). *Ergonomics.* Englewood Cliffs, NJ: Prentice-Hall.

Kryter, K. D. (1985). *The Effects of Noise on Man.* (2nd Ed.). New York: Academic Press.

Laubach, L. L. (1978). Human muscular strength. In *Anthropometric Source Book. Volume I:*

Anthropometry for Designers (pp. VII-1–VII-55). (NASA Ref. Pub. #1024). National Aeronautics and Space Administration, Scientific and Technical Information Office.

Miles, W. R. (1931). Correlation of reaction and coordination speed with age in adults. *American Journal of Psychology, 43*, 377–391.

Miles, W. R. (1933). Abilities of older men. *Personnel Journal, 11*, 352–357.

MIL-STD-1472D (Mar. 14, 1989). *Military Standard: Human Engineering Design Criteria for Military Systems, Equipment and Facilities.* Washington, DC: Department of Defense.

Mital, A., & Kumar, S. (1994). Special issue preface. *Human Factors, 36*, 195–196.

Nixon, C. W., & Berger, E. H. (1991). Hearing protection devices. In C. M. Harris (Ed.), *Handbook of Acoustical Measurements and Noise Control.* (pp. 21.1–21.24). New York: McGraw-Hill.

Olson, P. L., Cleveland, D. E., Fancher, P. S., Kostyniuk, L. P., & Schneider, L. W. (June 1984). *Parameters Affecting Stopping Sight Distance* (Report No. 270). Washington, DC: Transportation Research Board, National Research Council.

OSHA 3048 (1980). *Noise Control: A Guide for Workers and Employers.* Washington, DC: Occupational Safety and Health Administration.

Rodgers, S. H. (1985). *Working with Backache.* Fairport, VA: Perington.

Roebuck, J. (1995). *Anthropometric Methods: Designing to Fit the Human Body.* Santa Monica, CA: Human Factors and Ergonomics Society.

Roebuck, J., Smith, K., & Raggio. L. (1988). Forecasting crew anthropometry for shuttle and space station. In *Proceedings of the Human Factors Society—32nd Annual Meeting* (pp. 35–39). Santa Monica, CA: Human Factors Society.

Schaie, K. W. (1994). The course of adult intellectual development. *American Psychologist, 49*, 304–313.

Schieber, F., & Kline, D. W. (1994). Age differences in the legibility of symbol highway signs as a function of luminance and glare level: a preliminary report. In *Proceedings of the Human Factors and Ergonomics Society 38th Annual Meeting* (pp. 133–136). Santa Monica. CA: Human Factors and Ergonomics Society.

Sinaiko, H. W. (1975). Verbal factors in human engineering: Some cultural and psychological data. In A. Chapanis (Ed.), *Ethnic Variables in Human Factors Engineering* (pp. 159–177). Baltimore: The Johns Hopkins University Press.

Sloan, L. L. (1947). Rate of dark adaptation and regional threshold gradient of the dark-adapted eye: Physiologic and clinical studies. *American Journal of Ophthalmology, 30*, 705–720.

Stoudt, H. W., Damon, A., & McFarland, R. A. (1960). Heights and weights of white Americans. *Human Biology, 32*, 331–341.

von Gierke, H. E., & Ward, W. D. (1991). Criteria for noise and vibration exposure. In C. M. Harris (Ed.), *Handbook of Acoustical Measurement and Noise Control* (pp. 26.1–26.17). New York: McGraw-Hill.

Von Neumann, J. (1951). The general and logical theory of automata. In L. A. Jeffress (Ed.), *Cerebral Mechanisms in Behavior* (pp. 1–31). New York: John Wiley & Sons.

Wechsler, D. (1952). *The Range of Human Capacities.* Baltimore: Williams and Wilkins.

Wechsler, D. (1981). *WAIS-R Manual.* San Antonio, TX: The Psychological Corporation, Harcourt–Brace–Jovanovich.

White, R. M., & Churchill, E. (1971). *The Body Size of Soldiers: U.S. Army Anthropo-*

metry—1966 (Technical Report 72-51-CE). Natick, MA: U.S. Army Natick Laboratories, Clothing and Personal Life Support Equipment Laboratory.

Wisner, A. (1985). *Quand Voyages les Usines.* Mesnil-sur-l'Estrée, France: Société Nouvelle Firmin-Didot.

Wyndham, C. H. (1975). Ergonomic problems in the transition from peasant to industrial life in South Africa. In A. Chapanis (Ed.), *Ethnic Variables in Human Factors Engineering* (pp. 115–134). Baltimore: The Johns Hopkins University Press.

Chapter 6

Human Mental Characteristics

Chapter 5 was concerned primarily with those human characteristics that are structural and mechanical. As important as they may be, there is another part of human beings that is even more important for systems design, and that is human abilities to deal with the abstractions we call information. These are mental, or, as they are generally referred to these days, *cognitive* abilities. Do not be misled by my apparent division of human characteristics into structural and mechanical versus mental. The distinction is only one of convenience. There are complex interactions between the two—our bodies affect our minds, and vice versa.

In Chapter 5, I commented repeatedly on how difficult it is to get precise specifications of the structural and biological characteristics of human beings. Getting quantitative data about mental functions is even more difficult. Nonetheless, it is possible to get enough useful information about those functions to help in the design of systems we can use successfully. This chapter surveys some of those functions and reviews briefly what we know about them.

THE HUMAN SUBSYSTEM

Engineers have a machine model that they use in talking and thinking about systems (Figure 6.1). A machine receives inputs, transforms those inputs through some sort of function, and outputs the results. With only a change of terms this same model applies to human behavior. Human beings receive stimuli, process the information they receive, and output the results in the form of responses. This parallelism makes it possible and convenient to think of the human operator as a system component.

Although we may think of the human operator as a system component, we must not take the analogy too far, because all human activities are affected by a large

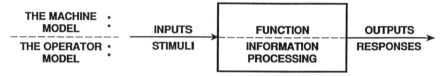

Figure 6.1 An engineering or machine model (*above*) and the corresponding operator model (*below*).

number of variables, some relating to the machines with which we interact and the jobs we do, some that are uniquely human. Among the former are the machine interfaces that operators use, the environments in which operators work, the instructions provided them, and the quality and quantity of training they have received. Among the distinctly human variables that affect performance, both physical and mental, are fatigue and boredom, stress, attitudes, motivation, and personality.

Human operators also have special needs and requirements if they are to function effectively or even survive. As I pointed out in Chapter 5, the environment has to be designed so that variables such as temperature, humidity, air pressure, oxygen content, and noxious gases and fumes are all within definite limits. Although some of these same factors have to be considered in the design of systems for the protection or functioning of hardware and software components, the tolerance limits for them are generally much greater than for humans. Machine systems can operate in the total vacuum of outer space, for example, whereas humans cannot survive unless they are provided with self-contained environments in the form of space suits. Finally, the human operator has both nutritional and psychosocial needs, and must be provided with periods of rest to recover from fatigue and to restore both mental and physical energies.

An Operator–Machine Model

Figure 6.2 is a more complete model of an operator–machine combination. Machine systems convey information to their operators by means of one or more displays. The displays may be any of a very large number of things: lights, dials, gages, bells, CRTs, buzzers, speech, vibration, or the tactile feel of controls. The operator senses this information, processes it, and makes a decision about what to do with the information. That decision is effected through any one or more of a large number of possible controls, such as switches, pedals, levers, cranks, and keys. The control action changes the behavior of the machine and the information presented on the displays, thus completing the cycle.

The operator–machine combination does not work in a vacuum; it works in an environment of some sort. Some of the constituents of that environment are shown in Figure 6.2, and were discussed in Chapter 5.

Examples Driving an automobile is a familiar example of a system that fits the model in Figure 6.2. The displays consist of (1) instruments—such as the fuel

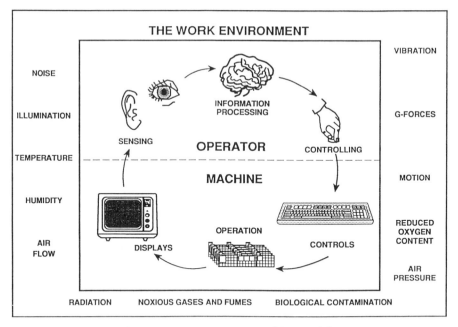

Figure 6.2 An operator–machine model.

indicator and speedometer on the instrument panel; (2) information about traffic—the road, signs, and obstacles that the driver senses through the windshield; (3) noises that convey information about the performance of the vehicle and traffic; and (4) tactual information about the "feel" of the vehicle and road surface. All this information is somehow integrated in the driver's head. The driver constantly makes decisions about it. One decision may be that everything is okay and that no action is required. Another decision might be that it would be a good idea to change lanes, in which case the driver carries out that decision by turning the steering wheel and perhaps pressing on the accelerator or brake. These actions change the movement of the automobile and the displays, thus completing the cycle. All these activities may be performed under extremes of temperature, humidity, and illumination. Vibration and noise are some of the other environmental factors that impact the driver.

Working with a computer fits the model in Figure 6.2. The displays are different, to be sure, for in this case they most likely consist of information on a computer screen. The operator reads that information, processes it, and makes a decision about appropriate actions to take. These actions may be typing into a keyboard, moving a "mouse," rolling a trackball, pointing, or sometimes even by speaking into the computer. These actions change the performance of the computer and the displays associated with them, thus completing the cycle. Although computer systems are generally in benign environments, some, such as those on factory floors, may be used under more severe conditions. Dirt, grease, noise, and glare are some of the environmental conditions that often hinder efficient computer operation.

These two examples should be enough to show the generality of the model in

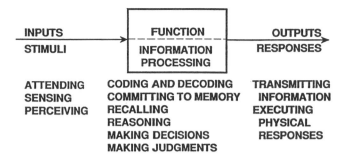

Figure 6.3 Some human functions that have been associated with various stages of the operator model.

Figure 6.2. Although it is a simple model, it fits many of the operator–machine combinations we encounter in our work and lives. The model also serves as a useful way of dissecting the human operator for purposes of analysis and discussion.

Figure 6.3 repeats the general model in Figure 6.1 with the addition of some of the specific kinds of human functions that are generally associated with each stage.

ATTENDING

All the information we receive comes to us through one or more of our senses. When we are awake and our senses are functioning properly, they have no way of avoiding the stimuli that constantly bombard them. Filtering is done by the brain in a process called attending. We "pay attention to," select out, or concentrate on those things we want to receive, and filter out the rest. Attention can also be focused inward toward our own thoughts. Sometimes we may concentrate so heavily on those thoughts that we are almost oblivious to external stimuli.

Three features of attending have design implications.

Selective Attention

The most direct way of selecting stimuli to attend to, is to orient your senses toward one set of stimuli and away from another. The principal way we select stimuli visually is by moving our eyes. Something catches our attention in the periphery and we almost involuntarily move our eyes so that the item of interest stimulates the fovea, the central part of the eye that is most sensitive and acute. For that reason things that we want to command our attention most readily should be positioned so that they will most naturally fall within that cone of most sensitive vision. (In this connection, see Figure 6.8.)

We can also physically orient other senses, but not as obviously or as effectively as with vision. We can turn our heads so that our ears will most readily pick up the sounds we want to hear, and we explore with our hands the location of keys on a keyboard and the shapes of controls on a control panel.

Because our ears are always open and receive information from all directions, it is difficult to direct attention to one source of sound when there are several in the environment, the "cocktail-party effect," or more formally, multi-channel listening. This occurs when several people are talking at the same time and you want to focus on, or listen to, one particular person. Such problems occur in some systems, for example, air-traffic-control towers, disaster control centers, and control rooms for some manufacturing plants. The problem has been extensively studied, and there are several things one can do to help operators attend to one source of speech: physically separate speakers over which information is coming, differentiate speech signals by frequency or intensity filtering, feed different channels into separate ears, and use a visual signal to show which channel is in use. For more information on these and other design options, see Kryter (1972).

Time Sharing

Humans have a very limited capacity for processing information, but many systems require operators to perform two or more different tasks simultaneously or in rapid alternation. This activity, known as time sharing, or dividing attention between activities, results in the degradation of performance of one or more of the tasks if their combined demands exceed the capacity of the individual. If you are an experienced driver, you have no difficulty driving and talking at the same time under ordinary conditions. But if you find yourself suddenly driving in difficult conditions, say in a thunderstorm, you will have difficulty doing both tasks and will most likely concentrate on driving rather than talking. Searching and navigating a helicopter in low-level flight is a more sophisticated example of the same kind of problem. From a design standpoint it is important not to have operators time-share activities that exceed their capacities or that may do so under stressful conditions. Unfortunately, we can estimate those capacity limits only approximately. If time sharing is likely to impose speed or load stress, it helps the receiver to know priorities so that attention can be given to the more important inputs first.

Capturing Attention

Many times it is important to capture an operator's attention when he or she is engaged in some other activity, for example, to warn personnel of impending danger, alert operators to a critical change in system or equipment status, remind an operator of a critical action or actions that must be taken, or provide advisory and tutorial information. Since it takes time for an operator to shift attention from a task at hand to an alarm, caution and warning displays should contain (1) an alerting signal, and (2) critical information concerning the event.

The first, and perhaps most important consideration in the design of an alerting signal is to pick the sense channel that is most appropriate for the job. Visual alarms are effective only if they can be seen and only if the operator is awake. Auditory alarms, on the other hand, are omnidirectional, and our ears are always open. They can be effective no matter where an operator is located and can even wake an operator from a sound sleep. If the work environment is already very noisy, however, a visual alerting system may be most effective. Of course, if the operator is already overloaded with visual information, an auditory alarm is probably still the best choice. In some situations a vibratory alarm may even be useful if the operator, for example, a pilot, is fully occupied with both visual and auditory inputs.

Having selected a sense channel, a number of other design considerations are important:

- Alarms should be compelling under all operating conditions, that is, they should be sufficiently different from other sources of sensory input in the environment so that they are immediately recognizable as alarms.
- Each abnormal condition should trigger a unique alarm, that is, it should be possible to identify rapidly the condition to which the alarm is calling attention.
- It should be easy to determine the source of the alarm.
- The alarm should convey information about the approximate level of the danger, hazard, or abnormality.
- In the case of multiple simultaneous abnormal conditions, alarms should be prioritized according to their importance or degree of hazard, that is, alarms should signal only the most important condition(s) and alarms for lesser problems should be suppressed.

The considerations listed above are very general ones. Detailed design recommendations can be found in a number of references, for example, NASA-STD-3000.

SENSING

We have more exact quantitative data about human sensing than about any other of the mental activities considered in this chapter. The encyclopedic work by Boff, Kaufman, and Thomas (1986) is an indispensable source of those basic data for researchers and specialists, but is much too esoteric and theoretical for engineers and designers. A companion three-volume work (Boff & Lincoln, 1988) is more practically oriented. It is an in depth treatment of human sensing and perception in terms of the variables that influence the human operator's ability to acquire and process information and make effective decisions. What follows here is a sketch of some of the most important features of human sensing that are covered comprehensively in those reference works.

Human senses respond to specific types and ranges of physical energy, called stimuli. Some sense organs respond to external energy, for example, the eye to luminous energy, while others respond to internal stimuli, for example, the kinesthetic receptors to limb movements. Table 6.1 lists the human sense organs, the types of energy that stimulate them, their corresponding sensations, and the origins of stimulation. Of all the senses listed in Table 6.1, only three are used routinely for

TABLE 6.1 Human Senses and the Energies That Stimulate Them*

Sensation	Sense Organ	Stimulation	Origin
Sight (vision)	eye	some electromagnetic waves	external
Hearing (audition)	ear	some pressure variations in surrounding media	extenal
Rotation	semicircular canals	change of fluid movement in inner ear	internal
	muscle receptors	muscle stretching	internal
Falling and rectilinear movement	otoliths	position changes of small bony particles in inner ear	internal
Taste	specialized cells in tongue and mouth	some chemical substances	external on contact
Smell	specialized cells in nasal cavity	some vaporized chemical substances	external
Touch	skin	surface deformation	on contact
Pressure	skin and underlying tissue	surface deformation	on contact
Temperature	skin and underlying tissue	temperature changes in surrounding media or objects, friction, some chemicals	external on contact
Pain	uncertain, thought to be free nerve endings	intense pressure, heat, cold, shock, some chemicals	internal or external on contact
Position and movement (kinesthesis)	muscle nerve endings	muscle stretching	internal
	nerve endings in tendons	muscle contraction	internal
	joints	unknown	internal
Mechanical vibration	no specific organ	variations of pressure on skin	external on contact

*Adapted from Van Cott & Warrick (1972) and from Mowbray & Gebhard (1958).

conveying information from machines and systems to operators: vision, audition, and touch.

The Absolute Threshold

A measure of practical importance is the smallest amount of energy that can be sensed, the absolute threshold. This is not a specific, precise value but is rather the average of a statistical distribution of energy values. Table 6.2 gives some stimulation-intensity ranges of human senses.

Young, healthy senses are extraordinarily acute. For example, fully dark-adapted

TABLE 6.2 Stimulation–Intensity Ranges of Human Senses*

Sensation	Smallest Detectable (threshold)	Largest Tolerable or Practical
Vision	10^{-6} mL	10^4 mL
Audition	2×10^{-4} dynes/cm^2	$<10^3$ dynes/cm^2
Mechanical vibration	25×10^{-5} mm mean amplitude at figertip (maximum sensitivity at 200 Hz)	varies with size and location. Pain likely at 40 dB above threshold
Touch (pressure)	at fingertip, 0.04 to 1.1 erg; "Pressure," 3 gm/mm^2	unknown
Smell	very sensitive for some substances, e.g., 2×10^{-7} mg/m^3 of vanillin	unknown
Taste	very sensitive for some substances, e.g., 4×10^{-7} molar concentration of quinine sulfate	unknown
Temperature	15×10^{-5} gm-cal/cm^2/sec for 3 sec. exposure of 200 cm^3 skin	22×10^{-2} gm-cal/cm^2/sec for 3 sec. exposure of 200 cm^2 skin
Position and movement	0.2–0.7 deg. at 10 deg./min. for joint movement	unknown
Acceleration	0.02 g for linear acceleration 0.08 g for linear deceleration 0.12 deg./sec^2 rotational acceleration for oculogyral illusion (apparent motion or displacement of viewed object)	5 to 8 g positive 3 to 4 g negative disorientation, confusion, vertigo, blackout, or redout

*Adapted from Van Cott & Warrick (1972) and from Mowbray & Gebhard (1958).

eyes can detect a candle flame 30 miles away on a clear dark night. If our ears were any more sensitive they would be able to hear the random collision of molecules in still air. This kind of sensitivity occurs only within very narrow limits and under special conditions, but it can be augmented by physical devices, for example, infrared viewing devices, binoculars, amplifiers, that detect energies outside human limits, or that increase sensitivities to levels that are more favorable for reception. When the outputs of such devices have been matched to the capabilities of human sense organs they can greatly increase human ability to receive information. To use human senses to their best advantage, external conditions, the energy source, its location, intensity, coding, rate, and/or manner of presentation, have to be selected or changed to match the capabilities of particular senses. A great deal of research yielding specific design implications and principles is summarized in a number of sources. (For example, MIL-STD-1472D, NASA-STD-3000, Sanders & McCormick 1993, Van Cott & Kinkade 1972.)

Signal Detection

Threshold measurements reflect not only a person's sensitivity—how well that person can see, hear, or feel—but also that person's response bias—how willing that person is to say, "I heard it," "I saw it," or "I felt it," when not entirely certain. The situation is illustrated in Figure 6.4, which shows four possible outcomes of a detection situation. If there really is a signal and the operator responds "Yes," we have a hit. If the operator responds "No," we have a miss. On the other hand, if there really is no signal and the operator responds "Yes," we have a false alarm, and if the operator responds "No," then we have a correct rejection.

For any fixed signal intensity, the proportions in the four cells of Figure 6.4 vary according to the person's mental set or response bias, and this response bias can be affected by attitudes and instructions. If an operator decides, or is told, to be absolutely sure about a signal before reporting it, the proportion of misses will be decreased, but so also will the proportion of hits and false alarms. On the other hand, if the operator decides, or is told, to report all signals that he or she thinks are sensed, even if he or she is not sure, the proportion of hits increases, and so also does the proportion of misses and false alarms. What happens as you vary response bias is shown in Figure 6.5, which is called a *receiver operating characteristic,* or ROC, curve (see, for example, Egan, 1975). If the signal is constant, the outcomes move up the curve as the response criterion is shifted from a strict one (reporting only those that are clearly signals) to a more liberal one (reporting all those that *might* be signals). In other words, to increase the proportion of hits, you have to be willing to accept a greater proportion of false alarms.

Operators often have to make decisions in situations resembling the one schematized in Figure 6.4. Is that a target on the radar screen, or isn't it? Is the ship beginning to drift off course, or isn't it? Does this item pass inspection, or doesn't it? Since operators can't always be correct in such situations, it is important for them to know whether it's more important to maximize hits or to minimize false alarms.

	Operator responds	
	"Yes"	"No"
Signal present	Hit	Miss
Signal absent	False alarm	Correct rejection

Figure 6.4 Four possible outcomes in a detection situation.

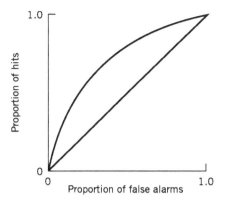

Figure 6.5 A receiver–operator–characteristic, or ROC, curve. Random responding results in the straight line from the lower left to upper right. Human responding typically results in a curve above the straight line varying in the amount of its concavity.

Upper Thresholds

For all senses there is an upper threshold, an intensity above which sense organs do not respond or which would be injurious. No one has tried to determine exactly the upper threshold for hearing, for example, because intensities above about 120 or 130 dB begin to cause permanent damage to the ears. Table 6.2 shows approximate upper-threshold values for some senses.

There are also limits in the frequency ranges to which our sense organs respond (Table 6.3). The lower limits for flicker (interrupted white light) and for mechanical vibration are the frequencies at which stimuli begin to be perceived as flicker or vibration, and not as separate pulses. Incidentally, both our tactile and auditory senses respond to pressure changes in the range of about 20 to 1,000 Hz.

TABLE 6.3 Approximate Frequency-Sensitivity Ranges for Four Stimulus Dimensions*

Sensation	Lower Limit	Upper Limit
Color (hue)	390 nm**	780 nm
Flicker (interrupted white light)	0.5 interruptions/sec	50 interruptions/sec at moderate intensities and on/off cycle of 0.5
Pure tones	20 Hz	20,000 Hz
Mechanical vibration	0.5 Hz	10,000 Hz at high intensities

*Adapted from Van Cott & Warrick (1972) and from Mowbray & Gebhard (1958).
**nanometers = 10^{-9} m.

Difference Thresholds

The magnitudes of sensations increase as physical energy increases, but the relationship is not linear. Equal increments of physical energy do not produce equal increments of sensation. For example, doubling sound energy does not double loudness; about a tenfold increase in energy is required for a sound to seem twice as loud. In general, the perceived magnitude of a sensation is proportional to some power of the physical magnitude, a power that is different for different stimuli.

The smallest change in a stimulus that can be detected is called the Just Noticeable Difference (JND), and is symbolized by Δ. Generally, ΔI is a JND in intensity, Δf a JND in frequency, and $\Delta \lambda$ a JND in wavelengths. Over typically useful ranges of some stimuli, the ΔI corresponding to a JND bears a roughly constant relationship to the size of the reference stimulus, that is, $\Delta I / I = C$, an equation called Weber's Law, after the man who first propounded it. The ratio $\Delta I / I$ is called the Weber fraction. Table 6.4 gives some representative middle-range values for the Weber fraction. An extension of Weber's Law is Fechner's Law, which states that the strength of a sensation increases as the logarithm of intensity, that is

$$S = k \log_{10} I$$

where S is the psychological magnitude, I the stimulus intensity, and k is a constant which depends on the sense stimulated and the stimulus dimension concerned, that is, whether intensity, frequency, or some other dimension. These equations are only *approximately* correct. Don't overestimate their precision.

Table 6.5 shows the number of JNDs (ΔI's or Δf's) within the sensitivity ranges of three sense organs. Although data such as these may be useful as guides for designing or evaluating displays, it is important to remember that they are ideal values. They were determined under carefully controlled laboratory conditions and with apparatus that typically presented the standard and comparison stimuli together and side-by-side, or at least close to each other. Practically useful detectable

TABLE 6.4 Representative Values of the Weber Fraction for Some Senses*

Sense	Weber Fraction $\Delta I / I$
Vision (brightness, white light)	1/60
Audition (tone of middle pitch and moderate loudness)	1/10
Kinesthesis (lifted weights)	1/50
Pain (thermally aroused on the skin)	1/30
Pressure (pressure on a skin "spot" receptor)	1/7
Smell (odor of India rubber)	1/4
Taste (table salt)	1/3

*From Geldard (1953). Reprinted by permission of John Wiley & Sons, Inc.

TABLE 6.5 Relative Discrimination of Physical Intensities and Frequencies*

Sensation	Number of JNDs
Brightness	570 discriminable intensities of white light.
Hues	128 discriminable wavelengths at medium intensities.
Flicker	375 discriminable interruption rates between 1 and 45 interruptions/sec. at moderate intensities and an on/off cycle of 0.5.
Loudness	325 discriminable intensities for pure tones of 2,000 Hz.
Pure tones	1,800 discriminable tones between 20 and 20,000 Hz at 60 dB.
Interrupted white noise	460 discriminable interruption rates between 1 and 45 interruptions/sec. at moderate intensities and an on/off cycle of 0.5.
Vibration	15 discriminable amplitudes in chest region with broad contact vibrator within an amplitude range of 0.05–0.5 mm.
Mechanical vibration	180 discriminable frequencies between 1 and 320 Hz.

*Adapted from Van Cott & Warrick (1972) and from Mowbray & Gebhard (1958).

changes in the intensity of warning lights, sounds, or vibrations, are usually several times greater than these ideal values. For example, easily discriminable contrasts on CRT screens, called *shades of gray,* are seven times a JND and increase by factors of $\sqrt{2}$.

Absolute Judgments

Many tasks require operators to make judgments about stimuli in the absence of a comparison. The pilot who has to judge separations visually, the navigator who has to determine the color of a distant beacon light, the maintainer who has to diagnose a difficulty from the sound of equipment, the vintner who has to judge the quality of a sample of wine by tasting it, all are making absolute judgments. There is an astonishing shrinkage in discriminability as one goes from relative to absolute judgments. Table 6.6, for example, gives data on the absolute identification of

TABLE 6.6 Number of Absolutely Identifiable Sensations*

Sensation	Number Identifiable
Brightness	3 to 5 with white light of 0.1–50 mL
Hues	12 or 13 wavelengths
Interrupted white light	5 or 6 rates
Loudness	3 to 5 with pure tones
Pure tones	4 or 5 tones
Vibration	3 to 5 amplitudes

*Adapted from Van Cott & Warrick (1972) and From Mowbray & Gebhard (1958).

intensities for some sensory dimensions. For many design problems, the values in Table 6.6 are more realistically usable ones than those in Tables 6.4 or 6.5. Data on several other sensory dimensions, for example, color, pitch, lengths of lines, flicker rates, and numbers of dots, are available in a number of sources (for example, Grether & Baker, 1972).

Vision

Of all human senses, vision is probably the most important for system design. Some investigators have estimated that as much as 80 percent of the information we receive from machines and systems comes to us by way of our eyes. Two of the most remarkable capacities of the visual system are its ability to discriminate forms and colors.

Visual Activity Visual activity refers to our ability to see spatial detail and the most useful measure of acuity is the visual angle, the angle subtended by the object viewed. For objects less than 10° in size and perpendicular to the line of sight, the visual angle is closely approximated by

$$\text{Visual angle (minutes of arc)} = (57.3)(60)L/D$$

where L = the size of the object measured perpendicularly to the line of sight and D = the distance from the front of the eye to the object.

Normal, healthy eyes are extraordinarily acute. They can detect objects subtending no more than about 0.5 seconds of arc (roughly a wire 1/4" in diameter at a distance of a quarter of a mile). Acuity measures vary greatly depending on the kind of target used to measure it and the task we are required to do. Figure 6.6 shows some common types of targets used in measuring acuity and S, L, W, C and α indicate the critical dimensions used in calculating the acuity measure. In that figure stereo acuity is different from the others. It is a measure of depth perception or the ability to detect that objects are at different distances.

Generally speaking, acuities decrease as the complexity of targets increases. They can be grouped in four categories (Table 6.7 and Figure 6.7):

1. Detection, merely detecting the presence of something;
2. Vernier, detecting misalignment;
3. Separation, detecting the separation or gap between parallel lines, dots or squares; and
4. Form, identifying shapes or forms.

It is important to keep in mind that all the measures in Table 6.7 and Figure 6.7 were obtained under ideal laboratory conditions with healthy eyes and so represent minimum values. Although the various kinds of acuity tend to maintain their same relative rankings in practical situations, their magnitudes may easily increase by a

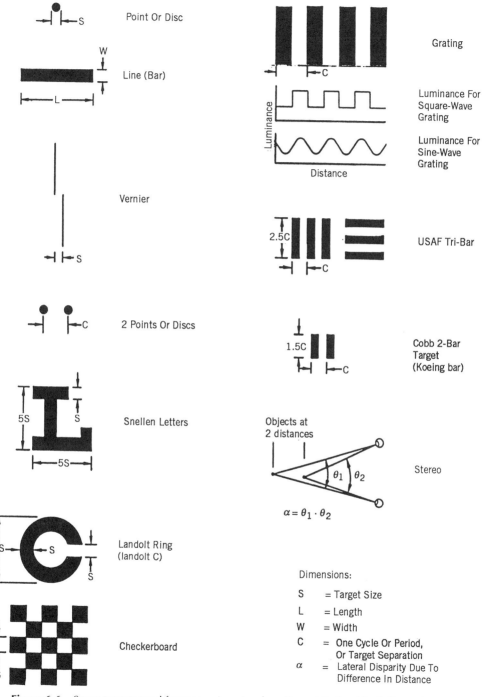

Figure 6.6 Some targets used for measuring visual acuity and their critical dimensions. (From Farrell & Booth, 1984)

TABLE 6.7 Visual Acuities Measured with Different Types of Targets*

Test target	Observer's Task	Acuity
Line bar 1 deg long	Detect presence	0.5 sec arc
Disc	Detect presence	30 sec arc
Vernier (misaligned lines)	Detect misalignment	2 sec arc
2 discs or 2 bars	See as two	25 sec arc
Gratings	Detect orientation or separation	35 sec arc/bar
Tri-bar	See as three, detect orientation	70 sec arc
Landolt ring	Detect orientation of gap	30 sec arc
Snellen letters	Read letters	0.8 min arc
Checkerboard	Detect the pattern	1 min arc

*Adapted from Boff & Lincoln (1988).

factor of ten or more when targets are less than sharp, operators have visual defects, or viewing conditions are less than ideal.

All these forms of acuity are affected by a number of physical and physiological factors. Among the former are illumination, contrast, time of exposure, and color. For dark targets on bright backgrounds, acuity increases as background luminance increases (Figure 6.7). For light targets on dark backgrounds, acuity increases initially then decreases with increasing target intensity due to an irradiation effect. Stroke widths of symbols on trans-illuminated displays must be thinner than for front-illuminated displays. Acuity also increases as contrast and viewing time increase. To some extent these physical factors can be traded-off against one another. For example, a low-contrast target can often be made more visible by increasing illumination, viewing time or both. Acuity for a target that must be seen quickly can usually be improved by increasing contrast, illumination, or both.

Acuity for colored targets varies greatly depending on the background against which the target is viewed. Bright orange is seen best against most backgrounds, and, as a general rule, one should avoid blue or red symbols if one is interested in good acuity.

Among the physiological factors that affect acuity, the most important one is the part of the eye stimulated (Figure 6.8). Acuity, however measured, is sharpest in the direct line of sight when images fall on the fovea of the eye. As images move away from the fovea or central line of sight, acuity drops off rapidly. Vibration and hypoxia, reduced oxygen content in the brain brought on by ascent to high altitudes or the inhalation of certain kinds of gases, also reduce acuity.

Nearly 50 percent of people have visual defects that decrease their visual acuity. The common ones are myopia (nearsightedness), hyperopia (farsightedness), and astigmatism (uneven curvature of the cornea). A visual defect that is becoming increasingly important these days is presbyopia, the farsightedness of old age, caused primarily by decreased elasticity of the lens as people age. Fortunately, all these defects are easily correctable with appropriate lenses. Eyeglasses do, how-

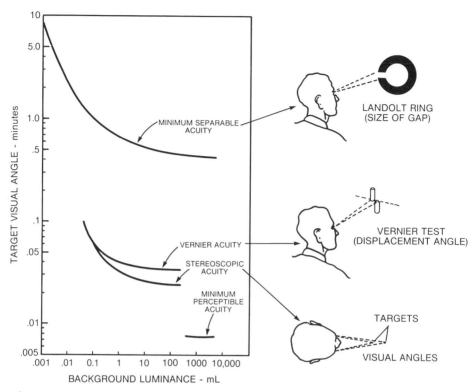

Figure 6.7 Acuity measures for four of the targets in Figure 6.6 as a function of background luminance. (Adapted from Figure 13–15 in Taylor, 1973)

ever, create problems in certain work situations. Computer operators who wear bifocals often have to tilt their heads back in awkward and tiring positions to view computer screens (Martin & Dain, 1988). Eyeglasses also make it difficult to don helmets and other kinds of protective equipment, and they may be difficult to keep in place in aircraft and other vehicles that engage in vigorous maneuvers.

Color Vision No one knows how many different colors the average person can discriminate, but that number has been estimated at several million. Of course, far fewer than that are usable in practical applications. (Remember what I said about absolute judgments on Page 217?) Nonetheless, the sheer number of available color sensations together with their vividness and distinctiveness make them ideal for a variety of practical purposes such as signalling, coding, and identifying and marking-off areas one from another.

The Psychological Dimensions of Color. The three principal psychological dimensions of color are hue, saturation, and lightness (or brightness) (Figure 6.9). Hue is what most people mean when they say color. It is the property of chromatic

222   6: HUMAN MENTAL CHARACTERISTICS

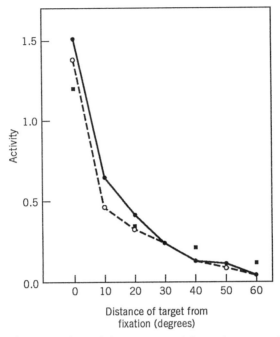

Figure 6.8 Visual acuity with Landolt rings at two different luminance levels (*open and closed circles*) and gratings (*squares*) as the separation from the fixation point—the direct line of sight—increases. (From Boff & Lincoln, 1988). Reprinted by permission of John Wiley & Sons, Inc.

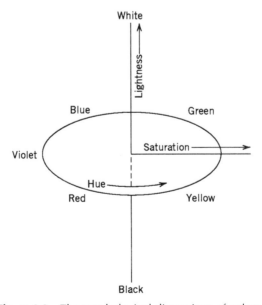

Figure 6.9 The psychological dimensions of colors.

colors that gives them their distinctive qualities, their blueness, greenness, yellowness, and so on. Hue is primarily dependent on wavelength or the combination of wavelengths constituting a color. Lightness is an intensity dimension, and is primarily dependent on the amount of light emitted or reflected by a color. Saturation is the "purity" of a color, the degree of difference between a color and a white or gray of the same lightness. All ideal neutral (achromatic) colors—white, gray, or black—have zero saturation.

The Specification of Color. I shall discuss three principal ways of specifying colors for engineering purposes: the CIE and Munsell systems, and Federal Standard Number 595a. The first two are described in detail in Kaufman and Haynes (1981).

The CIE System. The CIE (in English, ICI) system was devised by the Commission Internationale de l'Éclairage (in English, International Commission on Illumination), composed of representatives from standards organizations of the major countries of the world. The commission defined a Standard Observer (think of it as a hypothetical person having the visual characteristics of an average eye) and a procedure for calculating and plotting the location of colors in a two-dimensional diagram (Figure 6.10). The basis for the system rests on the fact that all colors can be matched by mixtures of three suitably chosen primary colors. So any color can be specified by the proportions of the three primaries in the mixture that matches it. In the CIE system the primaries are imaginary colors, but you will not be grossly in error if you think of them as red, green, and blue. In Figure 6.10 x is, roughly speaking, the proportion of red and y the proportion of green in mixtures. Only two dimensions are required to specify chromaticity since the proportions add to 1.0. Another dimension, the intensity of lights, the reflectance of surfaces, or transmittance of filters, has to be specified independently. For computational details, refer to Kaufman and Haynes (1981).

The curved line in Figure 6.10 is the locus of the spectrum colors and the straight line from 380 to 750 nm the locus of extraspectral purples (or violets), which do not appear in the spectrum. The purple and violet colors we see in the world around us are all made up of mixtures of red and blue wavelengths. White, gray, or black, being composed of roughly equal amounts of the three primaries, is located at the approximate center of the diagram. The dominant wavelength of a color (roughly speaking, its hue) is specified by the point on the spectrum locus intersected by a line from white through the color. Excitation purity (corresponding closely to saturation) is determined by the distance from white to the color.

The oddly shaped zones in the diagram indicate the tolerance limits of acceptable colors for marking physical hazards as defined in ANSI Z53.1-1979, and the small circles within those zones are the locations of the ideal safety colors. These colors have been chosen to provide maximum identification by both color-normal and most color-deficient persons.

224   *6: HUMAN MENTAL CHARACTERISTICS*

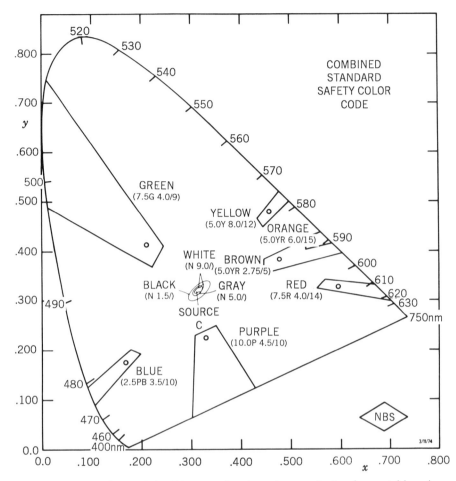

Figure 6.10   CIE Chromaticity Diagram, showing tolerance limits of acceptable colors for marking physical hazards. The small circles in each zone are the locations of the ideal safety colors. Munsell equivalents are given beneath each color name. (From National Bureau of Standards. Also reproduced in ANSI Z53.1-1979)

The Munsell System. The CIE system can be used to specify the colors of lights or surfaces. The Munsell system, another major system for specifying colors, can be used only for surface colors, but is much simpler to use. The complete version consists of a collection of over 1,000 color chips in an atlas (available from the Macbeth Division of Kollmorgen Instruments Corporation, P.O. Box 320, Newburgh, NY 12550) of charts, in each of which one of the three color variables is held constant. The chips have been selected to represent subjective scales of constant hue, saturation (called chroma in the Munsell system), and lightness (called value in the Munsell system), and are spaced in a constant number of multiples of JNDs of a color-normal person. To determine the specification of an unknown color, one merely has to find the chip that most closely approximates the unknown when both are viewed under a standard source of light.

The Munsell hue circle is shown in Figure 6.11. Figure 6.12 shows how chips at a constant hue (5.0R) are arranged and all the chips that are available at that hue. Colors are designated in hue value/chroma order, an example being 5.0Y 8.0/12. Tables are available showing the CIE equivalents of the Munsell chips. The CIE equivalent for this example is $x = 0.4562$, $y = 0.4788$. The small circles in Figure 6.10, the ideal safety colors, are the locations of the recommended Highway Traffic Sign Colors of the Federal Highway Administration of the Department of Transportation. Figure 6.10 also gives the Munsell designations for those colors.

Federal Standard Number 595a. There are a large number of local codes, national codes, federal standards, professional standards, and manufacturer's standards that designate colors in various ways. Most are highly specialized and apply only to specific products, for example, automobiles, fabrics, cosmetics, or pills. Federal Standard 595a is the most general of these other color standards. It is a

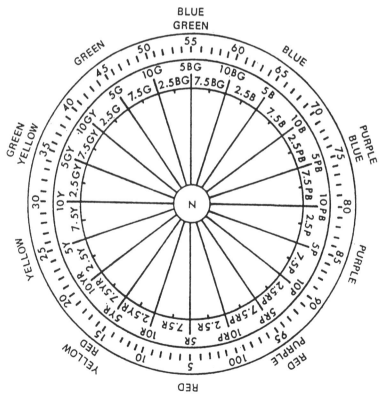

Figure 6.11 The Munsell System hue circle. (Reprinted with permission from EIA Standard RS-359)

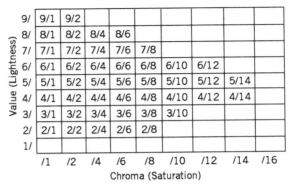

Figure 6.12 A sample page showing the arrangement of chips varying in value (*lightness*) and chroma (*saturation*) at constant hue (5.0R). Notice that it is possible to get highly saturated red colors only at medium levels of lightness.

collection of standard colors used by various departments of the government. Although these chips cannot be related to either CIE specifications or Munsell System chips, they have two classification features not available in either of the other two systems. Chips are classified into three categories, depending on their specular gloss and six categories of fluorescence. An arbitrary five-digit numbering system is used to identify individual chips. As with the Munsell System, paints or other material are identified by comparison with the chips in the standard.

Factors Affecting Color Appearance. The perception of colors is affected by the characteristics of the target (spectral content, luminance), the environment in which it is viewed (surround, contrast, illuminant), and the observer. Figure 6.13, for example, shows combinations of foreground and background colors that were rated good, average, or poor in an extensive series of tests by Lalomia and Happ (1987) on an IBM 5153 color display. Although the authors say they determined the luminance and CIE coordinates of the colors, they did not, unfortunately, report them in their article. Good ratings apply to the top 20 percent of combinations measured by both character legibility and subjective preference. Poor ratings apply to the bottom 20 percent and the remaining 60 percent are represented by dashes. As you can see, some backgrounds—black and blue—received a large number of good ratings; others—green and brown—a large number of poor ratings. You can also see some exceptions to these generalizations. Although blue is generally a good background color, it is poor when paired with a gray foreground. You can find other exceptions in the table.

Another important factor influencing the appearance of surface colors is the quality of the illumination. Two colors that appear identical in tungsten light may appear entirely different in daylight, and vice versa. High-pressure sodium lighting, in particular, tends to distort almost all colors. From a practical standpoint, this means that colors for coding or other purposes should be coordinated with the illuminations under which they will normally be used.

FOREGROUND	BACKGROUND							
	BLACK	BLUE	GREEN	CYAN	RED	MAGENTA	BROWN	WHITE
BLACK	X	–	–	GOOD	–	GOOD	–	GOOD
BLUE	–	X	–	–	POOR	–	–	GOOD
H.I. BLUE	–	–	POOR	POOR	–	–	POOR	POOR
CYAN	GOOD	–	POOR	X	–	–	POOR	–
H.I. CYAN	GOOD	GOOD	–	GOOD	GOOD	GOOD	–	–
GREEN	GOOD	GOOD	X	POOR	GOOD	–	POOR	POOR
H.I. GREEN	–	GOOD	–	–	–	–	–	–
YELLOW	GOOD	GOOD	–	GOOD	–	GOOD	–	–
RED	–	–	POOR	–	X	POOR	POOR	–
H.I. RED	–	–	POOR	–	–	–	–	–
MAGENTA	–	–	POOR	–	POOR	X	POOR	–
H.I. MAGENTA	GOOD	–	GOOD	–	–	POOR	–	–
BROWN	–	–	POOR	–	–	POOR	X	–
GRAY	–	POOR	–	–	POOR	–	POOR	–
WHITE	–	GOOD	–	POOR	–	–	–	X
H.I. WHITE	GOOD	–	GOOD	GOOD	–	–	–	–

(H.I. = HIGH INTENSITY)

Figure 6.13 Combinations of foreground and background colors that were rated good, average (*dashes*), or poor on an IBM 5153 color display. (Constructed from data in Tables 1 and 2 in Lalomia & Happ, 1987).

Color Vision Defects. About eight percent of otherwise normal men but less than one percent of women have some degree of color vision deficiency. Although this condition is popularly referred to as color blindness, that term is really incorrect. Almost all so-called "color blind" persons can discriminate some colors. True color-blind individuals, that is, those who can see only various shades of white, gray, or black (achromats), are very rare. Fewer than 200 have been reported in the scientific and medical literature.

There are various kinds and degrees of color deficiency. In the most common variety (deuteranopia), the individual is unable to discriminate, see the differences among, certain shades of red, yellow, and green. In addition, certain shades of blue-green and violet are confused with white or gray. Individuals with one other fairly common kind of defect (protanopia) have reduced sensitivity to colors in the red end of the spectrum.

Persons with defective color vision report a large number of problems when interacting with objects in their environments (Israelski, 1978). Some examples are: maintaining equipment with color-coded wire, resistors and components; adjusting color equipment in TV studios; titrating and identifying chemicals; reading color-coded symbols on drawings; correctly identifying color-coded controls on some household equipment and in automobiles; and discriminating traffic lights from street lights and lights on signs.

The foregoing notwithstanding, it is curious that some people with defective color vision go through life unaware that they do not see colors as well as most other people. There is enough redundancy in our visual world so that these individuals often get along reasonably well. A common rationalization is, "I just never learned all the color names." Their defects only become apparent in situations where there is little or no redundancy and the correct perception of colors is critical—seeing isolated colored signal lights, identifying colors on a computer screen, or matching colors on maps, charts, and other displays.

It is possible to find small sets of colors that almost all persons, color-normal and color-defective, can discriminate. The safety colors in Figure 6.10 are one such set.

Audition

Hearing is second only to vision in importance for human operators in systems, because it allows us to communicate by way of speech, and enables us to receive an enormous variety of non-speech sounds—bells, buzzers, beeps, horns, sirens, and sounds from the movements of people, vehicles, and machinery—all of which convey useful information. Unfortunately, our ears probably provide us with the greatest amount of annoying stimulation; we can close our eyes to block out visual stimuli, but cannot close our ears.

Because our ears are always open and can receive sounds from any direction, auditory signals can alert multiple operators simultaneously. They are also the best way to divert an operator's attention from current activities and so are useful for alerting operators to spoken messages that are to follow. In addition, they are effective in situations in which the operator is already overloaded with visual information, in darkness when vision is difficult or impossible, and when an operator may be suffering from hypoxia. Hypoxia, oxygen deficiency, occurs at high altitudes, when an operator is subjected to high positive g-forces, or when operators breathe certain kinds of gases, for example, even relatively small concentrations of carbon monoxide. Because of some evolutionary quirk, our visual systems are impaired with even moderate degrees of hypoxia, but our auditory systems are not. So, for example, the positive acceleration that pilots experience when they pull up sharply may drain the blood from their heads and produce hypoxia that leaves them essentially blind, but they can still hear.

Physical Parameters of Sound Sound is a physical disturbance in an elastic medium such as air or water. The physical parameters of sound are frequency (measured in Hertz or cycles per second), waveform, duration, amplitude (usually rms or root-mean-square sound pressure level expressed in dB, decibels), and phase relations. Waveform is the sound pressure of a wave plotted as a function of time and can also be described by its amplitude spectrum and its phase spectrum.

Variations in frequency are heard primarily as changes in pitch. Variations in amplitude are heard as changes in loudness. The correspondence is not perfect, however, since the perception of both pitch and loudness are determined in part by complex interactions between the two physical measures. Figure 6.14, for example, shows equal-loudness contours for pure tones. We are most sensitive to tones of about 3,000 Hz and the sound pressure level of frequencies higher and lower than that must be increased to produce the same subjective impression of loudness. This is the rationale behind the A correction to the decibel scale discussed in Chapter 5.

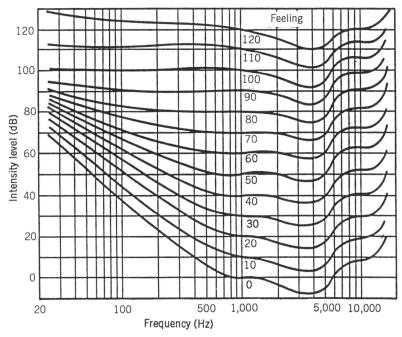

Figure 6.14 Equal loudness relations. The bottom curve is the threshold curve. The other curves show tones that sound equally loud. The number on each curve is the intensity level of a 1,000-Hz tone, and that number expresses the intensity level of any tone on the same curve. (From Chapanis, Garner, & Morgan, 1949, based on data from Fletcher & Munson, 1933. Reprinted by permission of John Wiley & Sons, Inc.)

Factors Affecting Auditory Sensitivity The sound pressure level necessary to detect a sound depends on the medium of transmission (thresholds are greater underwater than in air, particularly for high frequencies); frequency (see Figure 6.14); binaural versus monaural listening (binaural thresholds are about 3dB lower than monaural thresholds); signal duration (for signal durations less than 200 msec., intensity must be doubled every time duration is halved); bandwidth of complex tones (thresholds are raised when components of the tone are not in the same critical band); masking (the presence of other sounds may raise thresholds); age (thresholds increase from young adulthood into old age; refer again to Figure 5.7); and noise exposure (depending on the intensity and duration of exposure, increases in threshold may be temporary or permanent).

Auditory (Non-speech) Signals Auditory, non-speech, signals are appropriate when (a) the information to be transmitted is short, simple, transitory, and requires immediate or time-based responses; (b) it is desirable to warn, alert, or cue the operator to make a subsequent response; (c) custom or usage has created the anticipation of an audio signal; (d) voice communication is desirable, but needs to be cued

230   6: HUMAN MENTAL CHARACTERISTICS

by a nonverbal signal; and (e) operational or environmental factors limit vision but not hearing. To be effective an auditory signal must be heard, recognized, and understood.

Signal Audibility. Ideally, all intended listeners should hear an auditory signal on every occasion and under all operational conditions. It is virtually impossible to write specifications for sounds that will meet these requirements, but a reasonable auditory communication window has a lower limit of about 65 dB_A for sounds in quiet conditions or about 15 dB_A above the masked threshold in noise, and an upper limit of about 85 dB_A (intensities above this tend to produce undesirable startle reactions) within a frequency range of 300 to 3,000 Hz.

Signal Recognition. Auditory signals are effectively recognized if they capture an operator's attention and if the operator is able to discriminate the signal from all other sounds. Since abrupt onset and offset of intense sounds often produce startle reactions and disrupt mental activities, trains of rounded pulses, rather than abruptly rising, square-topped waveforms, are preferred. Sound levels should not rise more than 30 dB in 0.5 sec.

Subjective assessment of signal levels is unreliable and should not be used as the primary method of discrimination among signals. Discrimination between frequencies and temporal patterns is much more reliable.

Signal Meaning. Once signals are heard and discriminated, they must be understood, that is, operators should know what the signal means. To avoid ambiguity, systems should use no more than about six different auditory signals. If well-known sounds have been previously associated with specific meanings, the characteristics of those sounds should be incorporated in new signals. Urgency is usually associated with an increasing repetition rate of pulses.

Speech Communication   Speech communication problems fall into three fairly distinct categories, associated with: (1) unamplified, normally face-to-face, speech; (2) amplified speech; and (3) vocoded and synthetic speech. Design objectives for speech communication systems differ somewhat depending on the application, that is, where speech communication will be used and the purpose it will serve. For home telephone systems naturalness or voice quality is usually an important design objective. For air traffic control, communication, and most other systems, however, intelligibility is the single most important design objective.

Measuring Intelligibility. There are two general ways to measure speech intelligibility: by actual tests or by estimating it from computational formulas. In testing, standardized sounds (nonsense-syllables, phonetically balanced words, or sentences) are read under controlled conditions and the listener's responses are scored for accuracy. Since these tests are time consuming and expensive, computational alternatives are frequently used. One such measure is the Articulation Index (AI). Typically the differences between the levels of speech and noise in one-third-octave bands are weighted and summed to yield the AI. Details on procedures for using both methods can be found in a number of sources, for example, Kryter (1972).

Unamplified Speech. We engage in face-to-face speech almost constantly in offices, workshops, factories, vehicles, command and control centers, and many other places. The intelligibility of speech in such situations is determined by the background noise level and spectrum, reverberation, the hearing and speaking abilities of the communicators, the vocabulary used, and whether the language used is the native tongue of the communicators. Of these, background noise is by far the most important. The likelihood that a signal or speech will be accurately received is determined primarily by the signal-to-noise, or speech-to-noise, ratio—S/N. This ratio is the intensity of signals, or speech sounds, relative to the intensity of irrelevant background noise when both are measured at the same location, for example, a listener's ears, and for the same range of frequencies.

The Speech Spectrum. Vowel sounds contain power at frequencies as low as 50 Hz and the sibilants and fricatives, s and f, contain significant amounts of power at frequencies as high as 8,000 Hz. The bulk of the energy in speech, however, is in the low-frequency region around 250 to 500 Hz (Figure 6.15), and roughly 84 percent of total speech energy is at frequencies below 1,000 Hz. Speech comprehension is not in proportion to the intensities of sounds in various parts of the spectrum. Frequencies above 1,000 Hz contribute most to intelligibility, even though only 16 percent of the energy lies above that frequency (Figure 6.16). The

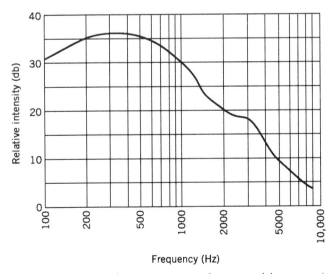

Figure 6.15 The normal speech spectrum. Note that most of the energy in speech is in frequencies below 1,000 Hz. This curve is the average of many measured spectra, male and female. (From Chapanis et al., 1949. Reprinted by permission of John Wiley & Sons, Inc.)

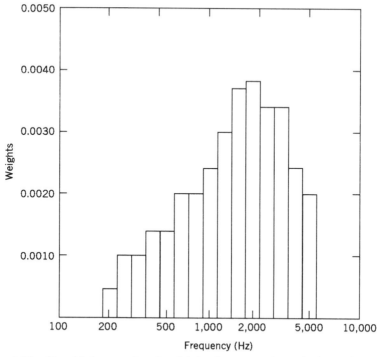

Figure 6.16 One-third-octave-band weighting factors used in calculating the Articulation Index. Frequencies above 1,000 Hz contribute more to the intelligibility of speech than do those below 1,000 Hz. (Constructed from data in Kryter, 1972)

region above 1,000 Hz contains the low-intensity but highly important consonant sounds and main vowel formats. These characteristics provide the rationale for the speech packaging frequently used to improve the intelligibility of speech in amplified systems.

Speaking in Noisy Surroundings. We automatically speak louder in noisy surroundings, but do not fully compensate for increases in background noise, because we are misled by our own sidetones. A sidetone is the acoustic feedback we normally get from our own voices when we talk. It is carried primarily from our mouths through the air to our ears, and secondarily by vibrations transmitted through our heads. Sidetones add to our perceptions of own speech levels with the result that we increase speech levels only about 5 dB for each 10 dB increase in background noise.

The reverse happens when operators wear hearing protection (earplugs, earmuffs, or the like) in noisy environments. Hearing protectors reduce noise at the talker's ear, but do not appreciably affect the sound of the talker's own voice which is transmitted to the talker's ears via headbone vibrations. Under these circumstances, talkers unconsciously speak more quietly since they can hear themselves above the noise perfectly well. The net result is often a reduction in intelligibility.

In reasonably quiet rooms, the long-term rms sound pressure level of normal conversational speech at a distance of one meter is approximately 65 dB. This level may go down as low as 40 dB when talkers speak confidentially. In rooms with moderate levels of noise, people still prefer to speak at a level of about 65 dB and do not normally raise their voices until the background levels reach values of around 50 to 55 dB$_A$. When noise levels exceed values of about 85 to 90 dB$_A$ communication becomes virtually impossible, even when speakers shout. A loud shout has a level of about 85 dB at a distance of one meter and a maximum shout, which can be maintained for only a few words, is about 90 dB.

Other Factors Affecting Intelligibility. The intelligibility of unamplified speech can be aided by selecting appropriate vocabularies, the phonetic composition of messages, the context of messages, and speakers. In general, intelligibility is improved by restricting the size of vocabularies, and by ensuring that the vocabularies used by speakers are known to listeners. Words can also be chosen to minimize confusion. The international word-spelling alphabet adopted by the International Civil Aviation Organization, for example, is designed to be recognizable when it is spoken by and to persons of different nationalities and accents (Table 6.8). Redundancy in spoken messages (for example, saying "Turn pressure valve 26 off" and then "Shut down the feedwater system") also helps to reduce errors. Speakers differ greatly in their ability to enunciate clearly and, for critical situations, they should be selected for their speaking abilities and then trained to improve their natural abilities even further.

Amplified Speech. Examples of amplified and electronically assisted speech are public address systems in buildings, telephones, radio links, and intercoms. The same factors that affect unamplified speech also apply to amplified speech. In addition, most speech-transmission systems distort speech by restricting bandwidth, by frequency and amplitude distortion, and by filtering. As with face-to-face communication, noise is usually the most disruptive factor. The noise may be background noise at the speaker's location, which is amplified along with the speaker's voice, noise in the communication channel itself, or background noise at the listener's location.

If intelligibility is more important than naturalness, some forms of distortion may be deliberately designed into a system to improve speech reception, particularly in noise. Peak clipping and reamplification of speech sounds, for example, packages speech by cutting off the peaks of low-frequency vowel sounds and amplifying the weaker but more important consonant sounds. (Refer to Figures 6.15 and 6.16 and the discussions about them.) Although speech packaged this way sounds "grainy," it is highly intelligible for listeners in noisy surroundings.

Vocoded and Synthetic Speech. Two principal methods are used to produce synthetic speech: (1) record, digitize, and play back an actual human voice; and (2) synthesize speech by concatenating a set of machine-generated phonemes or words. The former method requires that a human prerecord desired outputs before messages can be digitized and stored. This generally means anticipating all the words and messages that may be needed for all future uses. Speech synthesis is more flexible in the sense that no human verbal output is required to produce the speech. The principal difficulty with synthesized speech is that it does not simulate complex human speech characteristics such as inflection, emphasis, and stress very well.

For applications in which naturalness is important, digitized speech is generally preferable. Another consideration is that, although synthesized speech may be intelligible, studies show that when listeners are task loaded, accurate recognition of synthesized speech drops off much more rapidly than does the recognition of digitized speech. An analogy is the increased difficulty one has listening to a person with a pronounced foreign accent deliver a speech on a familiar topic.

Auditory Defects Many designers seem to assume that users of their systems will have adequate hearing. Unfortunately, hearing impairments are fairly common, particularly now that so large a proportion of our population is elderly. Nearly all people over the age of 60 have at least a mild hearing problem. Most of these problems are caused by damage to the inner ear or auditory nerves and cannot be improved medically. In addition, many people suffer from temporarily impaired hearing, either because of exposure to loud sounds or by the ingestion of certain drugs. For these, and perhaps other reasons, the most comfortable listening level, or preferred listening level, increases monotonically and rather dramatically with age. At age 20 it is about 55 dB, at age 50 about 67 dB, at age 70 about 76 dB, and at age 85 about 85 dB (Coren, 1994). Since designers usually have no control over who will use their systems, most systems should be designed with the expectation that they will be used by persons with hearing difficulties. Providing volume controls so that listeners can pick their own listening levels is one important design option.

Other Senses

Although they are not used as extensively as vision and audition for conveying information from machines to us, our other senses cannot be totally ignored in system design. In some cases these other senses can be harnessed usefully and, in still other cases, their functioning can seriously affect operator performance. A few examples follow.

Our cutaneous senses, feeling, touch, or vibration, have been used effectively in the design of tactual displays, the shape coding of keys and other controls, and as substitutes for hearing in the reception of coded messages, perception of speech, and localization of sound (See, for example, Sanders and McCormick, 1993.) Our sensitivity to smells and tastes is decreased in microgravity environments. As a result food that is judged to be adequately seasoned prior to flight usually tastes bland in space. Since the quality of food plays an important role in maintaining morale on extended missions, special attention needs to be devoted to this problem. A somewhat different set of problems involves the sense of smell. Particulate matter does not settle out in a weightless environment, so that odor in a space habitat may be more offensive than under similar conditions on Earth. Unpleasant odors in closed environments even on earth, for example, in tanks, bunkers, and command centers, may also cause nausea and result in decreased operator performance. These problems require special attention to the filtering and circulation of the ambient air.

Much more serious are the responses of our vestibular senses—the senses that tell us whether our bodies are tilted, right-side up, upside-down, or moving (refer to Table 6.1)—in certain machine systems. The most traumatic effects are those of seasickness, airsickness, or, more generally, motion sickness. Frequencies of 0.1 to 0.63 Hz are especially conducive to motion sickness and may almost totally incapacitate some operators. Particularly perplexing are the discomfort and sometimes nausea that people experience in some simulators, for example, automobile driving and aircraft simulators, even when the simulator platforms are stationary. (See, for example, McCauley, 1984.) Spatial disorientation and the space adaptation syndrome occur in about 50 percent of crew members in microgravity environments. Since the symptoms to these various vestibular disturbances generally take two to four days to subside, it is a good idea to avoid scheduling activities that depend on spatial orientation early in a mission.

Sensing Versus Perceiving

Although I have not explicitly called attention to the distinction, there is an important difference between sensing and perceiving. As you look at this page, you can see some black marks on a white background. That's sensing—the simple reception of stimuli. Recognizing that those black marks constitute letters and the letters make up words is a perceptual process. Similarly, hearing sounds is a simple sensing task. Recognizing that some sounds make words is a perceptual process. Perception, more precisely, is the temporal or spatial organization of simple sensory information into meaningful wholes.

Perception is a much more complicated process than simple sensing, because perception is highly dependent on learning, set, experience, and attitudes. You don't need to learn how to see or hear stimuli, but it took years of learning for you to read written words and understand speech.

The distinction between sensing and perceiving is an important one for design purposes. If an operator's task is merely to detect something—the presence of a plane in the sky, a sign along a highway, or the onset of an annunciator alarm—you have a simple sensing task. You have only to make sure that the objects to be sensed are above the absolute and difference thresholds and below the upper threshold. If the operator's task requires the identification or classification of those inputs—identifying the plane as USAir 780, United 563, or TWA 1156; reading the message on the highway sign; or understanding the significance of the annunciator alarm—you have a perceptual task and that perceptual task assumes a considerable amount of learning and more attention to the design of the things to be perceived.

REMEMBERING

A human ability vital to all system operations is the ability to remember things from the past and use them in the present. We rely on our memories when we almost instinctively reach for controls without looking at them, when we read and interpret symbols on a computer display, when we remember instructions for operating a VCR, and for thousands of interactions with the devices we use daily. All these things are possible because memory is intimately associated with learning and training. We are able to increase our store of knowledge and our skills only because we are able to retain in our memories things we have done, have experienced, or have been taught.

The complement of remembering is forgetting. People forget—they lose things from memory—and it is this aspect of memory that is perhaps of greatest concern to designers because users of equipment constantly forget things. If users didn't forget it would not be necessary to include checklists, warnings, and instructions with systems.

What are some of the principal features of this remarkable human ability? How are they used in systems?

Three Kinds of Memory

Psychologists generally agree that we have three kinds of memory: sensory, short-term, and long-term memory. We can't say exactly how they are bounded temporally, but, roughly speaking, our sensory memories hold incoming information for no more than a second, short-term memory for up to a minute, and long-term memory for up to a lifetime.

Sensory Memory This is a more or less exact copy of what you have just seen, heard, or felt—the mental picture, so to speak, of a collision you have just witnessed, the internal echo of words you have just heard, or the tingling that remains after vibration has stopped. Although items in sensory memory fade rapidly and are replaced with new inputs, the persistence of information, however brief, makes it available for further processing after a stimulus has ended or moved. There do not appear to be any design issues associated with this kind of memory.

Short-term Memory Short-term memory is a working memory. It is memory for things that have never left consciousness, that are needed temporarily, and are then forgotten. An air traffic controller keeps current aircraft data (identity, altitude, speed, heading) in mind, that is, in short-term memory, for immediate or nearly immediate use, and then discards or forgets them when they have been used. Hundreds, perhaps thousands of things we do every day require this kind of memory. The "throw-away" aspect of short-term memory is important to keep our minds from being cluttered with the countless bits of information that constantly bombard us and that are not needed for more than a few seconds.

"Chunking" and the Capacity of Short-term Memory. Short-term memory has a very limited capacity, and a brilliant insight by Miller (1956) defined that limit for us. He introduced the term "chunking" to refer to the organization of input information into coherent groups or chunks. It is, in short, an encoding strategy. A string of six random letters constitutes six chunks, but if letters are grouped into random

words or meaningful combinations, such as IBM, TWA, IRS, or ETC, each word or group of letters now constitutes a chunk. The important and curious finding is that the span of immediate memory is about 7, plus or minus 2 chunks, and is independent of the size of the chunks. You cannot expect a system operator to hold more than about seven pieces of information in short-term memory.

Recall. Recall from short-term memory is generally easy and effortless but items are not recalled equally well. Refer to Figure 5.3 in Chapter 5, which showed data from an experiment in which subjects saw a string of eight random numbers for five seconds, and were then given unlimited time to reproduce the numbers. Note that numbers at the beginning of the string were recalled with fewest errors, those at the end were recalled moderately well, and that the greatest number of errors was made with numbers closer to the end (at digit position 7). This kind of recall curve, called the serial position curve, has been found with enough other kinds of materials that we may consider it a stable characteristic of short-term memory.

One way to improve the capacity of short-term memory is to organize the items in some meaningful way. It is easier to remember the functions served by computer keys labeled ESC, DEL, and INS than those labeled PF1, PF2, and PF3. There are even significant differences in the meaningfulness of strings of digits (Chapanis & Moulden, 1990). For example, repetitive triplets—000, 111, 222—and triplets containing a pair—227, 933—are generally very easy to recall. Triplets that go down-up-down—264, 564, 683—or up-down-up—947, 759, 968—are harder to remember. You will notice that frequently used telephone numbers are often assigned easily remembered combinations of digits; unlisted telephone numbers more difficult combinations.

Maintenance Rehearsal. Although short-term memory decays in 15 to 20 seconds, items can be kept there almost indefinitely if you attend to them and rehearse them. After you look up a number in a telephone directory, repeating the number to yourself several times will keep it in short-term memory while you walk across the room to a telephone. If tasks require operators to keep items in short-term memory by rehearsing them, tasks should be designed so that there are no distractions to keep operators from that rehearsal. Checklists are another way of helping people to retain items in short-term memory. In this case, grouping items in sets of five, for example, will help operators to keep track of where they are.

Long-term Memory Long-term memory is a more or less permanent memory store. It contains items of information that have not been in consciousness for some time, but are recalled as needed. Unlike recall from short-term memory which is generally easy and effortless, recall from long-term memory requires an active search process which, in some cases, may be drawn out and frustrating. When items are found, they are transferred to short-term memory for conscious use.

The Capacity of Long-term Memory. No one really knows how much can be stored in long-term memory, but everyone agrees that it is very large—perhaps several billions of items. It holds our usable vocabularies (up to 50,000 or so words), names, facts, instructions, procedures, and skills—like driving or flying an airplane.

Retention of material in long-term memory is measured in two principal ways: by recall or recognition. In recall a person has to retrieve and reproduce, by speaking or writing, something that was once learned. Recognition only requires a person to acknowledge having seen, heard, or learned something previously. Recalling information is much more difficult than recognizing it. Although you probably cannot recall the exact location of all the keys on your computer terminal, you probably would have no difficulty recognizing the layout of your keyboard if you saw it displayed among others in a computer store.

Aids to Memory. Although memories in long-term storage may last a lifetime, most of what we learn is quickly forgotten. What can be done in a practical way to help operators remember? Perhaps the single most important thing is to have them practice or rehearse what they want to remember. "Practice makes perfect" is an adage that contains more than a germ of truth. Overlearning—practicing or rehearsing beyond that point where you can correctly recall something once—helps to stamp things firmly in long-term storage.

Perhaps the second most important thing is to organize material to be remembered in some meaningful way, or to associate it with things already known or remembered. For example, using more English-like commands in computer programs rather than arbitrary symbols facilitates both learning and retention. (See, for example, Chapanis, 1988, Ledgard, Singer, & Whiteside, 1981.)

Finally, designers can and should use memory aids: well-designed checklists, prompts and HELP messages on computer screens, and instruction manuals are all valuable and, in many cases, indispensable. In fact, one way of quickly spotting poorly designed equipment is to observe notes and reminders that operators themselves attach to their equipment with thumbtacks or adhesive tape. These additions are testimony both to the usefulness of written memory aids and to the failures of design.

DECISION-MAKING

Decision-making is choosing a course of action from among two or more alternatives. Mathematicians, statisticians, and economists have devised rigorous techniques, for example, multi-attribute utility analysis, decision analysis, and Bayesian statistics, for arriving at the best solutions to a variety of problems that require decisions, such as the allocation of a fixed amount of resources to various activities, or procedures for testing products on a production line. These mathematical models tell us what a *rational* decision-maker should do under a specific set of circumstances.

Although these rigorous strategies may seem useful, they fail to consider some important factors that characterize real-world decisions. Classical strategies degenerate when confronted with time pressures. They simply take too long. Operators in many systems have to make vital decisions fast. They do not have the luxury of securing all the relevant information, assuming that it could be obtained, and evaluating that information in a leisurely way. Even under low time pressure, these rigorous methods still require extensive work and lack the flexibility needed to deal with rapidly changing conditions. Moreover, it is difficult to factor in ambiguity, vagueness, and inaccuracies when applying analytical methods. Most problems in real life have to be solved in the context of uncertainty. These are problems for which we do not have all the information, and for which we are not certain of the accuracy of our premises.

Perhaps the greatest difficulty with these analytical methods is that people are not rational decision-makers. We tend to make decisions based on past experience and on what we perceive to be the best choice among possible alternatives. Although human decision-making may be "irrational" from a logical or mathematical point of view, it is not willy-nilly. There are consistencies in the way we make decisions.

Heuristics To solve most problems, people use intuitive strategies called *heuristics*, informal rules of thumb or general guidelines. Some people, through years of training and experience, have become so successful at solving some types of problems that they have become experts in particular areas. Examples of such expertise can be found in almost any profession. Some experts using heuristics can diagnose equipment malfunctions rapidly, others can conduct brilliant military campaigns, and still others discover new scientific principles. Expertise of that caliber is, unfortunately, in short supply because it is a knowledge-intensive skill that requires years of schooling and on-the-job experience to develop. The use of heuristics is often so complex and so vaguely defined that even experts cannot provide a simple explanation of how they do it.

Expert Systems It is unlikely that we shall ever see a simple how-to manual that can substitute for expertise. Nonetheless, some computer programs, called expert systems, can do almost as well as corresponding human experts and often much better than someone who has been trained to do a job, but who is not an expert. These computer programs solve problems by using knowledge laboriously extracted from experts together with certain heuristics. Although expert systems are not yet good enough to replace human experts, they are often useful as knowledgeable assistants.

Expert Human Decision-Making Studies of experienced decision-makers (Klein & Klinger, 1991) show that they tend to classify tasks into familiar or prototypical categories. They generate options serially, rather than concurrently, and the typical course of action is the first one considered. Experienced decision-makers are able to respond quickly, by using experience to identify a plausible course of action rather than having to generate and evaluate a large set of options. Under time pressure they are poised to act while evaluating a promising course of action rather than remaining paralyzed while waiting to complete an evaluation of different options. Their focus is on action rather than analysis.

Sources of Bias For most of us heuristics are often efficient, practical and effective. At the same time people sometimes misuse heuristics in systematic, predictable ways with the result that both novices and experts make notoriously poor judgments. One source of bias has been called the *availability heuristic*. It refers to decisions that are biased by things most readily available in memory at the time the decision is made. Recent vivid experiences, for example, may overwhelm all other evidence. This heuristic probably accounts for errors in the estimation of certain kinds of "everyday" knowledge, for example, in the causes of death. Because dramatic accidents, such as those involving aircraft and nuclear power plants are so well publicized, accident frequencies attributable to these sources are generally grossly overestimated.

Another common source of bias has been labeled the *gambler's fallacy,* a misconception about the fairness of the laws of chance. This is the belief that a series of events in one direction, for example, a series of heads in tosses of a coin, increases the probability that the next event will be in the other direction. Chance, contrary to common belief, is not self-correcting.

Recommendations What can be done to improve the quality of decision-making? Our understanding of this process is so limited that most human-factors textbooks never even mention it. In any case, we can offer only general advice. One thing seems clear, however: There are no shortcuts. Understanding the kinds of biases that we are prone to, abundant practice, clear-cut feedback on the correctness of our decisions, thorough review of alternative courses of action that might have been taken with their consequences, and reinforcement for correct performance, is the best we can recommend.

From a design standpoint, decision-making can be aided by providing an operator or maintainer with up-to-date, complete, and easily understood information about the state of a system and the consequences of taking alternative actions. In some cases, it is also useful to provide the operator with information about past history—what the system was doing in the immediate past that led to its current status. A recommendation that is easy to make, but difficult to determine: it is important not to overload the operator with too much information or information that is irrelevant to making a decision. In some systems—aircraft, nuclear power plants, ships—predictor displays can aid decision-making. These displays show operators what the state of the system will be at some future time if no action, or if certain kinds of actions are taken.

LEARNING

Once systems have been built, the people who are to operate them have to learn how to do it. Even the most highly automated systems require some learning, if for no other reason, so that users will know where the START button is located. As systems increase in complexity, so also does the amount of learning required. In the case of some jobs (air traffic control, computer programming, piloting) it may take years of learning before operators are able to perform satisfactorily on their own. In fact, for most systems it is safe to say that operators never stop learning. Studies show that to attain a very high level of performance in most major domains requires about ten years of essentially full-time learning (Ericsson & Charness, 1994).

Learning Defined

Learning refers to relatively permanent changes in behavior (or the potential for them) produced by experience and not by fatigue, drugs, or sensory or motor changes. The word "potential" is important in the definition because learning may sometimes occur but not be evident until much later. Although the distinction between learning and training is by no means sharp, learning refers to the principles by which we acquire skills and knowledge. The application of those principles to the learning that goes on in schools, business, industry, or military establishments is generally referred to as training, and will be discussed in Chapter 7.

The Course of Learning

Quantitative changes in behavior caused by learning are usually shown in the form of learning curves that plot some measure of performance (productivity, time to perform tasks, errors) against time on the job. Figure 6.17 is an example of such data. Notice especially that improvement continues for at least three or four years. Crossman (1959) gives a learning curve for women operating cigar-making machines that shows improvement over a period of more than two years, during which

Figure 6.17 Improvement in performance as a function of time on the job for bank proof machine operators in one bank. (From Klemmer & Lockhead, 1960). Reprinted by permission from International Business Machines Corp.

time the women had performed about three million cycles! See de Jong (1958) for still other examples of learning that takes place over long periods of time in a number of practical work situations.

Although harder to show, the quality of work also improves as learning progresses. An example appears in a study by Lindahl (1945), who described an industrial task in which thin discs were sliced from the end of a tungsten rod by means of a cutting machine. Successful operation required operators to perform complex hand and foot actions. Failure to apply foot pressure properly resulted in damage to the discs, excessive breakage and use of wheels, and wastage of material. Figure 6.18 shows the ideal pattern of foot movements and Figure 6.19 shows the patterns made by a trainee at various times after the start of learning. Notice the gradual refinement in the movement patterns as learning progressed. After 239 hours of training, the movement patterns of this operator closely approximated the ideal.

You can notice qualitative changes of this kind yourself in such common situations as learning to drive, type, or ride a bicycle. As learning progresses, waste motions are gradually eliminated and movement patterns become smooth and efficient so that they require less energy. Supporting these observations, Fleishman and Quaintance (1984) have shown that different abilities are required for early stages of learning than for later ones.

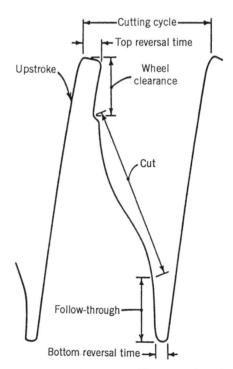

Figure 6.18 Ideal disk-cutting cycle (From Wolfle, 1951, based on data from Lindahl, 1945. Reprinted by permission of John Wiley & Sons, Inc.)

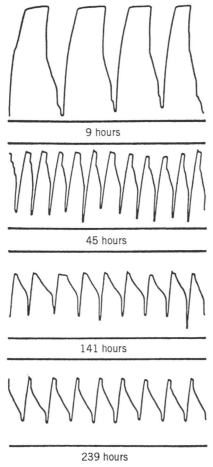

Figure 6.19 Foot movement patterns of one operator learning the pattern in Figure 6.17. (From Wolfle, 1951, based on data from Lindahl, 1945. Reprinted by permission of John Wiley & Sons, Inc.)

I have emphasized the length of time over which learning occurs because most human-factors studies, and that includes most studies of human performance in simulated systems, are conducted for very short periods of time—frequently hours, or at most a few days. Measures of human performance, obtained from studies of such short duration, may not be good predictors of task performance that might be expected of highly skilled and trained operators. Use caution in extrapolating from laboratory or simulator studies of performance.

Factors Affecting Learning

The principal task variables that affect the rate of learning are feedback, the amount of material to be learned, its meaningfulness, and difficulty. Of these, feedback, or knowledge of results, is probably the most important single factor. In order to learn, you have to know how you are doing and whether what you are doing is correct. To be most effective, feedback should be quantitative and immediate. Some systems, such as driving an automobile, provide their own feedback, but, for many other systems, feedback must be designed into the system. The successfulness of teaching machines is due largely to the immediate feedback they provide. In general, learning is also more rapid for smaller amounts and for more meaningful and easier material.

Important human variables that affect learning are age, motivation, and intelligence. Studies show a rather consistent improvement in learning from childhood to maturity, but it is difficult to know whether this reflects genuine learning ability or the effects of other variables, such as amount of prior learning and increased motivation. Most studies show a slow but continuous decline in learning ability from maturity to old age.

Motivation and intelligence are probably the most important human variables in learning. People will not learn unless they want to, nor will they learn if they are incapable of doing so. Figure 6.20, for example, shows the performance of three groups of Army recruits in learning how to report the bearings and ranges of targets appearing on a combat plotting board such as that shown in the figure. All the subjects in the highest category of mental ability were able to perform correctly at the end of three trials. It took seven trials before all subjects in the middle category could master the task, but the average subject in the lowest category had not mastered the task at the end of ten trials.

Anyone reading this book will find it difficult, perhaps impossible, to appreciate how difficult it is for many people to learn what may seem to be easy tasks. One of the principal obstacles to good system design is the inability of engineers, designers, and programmers—or, for that matter, even human-factors professionals—to recognize that what seems easy to them may not be so for typical users. Merely saying "Design for the typical user" is not good enough. Designs have to be tested with a full range of typical users and modified in the light of difficulties those users experience.

Motor Learning and Skill Acquisition

The acquisition of motor skills, such as the skilled movements made in driving, flying, postal sorting, entering data into a computer, or aiming and firing weapons, is a kind of learning that is essential for the operation of many systems. When skills are highly trained they can:

- Reduce energy demands on the body;
- Reduce stress on muscles and joints, and so reduce susceptibility to injury;
- Reduce the time needed to perform tasks;
- Improve performance and the quality of work;
- Decrease errors; and
- Decrease feelings of exertion.

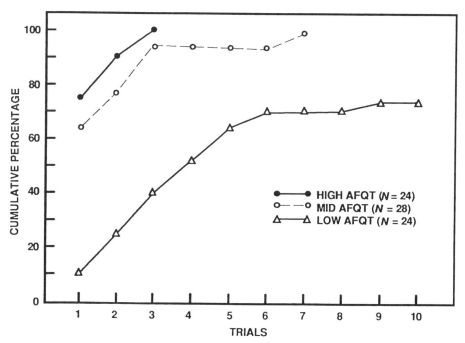

Figure 6.20 Learning curves for three groups of men. (AFQT = Armed Forces Qualification Test) (Adapted from Fox, Taylor, & Caylor, 1969).

Motor skills appear to be controlled by some kind of internal neuromuscular program. During initial stages of learning, feedback necessary to establish the program is provided by information from proprioceptors in the muscles and from the visual and auditory systems. With repetition, tasks eventually become routine and are performed almost automatically. Highly skilled operators work without consciously thinking about the movements they make. In fact, actively thinking about those movements usually results in a disruption of performance and an increase in errors. The next time you tie your necktie or your shoelaces, think in advance about the movements you have to make and note what happens.

Two kinds of feedback are necessary for an operator to learn new skills: (a) knowledge of results, which produce information about the outcome of movements; and (b) knowledge of performance, which provides information about the movements themselves. Of the two, the former is the more important. Imagine trying to learn how to play tennis, golf, or ping-pong if you could never see where the ball went after you hit it. Although of lesser importance, knowledge of performance helps learning by providing information about the correct timing, positioning, and sequencing of movements, and the forces that should be exerted. Feedback of this kind can be given by demonstrations, descriptions of movement patterns, or by videotaped replays.

Transfer of Training

From birth on, everything we learn builds on prior learning. Although we rely heavily on things we have already learned to help us learn new things, we sometimes find that old habits interfere with the acquisition of new ones. The ways in which previous learning either aids or impedes new learning is generally discussed under the heading of transfer of training.

Positive Transfer   Positive transfer is said to occur when previous learning helps us learn new material. Inputting text into a computer can be learned faster by skilled typists than by people who have never used a keyboard. The typist already knows the location of the alpha keys and can most likely key text without even looking at the keys.

Positive transfer provides the rationale for training simulators and training devices. It is cheaper and safer, and does not interfere with operations to train operators on simulators. Simulators can also provide training for systems that have not yet been built. But the construction and use of simulators can be justified only if the training they provide actually reduces the time needed to learn how to operate the equipment they simulate. There is no guarantee that simulators will result in positive transfer, and in the past some have even resulted in negative transfer.

Negative Transfer   Negative transfer is said to occur when previous training impedes the learning of new material. Skilled typists who have learned touch typing on a QWERTY keyboard find it difficult to learn to use an alphabetic, Dvorak, or AZERTY keyboard, because they have to unlearn deeply rooted movement patterns (see Michaels, 1971, for an example). Millions of people have been trained with the QWERTY arrangement, and this widespread skill makes retraining a formidable obstacle to the introduction of alternative keyboard arrangements.

Habit interference that results in negative transfer is one reason why standardization of displays, controls, and procedures is such an important human-factors principle. Lack of standardization, even in apparently minor details, can be disruptive.

For example, I have been using an IBM PC and an IBM Wheelwriter (essentially a word processor), each in different work locations. Although the basic keyboards for both machines are similar, they differ in several important respects. For one thing, there are two keys on the left of both keyboards, the lower one of which shifts type into uppercase momentarily, the upper one locks the keyboard in uppercase. To unlock the keyboard on the PC, one has to push the LOCK (upper) key again, but on the Wheelwriter one unlocks the keyboard by pushing the lower key. This seemingly trivial discrepancy has caused me great annoyance when I go from working with one machine to the other. Such lack of standardization is, unfortunately, all too common in systems today. Compare the arrangement of the numerical keys on a push-button telephone with those on hand-held calculators.

Factors Affecting Transfer Transfer of training is facilitated if the learners are taught general principles that apply to the new learning. Without such principles, habits tend to be highly specific to the situation in which they were originally learned. Knowing the principle behind operations is one reason why having a "mental model" of an operation is so important. General principles can be taught by lectures or instructional manuals.

Positive transfer is also greater the greater the similarity between the originally learned task and the one being learned. In simulators and training devices, this issue of similarity is generally referred to as "fidelity of simulation." Although the general principle seems valid, years of research on fidelity of simulation have failed to produce a metric by means of which we can predict it precisely before a simulator has been built.

RESPONDING

Once information has been received and processed, often in combination with stored information, the operator is ready to respond. Although it is theoretically possible to record human responses by tapping into nervous activity, for all practical purposes all human responses, or motor responses, are the results of some form of muscular activity. In fact, the only way we know that information has been correctly received and processed is by what a person does—the actions he or she performs. Two major forms of responding are speaking, which is a specialized kind of muscular activity, or making some sort of overt movement.

Perhaps the most important thing we can say about human responding is that it is extremely versatile. In speaking, we can form an almost infinite number of different utterances. Although speech has even been used to activate controls, movements of the fingers, hands, elbows, feet, knees, shoulders, head, and eyes are most commonly used for that purpose. What is more impressive is that people who have lost one hand can learn to perform with the other. People with no hands have learned to perform complicated activities with their feet and toes, and quadriplegics have even been able to write and to paint by holding writing implements in their lips. This extraordinary human adaptability provides designers with many options in designing controls for machines and systems. Still, some forms of responding are more efficient than others, and better suited to certain tasks.

Speaking

Speaking is so easy and convenient that most people think it must also be very simple. In fact, speaking is one of the most complicated and intricate skills we ever acquire. Communication is the primary function served by speaking and one thing that makes this form of communication so important is its universality. Everyone, except a few severely handicapped persons, can speak. Although the creation of sounds is an inborn ability common to all children, everything else about speaking must be learned. Nonetheless children learn to do it—and very quickly. By the age of two most children are able to speak simple sentences. By the age of five they usually understand about 2,000 words, and by the age of six they have learned virtually all the rules of grammar, can string words together to create meaningful new sentences of their own, and understand the meaning of sentences they have never heard before.

Our utterances, except for those that are deliberately playful or frivolous, are meaningful and serve diverse purposes. They can inform listeners, warn them, order them to do something, or question them. Spoken language is symbolic, that is, it can convey information about objects and events that are not immediately present. It is also innovative, that is, we have the ability to create new sentences or combinations of utterances that we have never encountered before. Not only can we generate such expressions effortlessly, but other people can understand and respond to them as well.

Finally, our utterances are arbitrary, in the sense that the sounds we use in speaking do not have any intrinsic relationship to the objects they symbolize. Moreover, completely different sounds may be used to refer to the same objects. For example, *input* and *output* are *entrée* and *sortie* in French, and *Eingabe* and *Ausgabe* in German. One implication of this arbitrariness is that for effective communication both speakers and listeners must use and understand the same language, a difficulty often encountered when speakers who speak one language must communicate with persons who speak some other language. These difficulties occur even in those situations where one language is supposed to be spoken by all operators. For example, by international agreement, English is the common language for all international aircraft operations. Unfortunately, there are marked differences in the pronunciation of English spoken by persons of diverse nations, or even from different regions of a single country, say, the United States. Indeed, difficulties in communication between persons of different nationalities have been implicated in aircraft accidents (Ruffell Smith, 1975).

When communication links are poor because of distance, atmospheric conditions, or limited bandwidth, or the noise is excessively great, it is sometimes necessary to spell-out each word you are trying to communicate. Although this procedure is tedious and time consuming, it is very effective. Its effectiveness can be increased by making each letter into a word, a technique known as word spelling. Table 6.8 shows the international word-spelling alphabet adopted by the International Civil Aviation Organization (ICAO). This set of words was chosen after a considerable amount of research to satisfy a number of criteria: the words must be easy to pronounce, readable by persons of diverse educational backgrounds, easily distinguishable from one another, and recognizable when spoken by and to persons of different nationalities and accents. I use this technique constantly when I have to spell my name over the telephone. The consonants *p* and *t* are virtually indistinguishable because of bandwidth limitations of telephone circuits. So, to avoid having my name come back as *Chatanis,* I always spell it out: C–H–A–P–(as–in–Papa)–A–N–I–S. The same applies to my address, *Bellona,* because *b* is so often heard as *v*.

Speech and Language Ability   Speech may be thought of basically as a way of manipulating abstract ideas and symbols. For that reason, it is not surprising to find that verbal ability is highly correlated with general mental ability. A practical implication is that, in designing systems that make use of speech and language, engineers and human-factors professionals need to keep in mind that they have verbal skills that are probably more advanced than most people's.

Some Advantages of Speech Output   In addition to what I have already said, speech has a number of other advantages as a form of output. It is fast. People can speak much more rapidly than they can write, type, or communicate by signs. At a comfortable rate of speaking the average person can utter about 170–180 words per minute, whereas skilled typists average only about 65 words per minute. Speech is also very effective. A number of research studies (Chapanis, 1981) have shown in several ways that people can solve problems and communicate better when they can

TABLE 6.8 The International Word Spelling Alphabet*

A	Alpha	N	November
B	Bravo	O	Oscar
C	Charlie	P	Papa
D	Delta	Q	Quebec
E	Echo	R	Romeo
F	Foxtrot	S	Sierra
G	Golf	T	Tango
H	Hotel	U	Uniform
I	India	V	Victor
J	Juliet	W	Whiskey
K	Kilo	X	X-ray
L	Lima	Y	Yankee
M	Mike	Z	Zulu

*From Moser & Bell (1955).

speak than when they have to use communication channels without speech. That finding is all the more impressive because analyses of spoken communication show that it is extremely unruly and often seems to follow few grammatical, syntactical, or semantic rules. That unruliness can be tamed, in part, by carefully selecting vocabularies. People can communicate very effectively with highly restricted but carefully chosen vocabularies (Kelly & Chapanis, 1977). Another advantage of restricting vocabularies is that they are much more comprehensible in noisy or degraded situations.

Speech can be used to communicate to a large number of people at the same time, and listeners do not have to be looking at, or even be visible to, the speaker. Because it is so rapid and immediate, it lends itself to rapid interchanges. Finally, it can convey subtle but important overtones, such as urgency, in communications.

Some Disadvantages of Speech Output One important limitation is that it is transient—once spoken the words are gone and cannot be retrieved. Because it is so easy to miss a telephone number spoken in television or radio commercials, announcers invariably repeat them. Speech is also a poor tool for expressing quantitative relations. Finally, spoken messages can be ambiguous, which is one reason why it is so hard to use natural speech as inputs to computers. Speech-recognition devices are continually being improved, and some current ones can recognize 1,000 or more discrete utterances (words or short phrases) with a satisfactory level of accuracy when speakers have been trained, are unhurried and speak distinctly. However, accuracy drops off rapidly if speakers are stressed, have respiratory ailments, or are in environments different from those at ground level.

Speech Intelligibility The intelligibility of speech can be measured by testing talkers and listeners under carefully controlled conditions and with specially selected speech samples: (a) nonsense syllables, (b) monosyllabic Phonetically Balanced (PB) words, (c) Modified Rhymes, or (d) sentences. Since that kind of intelligibility testing is usually much too difficult and time consuming for practical design purposes, human-factors professionals and engineers are more likely to use predictive measures of intelligibility, for example, the Articulation Index or Speech Interference Level (SIL). For a good introduction to these methods, see Kryter (1972) or Boff and Lincoln (1988).

Speech can be shaped and packaged in various ways to increase its intelligibility in noise, with headsets, for hearing aids, in unusual environments (high altitudes, underwater) or even for face-to-face communication. Peak clipping, speech compression, and automatic gain control, properly applied, are some of the techniques that can measurably increase intelligibility (see Kryter, 1972, and Boff & Lincoln, 1988). Successful communication in systems also depends on talkers and listeners. There are large differences in fundamental intelligibility among individuals and these differences tend to persist, even with practice and training, although proper training does, of course, improve performance. Finally, a number of language factors affect speech intelligibility: the information content of words, the size of the vocabulary, sentence or phrase structure, situational constraints, and the fundamental speech sounds used. Refer to Kryter (1972) or Boff and Lincoln (1988) for a good introduction to these and other related topics.

Motor Responses

Human motor responses can be classified in various ways (Fleishman & Quaintance, 1984). One such classification follows:

- *Positioning* movements are those in which a hand or foot moves from one position to another, as in reaching for the ON–OFF switch on a computer.
- *Continuous* movements are those that require muscular control movements of the same kind during the movement, as in moving a mouse or trackball to control the position of a cursor.
- *Repetitive* movements are those in which the same movement is repeated, as in using a handsaw or pedaling a bicycle.
- *Manipulative* movements involve handling parts, tools, or control mechanisms, typically with the fingers or hands.
- *Serial* movements are several relatively separate, independent movements in a sequence, as inserting a diskette into a slot, turning a computer on, and keying information into the computer.
- *Static* responses involve no movement, but consist of holding a body member stationary for a period of time, as in applying steady pressure on the accelerator of a car.

Classification systems such as this one are primarily useful for basic researchers, because most skilled activities, such as driving or working with a computer terminal, are extraordinarily complex and involve several kinds of movements that blend into one another. In system design our interest is in integrated performance and not in the components that constitute it.

Principles of Motion Economy Early in this century most work in industry was manual, and time-and-motion engineers were among the first to study systematically ways of improving performance at work, by which they usually meant productivity. Among other things, these early pioneers developed a number of principles of motion economy. Although computers and automation have changed the nature of work dramatically in the intervening years, there is still a considerable amount of manual work done in offices, stores, and factories, and the principles of motion economy developed by those engineers are still valid today. Table 6.9 lists some of those principles.

Reaction Time Data of interest in many design applications are the times taken by operators to respond to some external event or input, a change in a traffic light, an annunciator warning signal in a power plant, or a jolt when our automobile is hit from behind. The time from the onset of such a stimulus to the beginning of a

TABLE 6.9 Some Principles of Motion Economy*

- Both hands should begin and complete their motions at the same time.
- Motions of the arms should be simultaneous and in opposite and symmetrical directions.
- The best motion sequences are those that have the fewest stops or breaks.
- Relieve hands of all work that can be performed advantageously by the feet, or other parts of the body.
- Where possible, hold work by jigs, clips, or clamps to free the hands for operations.
- Pre-position tools and materials to eliminate searching and selecting.
- Arrange the heights of the workplace and chair so that the operator may sit or stand at his or her option.
- Use continuous curved movements, rather than straight-line motions involving sudden and sharp changes in direction.
- Use ballistic movements rather than restricted or controlled movements.
- Design work so that the operator can use an easy and natural rhythm.
- Arrange successive movements so that one movement passes easily into the next, each ending in a position favorable for the beginning of the next movement.
- Arrange movements so that they do not have to be made against gravity.
- If forcible strokes are required, arrange movements and material so that the stroke is delivered when it has reached its momentum.
- Eliminate hesitations, or temporary cessations of motion.
- When a definite best combination of movements has been arranged, conduct training without exception from beginning to end. That is, emphasize form rather than accuracy, even if this results in poor performance at the beginning of learning.

*Adapted from Barnes (1949). Reprinted by permission of John Wiley & Sons, Inc.

response, variously called the reaction time, response time, or response latency, has probably been studied more than any other aspect of responding.

Reaction time tasks are of three types: *simple, disjunctive,* or *choice.*

Simple reaction time refers to those situations in which there is only one stimulus to which the person makes only one kind of response. This yields the shortest times, but is influenced by a number of factors: the stimulus modality, whether visual, auditory, or tactile; stimulus characteristics, such as frequency, wavelength, intensity, duration; location of the stimulus with respect to the sensing organ; part of the body making the response; and kind of response required. Table 6.10 gives some typical simple reaction times for stimuli at moderate intensities.

In a disjunctive task there is also only one stimulus and one response required, but the stimulus appears in the presence of other distracting stimuli. These times are almost universally longer than simple reaction times.

Choice reaction times refer to those situations in which there are multiple stimuli and multiple responses. These generally yield the longest reaction times.

Reaction times vary greatly from person to person, and even between successive trials by the same individual. Some people decrease their reaction times with practice; others do not. Reaction times are generally fastest for people between 15 and

TABLE 6.10 Simple Reaction Times (in seconds) to a Light, Buzzer, and Tactile Pulse of 200 pps Delivered to the Palm, Singly and in Combination*

Stimulus	Mean	Standard Deviation
Light	.24	.11
Buzzer	.23	.092
Tactile pulse	.21	.085
Light and buzzer	.20	.063
Light and tactile pulse	.19	.057
Buzzer and tactile pulse	.18	.043
Light, buzzer and tactile pulse	.18	.043

*Adapted from Swink (1966). Copyright 1966 by the Human Ergonomics Society. All rights reserved.

60 years of age, slower for people over age 60, and slowest for children under the age of 15. Sleep deprivation, fatigue, time of day, environmental stresses, drug use, and disease, all generally add to both an individual's mean reaction time and its variability.

Perhaps the most important thing one can say about the many reaction times you can find in textbooks and handbooks is that they should be applied to practical situations only with great caution. Most reaction time data have been collected under ideal laboratory conditions. Subjects had only one task, knew within reasonably close limits when the stimulus would appear, and were prepared and motivated to respond. In operational situations, by contrast, warning signals or emergencies may occur at unexpected times, at infrequent intervals, and when the operator is engaged in other activities. Reaction times under these conditions are invariably much longer than those collected in a laboratory and may sometimes yield results that are contrary to conventional wisdom.

Lerner (1993), for example, measured driver perception–reaction times, the time required to perceive, interpret, decide, and initiate a response to some stimulus, under highly realistic conditions. Drivers drove their own cars in their normal manner, and drove on actual roadways. By virtue of extended preliminary driving they were at ease, and not expecting any unusual event at the time of braking, and the stimulus occurred at a location without any features (curves, crests, driveways) that might enhance alertness. The stimulus was a large crash barrel that was remotely released from behind a bush on a berm and rolled toward the driver's path.

The mean reaction time was 1.5 s, more than twice the braking reaction time one typically finds in laboratory studies. More interesting, however, was the finding that older drivers (65–69 and 70+ years of age) were not substantially slower than younger ones (20–40 years of age). Similar findings have been reported by other investigators—Olson and Sivak (1986), for example. It may be that factors of experience and compensation in complex skills enable older persons to maintain performance, even in the face of reduced elemental capabilities.

Differences between laboratory and real-life findings are dramatically illustrated in a study by McClelland, Simpson, and Starbuck (1983), who tested a device to warn maintenance crews working on rail lines about approaching trains that might not be visible. In laboratory tests, the average time to detect the signal by crews performing a task that was both physically demanding and required decision making was 3.6 sec, but in the field, under genuine working conditions, the median time was 38.8 sec!

Even larger discrepancies were found by Beare and colleagues (1982) between field and simulator performance of nuclear power plant operators. The performance measure was the time for operators to initiate the first correct manual action in response to an abnormal or emergency event. The field data for seven events came from a total of 126 incidents at five separate boiling-water reactor plants. Analogous data were collected from refresher-training exercises performed on a boiling-water reactor simulator. The median response time for the seven events in the simulator was 47 sec, and for the field data over 200 sec—over 3 min! Even more impressive are the data for the 95th percentile—the response times by which 95 percent of individuals would have responded. These times were 141 sec for the simulator, and a whopping 2,248 sec (over 37 min) for the field data!

Although less impressive, findings consistent with those above, were found by Warrick, Kibler, and Topmiller (1965), who had secretaries, while typing, respond to a buzzer that was sounded without forewarning at irregular intervals once or twice a week over a period of seven months, and by Johansson and Rumar (1971), who measured braking reaction times of drivers who have to brake suddenly and completely unexpectedly in traffic situations compared with braking times to anticipated signals.

These are sobering data. All too often we see decisions and designs based on reaction time data obtained in the laboratory or in simulators. The fact of the matter is that, despite hundreds, perhaps thousands, of studies on human reaction times, we have few reliable data for accurately predicting response times in most practical work situations.

Data Entry Human outputs, motor responses, are typically inputs to systems. Because computers and computerized systems are so ubiquitous, an important class of motor responses are those used for data entry. Data entry may consist of simple operations, such as keying from clear unambiguous hard copy, or of more complex operations such as those that are involved in interactive problem solving. The motor responses used in data entry are not a homogenous set of activities because of the diverse forms the data may take and the diversity of devices, keyboards, trackballs, pointing devices, joysticks, mice, microphones, used for data entry. All have advantages and disadvantages that make them suitable for some applications, but not others.

Figure 6.21 shows some representative data entry rates for a variety of conditions. Notice first the enormous range of performance represented here, from about 18 characters per minute for marking letters to about 850 strokes per minute for top keying tasks. Then study this figure carefully and you will see that data-entry rates depend on (a) the task, for example, handprinting letters versus handprinting num-

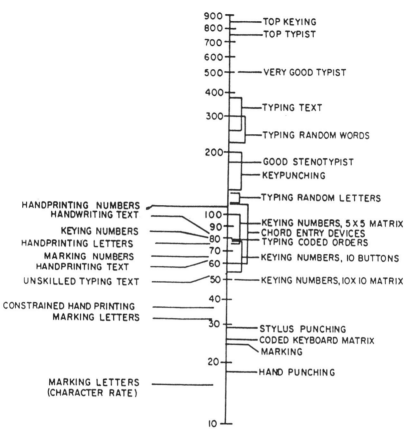

Figure 6.21 Data entry rates for diverse tasks, data-entry devices and operator skill levels. (From Devoe, 1967). (©1967 IEEE).

bers; (b) the data-entry device, for example, keying numbers with a 10×10 matrix versus a 5×5 matrix; and (c) the skill level of the operator.

SUMMARY

This chapter examined a few human characteristics involved in our ability to deal with those abstractions called *information*. Human operators receive information, process it in various ways, and output the results of that processing. To receive information we must first attend to the source of that information, sense it and, most typically, perceive what that information means. Although we have a number of sense channels, only three—vision, hearing, and touch—are used for transferring information from machines to people. Some of the main features of those sense channels are described in the chapter.

Information processing is thought to consist of a number of activities. The ones discussed in this chapter are remembering, decision making, and learning. Finally, processed information is conveyed to machines, systems, or other people through some form of responding, either by speaking or some other form of muscular activity.

We know a great deal about some aspects of this sequence of activities, much less about others. Whole books, full of basic quantitative data, have been written about human sensing and psychomotor activities, and this chapter has touched on only the highlights. Although there are also books about human information processing, they contain little quantitative data that can be directly applied to system design.

Finally, in several places throughout the chapter I have cautioned against the uncritical use of basic quantitative data in system design. Most of the data you will find in textbooks and handbooks were obtained under highly artificial and sterile conditions that do not duplicate the realities of human performance in operational systems.

REFERENCES

ANSI Z53.1-1979 (1979). *American National Standard Safety Color Code for Marking Physical Hazards.* New York: American National Standards Institute.

Barnes, R. M. (1949). *Motion and Time Study.* (3rd ed.). New York: John Wiley & Sons.

Beare, A. N., Dorris, R. E., & Kozinsky, E. J. (1982). Response times of nuclear power plant operations: Comparison of field and simulator data. In *Proceedings of the Human Factors Society 26th Annual Meeting* (pp. 669–673). Santa Monica, CA: Human Factors Society.

Boff, K. R., Kaufman, L., & Thomas, J. P. (Eds.) (1986). *Handbook of Perception and Human Performance. Volume I: Sensory Processes and Perception. Volume II: Cognitive Processes and Performance.* New York: John Wiley & Sons.

Boff, K. R., & Lincoln, J. E. (Eds.) (1988). *Engineering Data Compendium: Human Perception and Performance.* Wright-Patterson Air Force Base, Ohio: Harry G. Armstrong Aerospace Medical Research Laboratory.

Chapanis, A. (1981). Interactive human communication: Some lessons learned from laboratory experiments. In B. Schackel (Ed.), *Man–computer Interaction: Human Aspects of Computers & People* (pp. 65–114). Alphen aan den Rijn, The Netherlands: Sijthoff & Noordhoff.

Chapanis, A. (1988). 'Words, words, words' revisited. *International Review of Ergonomics, 2,* 1–30.

Chapanis, A., Garner, W. R., & Morgan, C. T. (1949). *Applied Experimental Psychology: Human Factors in Engineering Design.* New York: John Wiley & Sons.

Chapanis, A., & Moulden, J. V. (1990). Short-term memory for numbers. *Human Factors, 32,* 123–137.

Coren, S. (1994). Most comfortable listening level as a function of age. *Ergonomics, 37,* 1269–1274.

Crossman, E. R. F. W. (1959). A theory of the acquisition of speed skill. *Ergonomics, 2,* 153–166.

de Jong, J. R. (1958). The effects of increasing skill on cycle time and its consequences for time standards. *Ergonomics, 1,* 51–60.

Devoe, D. B. (1967). Alternatives to handprinting in the manual entry of data. *IEEE Transactions on Human Factors in Electronics, HFE-8(1),* 21–32.

Egan, J. P. (1975). *Signal Detection Theory and ROC Analysis.* New York: Academic Press.

Ericsson, K. A., & Charness, N. (1994). Expert performance: Its structure and acquisition. *American Psychologist, 49,* 725–747.

Farrell, R. J., & Booth, J. M. (1984). *Design Handbook for Imagery Interpretation Equipment* (Report Number D180-19063-1). Seattle, WA: Boeing Aerospace.

Federal Standard No. 595a (1984). *Colors.* Washington, DC: Office of Engineering and Technical Management, Chemical Technology Division, Paints Branch, General Services Administration.

Fleishman, E. A., & Quaintance, M. K. (1984). *Taxonomies of Human Performance.* New York: Academic Press.

Fletcher, H., & Munson, W. A. (1933). Loudness, its definition, measurement and calculation. *Journal of the Acoustical Society of America, 5,* 82–108.

Fox, W. L., Taylor, J. E., & Caylor, J. S. (May 1969). *Aptitude Level and the Acquisition of Skills and Knowledges in a Variety of Military Training Tasks* (Technical Report Number 69-6). Alexandria, VA: Human Resources Research Office.

Geldard, F. A. (1953). *The Human Senses.* New York: John Wiley & Sons.

Grether, W. F., & Baker, C. A. (1972). Visual presentation of information. In H. P. Van Cott & R. G. Kinkade (Eds.), *Human Engineering Guide to Equipment Design* (Rev. ed.) (pp. 41–121). Washington, DC: Government Printing Office.

Israelski, E. W. (1978). Commonplace human factors problems experienced by the colorblind—a pilot questionnaire survey. In *Proceedings of the Human Factors Society 22nd Annual Meeting* (pp. 347–351). Santa Monica, CA: Human Factors Society.

Johansson, G., & Rumar, K. (1971). Drivers' brake reaction times. *Human Factors, 13,* 23–27.

Kaufman, J. E., & Haynes, H. (Eds.) (1981). *IES Lighting Handbook: Reference Volume.* New York: Illuminating Engineering Society of North America.

Kelly, M. J., & Chapanis, A. (1977). Limited vocabulary natural language dialogue. *International Journal of Man–Machine Studies, 9,* 479–501.

Klein, G., & Klinger, D. (Winter 1991). Naturalistic decision making. *CSERIAC Gateway, 2(1),* 1–4.

Klemmer, E. T., & Lockhead, G. R. (1960). *An Analysis of Productivity and Errors on Keypunches and Bank Proof Machines* (Report Number RC-354). Yorktown Heights, NY: International Business Machines Corporation.

Kryter, K. D. (1972). Speech communication. In H. P. Van Cott & R. G. Kinkade (Eds.), *Human Engineering Guide to Equipment Design* (pp. 161–226). Washington, DC: Government Printing Office.

Lalomia, M. J., & Happ, A. J. (1987). The effective use of color for text on the IBM 5153 color display. In *Proceedings of the Human Factors Society 31st Annual Meeting* (pp. 1091–1095). Santa Monica, CA: Human Factors and Ergonomics Society.

Ledgard, H., Singer, A., & Whiteside, J. (1981). *Directions in Human Factors for Interactive Systems.* New York: Springer-Verlag.

Lerner, N. D. (1993). Brake perception-reaction times of older and younger drivers. In *Proceedings of the Human Factors and Ergonomics Society 37th Annual Meeting* (pp. 206–210). Santa Monica: Ca: Human Factors and Ergonomics Society.

Lindahl, L. G. (1945). Movement analysis as an industrial training method. *Journal of Applied Psychology, 29,* 420–436.

Martin, D. K., & Dain, S. J. (1988). Postural modifications of VDU operators wearing bifocal spectacles. *Applied Ergonomics, 19,* 293–300.

McCauley, M. E. (Ed.) (1984). *Research Issues in Simulator Sickness: Proceedings of a Workshop.* Washington, DC: National Academy Press.

McClelland, I. L., Simpson, C. T., & Starbuck, A. (1983). An audible train warning for track maintenance personnel. *Applied Ergonomics, 14,* 2–10.

Michaels, S. E. (1971). Qwerty versus alphabetic keyboards as a function of typing skill. *Human Factors, 13,* 419–426.

Miller, G. A. (1956). The magical number seven, plus or minus two: Some limits on our capacity for processing information. *Psychological Review, 63,* 81–97.

MIL-STD-1472D (1989). *Military Standard: Human Engineering Design Criteria for Military Systems, Equipment and Facilities.* Washington, DC: Department of Defense.

Moser, H. M., & Bell, G. E. (1955). *Joint United States–United Kingdom Report* (Report Number AFCRC-TN-55-56). Cambridge, MA: Air Force Cambridge Research Center.

Mowbray, H. M., & Gebhard, J. W. (May 1958). *Man's Senses as Informational Channels* (Report No. CM-936). Silver Spring, MD: The Johns Hopkins University, Applied Physics Laboratory.

NASA-STD-3000 (Oct. 1989). *Man–Systems Integration Standards. (Revision A).* Houston: National Aeronautics and Space Administration, Lyndon B. Johnson Space Center.

Olson, P. L., & Sivak, M. (1986). Perception–response time to unexpected roadway hazards. *Human Factors, 28,* 91–96.

Ruffell Smith, H. P. (1975). Some problems of voice communication for international aviation. In A. Chapanis (Ed.), *Ethnic Variables in Human Factors Engineering* (pp. 225–230). Baltimore: The Johns Hopkins University Press.

Sanders, M., & McCormick, E. J. (1993). *Human Factors in Engineering and Design.* (Seventh ed.). Hightstown, NJ: McGraw-Hill.

Swink, J. R. (1966). Intersensory comparisons of reaction time using an electro-pulse tactile stimulus. *Human Factors, 8,* 143–145.

Taylor, J. H. (1973). Vision. In J. F. Parker, Jr., & V. R. West (Eds.), *Bioastronautics Data Book* (pp. 611–665). Washington, DC: Scientific and Technical Information Office, National Aeronautics and Space Administration.

Van Cott, H. P., & Kinkade, R. G. (Eds.) (1972). *Human Engineering Guide to Equipment Design.* (Rev. ed.). Washington, DC: Government Printing Office.

Van Cott, H. P., & Warrick, M. J. (1972). Man as a system component. In H. P. Van Cott &

R. G. Kinkade (Eds.), *Human Engineering Guide to Equipment Design* (Rev. ed.) pp. 17–39. Washington, DC: Government Printing Office.

Warrick, M. J. Kibler, A. W., & Topmiller, D. A. (1965). Response time to unexpected stimuli. *Human Factors, 7,* 81–86.

Wolfle, D. (1951). Training. In S. S. Stevens (Ed.), *Handbook of Experimental Psychology* (pp. 1267–1286). New York: John Wiley & Sons.

Chapter 7

Personnel Selection and Training

No system, however well designed, is usable until people have been selected and trained to operate it, a step that was mentioned briefly in connection with the preparation of system specifications (see pp. 71 and 72). These "people considerations" constitute what is often referred to as the *personnel subsystem*. Human-factors personnel involved in system development are not normally concerned with selection, since that is the responsibility of the customer for whom the system is being built. Customers will, however, almost certainly want to obtain from system developers and human-factors personnel information about the skills, knowledge, and other characteristics system users must have. This information is needed to select or design appropriate selection devices.

Training is another matter. Since training may be, and often is, delegated to a training group separate from the system-development team, especially for large systems, training system design may be considered separately from system development. Still, system development cannot ignore training issues, because the way a system is designed has a strong and direct influence on the selection and training process. In this chapter I shall not discuss all the complexities of training system design—a major topic itself (see, for example, Gordon, 1994)—but shall concentrate only on the principal ways in which system design impacts and contributes to selection and training. Figure 7.1 is our point of departure.

RELATIONSHIPS BETWEEN SYSTEM DESIGN, SELECTION, AND TRAINING

The operability of a system is a direct function of its design, and so impacts personnel requirements in two ways: (a) by the numbers and kinds of people (users,

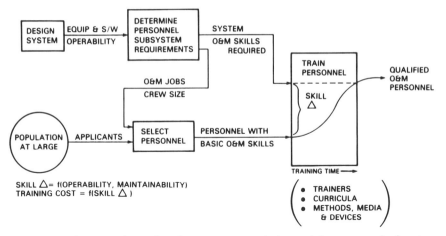

Figure 7.1 The interrelationships between system design and the personnel subsystem. (From Chapanis & Shafer, 1986. Reprinted with permission of IBM.)

operators, maintainers) who use or who will operate and maintain the system; and (b) by the skills those users, operators, and maintainers must have.

Meanwhile, applicants from the available population come with basic skills. Some of these skills are general ones, for example, language comprehension and motor skills; others are specific, for example, operating a motor vehicle, programming a computer, that have been learned with other systems. The kinds of skills new recruits have is partly a function of the time and effort that have gone into selection. Selecting highly skilled people usually requires greater recruiting efforts, more sophisticated selection devices, and more highly trained interviewers and testers, all of which add up to more dollars.

Even the best selection procedures typically do not yield personnel with the exact skills needed to operate new systems. That discrepancy, the discrepancy between the basic skills of newly selected personnel and the specific skills needed to operate the system, the skill Δ, has to be reduced or eliminated by training. The greater the discrepancy between the skills required and those available, the longer the training time. Training involves trainers, curricula, training methods, media, and devices, all of which add up to still more dollars. In other words, training costs are a direct function of the skill Δ. If you consider that, up to two or three years may be required for nuclear power plant operators, air traffic controllers, and airline pilots to become fully proficient, you will appreciate that training costs can run into hundreds of thousands of dollars.

To complete the circle, the discrepancy between skills needed and those available comes back to system design. Simplifying the design of a system, making it easier to operate and maintain, reduces the numbers of personnel who will be needed to operate and maintain it, the skill levels those personnel must have, and thereby selection and training costs. In many cases those costs are a substantial part of system life cycle costs. In addition, the supply of highly skilled people is limited by the size of the population, the basic knowledge and skills available in the population, and competition from other employers. These, collectively, are the reasons system development cannot ignore personnel considerations.

To be sure, many products, particularly consumer items, try to avoid or minimize selection requirements. Microwave ovens, washing machines, power lawnmowers, typewriters, television sets, VCRs, automobiles, home computers, and hundreds of other products aim at a broad cross-section of the population and, by and large, they are successful. Still, not everyone can use some microwave ovens, operate some power lawnmowers, program some VCRs, drive an automobile, or use some home computers. And even the most carefully designed products cannot avoid training. All come equipped with instructions, in some cases, whole books. The basic problems of selection and training cannot be entirely avoided. They only vary in degree, according to the complexity of the tool, machine, or system.

MANDATORY PERSONNEL REQUIREMENTS

The perspective I have presented in Figure 7.1 is not one you can find in most textbooks of systems engineering. For example, neither Blanchard and Fabrycky (1990) nor Cushman and Rosenberg (1991) mention selection and training at all. Wymore (1976) and Kirk (1973) both mention training, but only as something that must be done after a system has been designed. Neither sees it as an integral part of system development. Despite this apparent lack of concern about personnel requirements by some textbook writers, these requirements figure heavily in many industry and government standards. I have selected two of the latter as illustrations.

MANPRINT

MANPRINT (1987), an acronym for *Man*power and *Per*sonnel *Int*egration, is an Army program designed to ensure that personnel resources, among them, personnel and training, are continuously integrated into all phases of systems development. Before a development program for the Army can even begin, the developer must conduct studies, analyses, and evaluations to determine initial MANPRINT requirements and show how those requirements will be dealt with during system development. Although it is limited to Army equipment, the MANPRINT program has been receiving a considerable amount of attention recently, both in the United States and abroad, as a model for other development projects.

DI-CMAN-80008A

The other document is one of a whole series called Data Item Descriptions (or, as they are usually referred to, DIDs). These have been discussed earlier in this book. DIDs amplify the more general requirements contained in standards. For example, they contain information about the format of specifications and other work products and detailed instructions for the preparation of those work products. The particular DID I have selected to write about, DI-CMAN-80008A (1988) presents instructions for the preparation of a system/segment specification (see Pages 71 and 72).

In writing a system/segment specification, this DID says that one paragraph of the specification shall (note the use of the mandatory work *shall*) specify personnel requirements that must be integrated into system design. These requirements are:

- The numbers and skills of personnel who will be used in the operation, maintenance, and control of the system; and
- The numbers and skills of support personnel who will be required by the system.

Another paragraph states that the specification shall include training requirements covering, specifically;

- Who will be responsible for training, the contractor or government; and how training will be done, for example, in a school or on-the-job;
- Training devices that must be developed, the characteristics of those training devices, and the skills to be trained with those devices;
- Training time and the locations available for a training program; and
- Source material and training aids to support training.

As is evident, these make up a rather impressive list of requirements. Although this is a military document, it serves as a model for personnel requirements of commercial systems as well.

REDUCING SYSTEM COMPLEXITY

Since personnel requirements are so heavily influenced by system complexity, the most important way human factors can contribute to selection and training programs is to reduce complexity. Quickly review some design factors that will do that. They are not new; they are good general design principles that should already have been followed in earlier stages of system development. However, they are worth identifying here, in the present context.

To reduce operational complexity,

- reduce the number of operating modes;
- reduce the number and complexity of decisions required;
- reduce the number and complexity of displays;
- increase the effectiveness of automated functions; and
- reduce workloads.

To reduce maintenance complexity,

- increase built-in-test (BIT) coverage to reduce the amount of manual troubleshooting required;
- design and provide appropriate tools and test equipment; and
- increase accessibility to maintenance areas.

HUMAN-FACTORS CONTRIBUTIONS TO TRAINING PROGRAMS

A training program is a systematic series of learning events that focus on work that needs to be done. As shown in Figure 7.1, the aim of such a program is to have trainees acquire the skills, rules, concepts, and attitudes that are necessary for system operation. Perhaps because of the pervasiveness of gadgets available these days, all too often people who are developing training programs tend to become involved with technological issues, such as computerized instruction, a library well stocked with video tapes, or elaborately designed classrooms, instead of focusing on the really important question of *what* has to be learned. Work products that human factors will have completed as part of system-development provide the proper basis for a training program.

The most important objective of a training program is to train operators and maintainers for the work that needs to be done—to meet operational needs—the reasons why the system was designed in the first place. Meeting that objective means that trainees must learn how to perform in the various operational scenarios that were determined during preparation of the operational concept early in the development process. It goes without saying that training should be realistic, but the degree of realism must, of course, be traded-off against costs. Operators also need to be trained how to respond to emergencies and how to recover from them. The specific procedures and skills that should be trained are derived from task analyses that would have been made during system development. Finally, the kinds of equipment on which that training should be conducted would also have been specified earlier.

Performance, in the case of maintainer training, does not mean performance in using the system, but rather assuring that the system is available when it is needed. An additional requirement for maintainer training is that it should focus on maintenance activities for which there is no built-in-test (BIT) equipment, because those are the maintenance problems that are most time consuming and labor intensive. That training should also cover both soft and hard, as well as frequent and infrequent failures. A "soft" failure is one that does not completely render a system inoperable (for example, a flat tire). In contrast, a "hard" failure (for example, a broken axle), does.

Most of the ingredients for the specification of training requirements are available from human-factors work done during system development. They only need to be compiled and assembled into appropriate requirements.

TRAINING EQUIPMENT

In choosing equipment on which to conduct training one option is to use operational equipment, the actual system. This has advantages and disadvantages. The principal advantage of that approach is that fidelity, realism, is maximized but that benefit has to be traded-off against a number of disadvantages. Using operational equipment:

- Competes for resources, that is, trainees may actually interfere with operators;
- Often costs more than other alternatives;
- May not train effectively because of the lack of scoring devices;
- May be dangerous in case emergencies or failures occur; and
- May be too much, too soon.

To clarify the last point, bringing a novice face-to-face immediately with the barrage of indicators, controls, and other items in the cockpit of a B-747 aircraft, or the control room of a nuclear power plant, would be so overwhelming that it would actually hinder, rather than abet training. For the above reasons, it is usually more cost effective to conduct training in classrooms or with training devices.

Training Devices

Training devices fall into two main classes: *training aids* and *training equipment*. Training aids are things like films, video tapes, graphic materials, mockups, and equipment components that are used by an instructor to help present subject matter. Training equipment, on the other hand, gives the trainee "hands-on" experience, that is, the opportunity to practice skills actively.

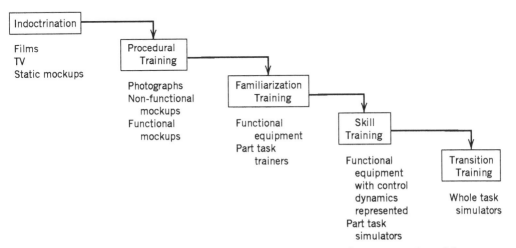

Figure 7.2 Training devices appropriate at various stages of training. (Adapted from Kinkade & Wheaton, 1972)

Stages of Training Different kinds of training devices are appropriate at various stages of training, and the sophistication and complexity of devices increases as training progresses (Figure 7.2). Training generally begins with indoctrination, during which time the trainee learns what tasks consist of and how they should be performed. Novice drivers often learn these things through lectures in a classroom; pilot ground school training is often done at a computer terminal.

Procedural training consists of teaching the trainee the names and locations of things, and the sequences in which tasks have to be performed. Student drivers learn "This is the accelerator. It regulates the speed of the vehicle. This is the directional turn indicator. . ." And so on. Figure 7.3 shows a much more complicated device, used as a bridge between pilot ground school training and simulator training. It's

Figure 7.3 A B747-400 cockpit methods trainer. (Courtesy of United Airlines)

Figure 7.4 An A-320 simulator. (Courtesy of United Airlines)

important for the student pilot to know where to reach for various cockpit switches and controls and where to look for flight information. Since there's no need to tie-up a working simulator to teach these things, the switches and controls on this device don't do anything.

During familiarization training, the trainee begins to practice skills and learn something about task dynamics, that is, get a "feel" for what the task is like. Skill training permits the trainee to become proficient, and transition training is the final stage that prepares the trainee to go to the system. Figure 7.4 shows an operational simulator, on which pilots receive skill training before being assigned to an actual aircraft.

Training Devices Are Systems Perhaps the most important thing we can say about training devices is that the more sophisticated ones are systems that need to go through the same development process as any other system. To be sure, they vary greatly in complexity. Thus the amount of development effort required must be tailored to the device. Nonetheless, most complex training devices involve hardware, software, and personnel—instructors, students, operators, and maintainers. In addition, most training devices have faults and emergencies deliberately built or programmed into them, and scoring devices for assessing student performance throughout training. Because of these additional requirements, the development of sophisticated trainers may actually be more complex than the development of the systems for which they are to train. They may greatly increase life cycle costs and their development needs to be started well before a system can actually be operated. As an example, the average control room simulator for an NRC-licensed nuclear power plant costs $10 to $12 million in 1994 dollars, and takes two to three years to put into operation. The simulator illustrated in Figure 7.4 costs between $15 and $20 millions in 1994 dollars. Clearly we are not dealing with minor pieces of equipment.

SUMMARY

All the good work that has been done in system development will come to naught if people have not been selected and trained to operate the system when it is built. In this chapter I have shown that a new system impacts selection and training—the personnel subsystem—by the numbers and kinds of people who will be required to operate and maintain it, and the skills that these people must have. Although personnel selection is typically the responsibility of the customer, system development must consider training. Training systems are often developed separately from systems, but training requirements are derived from work products (for example, statements of operational need, operational-concept documents, and task analyses) that would have been performed during system development. Training for complex systems is usually best done with training devices rather than on operational systems. Training devices vary greatly in sophistication and complexity and the more complex ones are themselves systems that have to be developed just like any other system.

REFERENCES

Blanchard, B. S., & Fabrycky, W. J. (1990). *Systems Engineering and Analysis*. (Second ed.). Englewood Cliffs, NJ: Prentice-Hall.

Chapanis, A., & Shafer, J. B. (1986). Factoring humans into FSD systems. *IBM Technical Directions, 12(1),* 15–22.

Cushman, W. H., & Rosenberg, D. J. (1991). *Human Factors in Product Design*. Amsterdam: Elsevier.

DI-CMAN-80008A (Feb. 29, 1988). *Data Item Description: System/Segment Specification*. Washington, DC: Department of the Air Force.

Gordon, S. E. (1994). *Systematic Training Program Design: Maximizing Effectiveness and Minimizing Liability*. Englewood Cliffs, N.J. Prentice-Hall.

Kinkade, R. G., & Wheaton, G. R. (1972). Training device design. In H. P. Van Cott & R. G. Kinkade (Eds.), *Human Engineering Guide to Equipment Design* (pp. 667–699). Washington, DC: Government Printing Office.

Kirk, F. G. (1973). *Total System Development for Information Systems*. New York: John Wiley & Sons.

MANPRINT (Apr. 17, 1987). *Manpower and Personnel Integration (MANPRINT) in Materiel Acquisition Process* (Army Regulation 602-2). Washington, DC: Department of the Army.

Wymore, A. W. (1976). *Systems Engineering Methodology for Interdisciplinary Teams.* New York: John Wiley & Sons.

Chapter 8

System Requirements

At various places throughout this book I have pointed out the contributions that human factors can make to system development. Let's review them quickly.

In Chapter 2 I defined and described systems, systems engineering, system life cycles, and the systems engineering process. In discussing the systems-engineering process, specific contributions that human factors could make at each stage of that process were identified (see pp. 47–52).

Chapter 3 stated that systems engineers are driven by standards, specifications, and codes that are imposed on them, and with which they must comply. Since many of those governing documents contain human-factors requirements, they drive human-factors professionals as well. By *drive* I mean that the human-factors requirements in those documents must be met. Ensuring that they are met is a human-factors responsibility. That last statement is not literally correct because, for virtually all projects, it is the system engineer's responsibility to ensure that these requirements are met. The situation is similar to that of a large ship, military organization, or industry, in which the captain, senior officer, or CEO is technically responsible for everything that goes on under his command. As a practical matter, however, subordinates do the real work. So, in the case of a systems-development project, while the systems engineer may be nominally responsible for seeing that human-factors requirements are met, the engineer must depend on someone with human-factors expertise to be sure that has happened.

Chapter 3 pointed out that the designer has to prepare work products—for example, specifications, design documents, rationale reports—and that many of those work products require human-factors inputs. The chapter then identified those inputs for each work product. (pp. 69–76). One work product, the Human Engineering Program Plan (HEPP), is specifically a human-factors responsibility. The contents of such a plan were described (pp. 64–69).

Chapter 4 was concerned with some of the principal methods that are used to obtain the human-factors inputs needed to satisfy the human-factors requirements in standards, specifications, and codes, and for the work products described in Chapter 3. In discussing each method I listed its products, that is, the usefulness of each method for system development.

Chapters 5 and 6 were concerned with human characteristics, the raw material on which the human-factors methods described in Chapter 4 are applied, to derive the human-factors inputs needed for the various documents described in Chapter 3. My treatment of this topic is highly selective, since it ignores theoretical considerations and concentrates on those human characteristics that are relevant to design.

Chapter 7 showed that the way a system is designed has a direct impact on personnel requirements—the numbers and kinds of personnel that will be needed to staff the system when it is put to use and the kinds of skills those personnel must have. Personnel selection is typically not a human-factors responsibility, but training very often is. I mentioned some of the important contributions human factors can make to the training process.

This chapter, 8, is perhaps the most important one in the book, because it is here that we come to grips with exactly how we specify design requirements so that the system, when it is built, will be consonant with the human characteristics discussed in Chapters 5 and 6.

SPECIFYING HUMAN–SYSTEM REQUIREMENTS

The proper role of the human-factors professional in system development is to write the system requirements necessary to ensure that human-factors goals will be met. More precisely, it is to write requirements of the externals of the interfaces between users and the system, with *what* interfaces should be and *how well* they should serve their intended functions. *How* those requirements are met, on the other hand, is left to the creativity of the designer. *Externals* refer to the surfaces (panels, displays) from which operators receive information, the environment in which they work, and the controls they operate.

What goes on behind those surfaces, the wiring, cables, pneumatic tubing, and computer codes that make the displays and controls work, and maintain the work environment at specified levels are *internals*. These internals are not human-factors responsibilities.

Requirements versus Design

Human-factors professionals generally do not design anything. That is not entirely true because, in rare instances, talented human-factors specialists may, in fact, design something. But that is the exception rather than the rule. What most of us do is write requirements. For example, the following is a WHAT requirement:

> Errors made in typing an input shall be correctable by positioning the cursor under the incorrect entry with one or more of the four cursor control keys, and retyping the correct entry.

It specifies what a particular system will do. That requirement may be met by diverse computer hardware, computer programs, programming languages, and algorithms. *How* the requirement is met is the province of the designer.

The following is a HOW WELL requirement:

> The system response time to a menu selection shall not exceed 0.2 seconds.

It specifies *how well* a given function shall be performed. *How* that requirement will be met is once again the province of the designer, and may be met in diverse ways.

In other words, our job is to tell designers exactly what requirements must be to make the system usable, and to state those requirements with sufficient detail and precision so that the designer and an inspector can determine whether they have been met. For an example of the application of these ideas to human–computer interfaces, see Chapanis and Budurka (1990).

System versus Human Requirements

That the proper role of human-factors professionals is to write *system* requirements is a perspective that is quite different from what can be read in most human-factors textbooks. Human-factors publications typically focus on *user* requirements, not on *system* requirements (Remember the discussion of this point on pp. 44–45). Designers cannot design to meet *user* requirements or usability goals. There is nothing wrong with those requirements or goals from a human-factors standpoint, but, to be usable by designers, those requirements and goals have to be translated into *system* requirements covering specific displays, controls, workplace configurations, and all the other design variables that affect usability. (See, for example, Wulff, Westgaard, & Rasmussen (1994.)

Human factors also attempts to define criteria for good human-factors design, and has done much to define those criteria operationally. But typically the problem is viewed from the standpoint of the human, not from the standpoint of what needs to be done with the hardware and software so that the human will find the system usable. That additional step of translation produces what Smith (1986) calls *design rules*, but it is a step that human-factors professionals have generally not taken.

What this means is that decisions about human–systems interfaces are typically left to designers. Most human-factors professionals who write for system designers appear to assume that designers deliberately seek out human-factors literature. [They mostly don't. See, for example, Meister (1989), p. 81; Burns & Vincente (1994)]. They assume that the designers read it, and thereby become qualified to make decisions about human-factors issues. This imposes responsibilities on designers they have not been trained to handle, and adds greatly to their burden of

SPECIFYING HUMAN-SYSTEM REQUIREMENTS 273

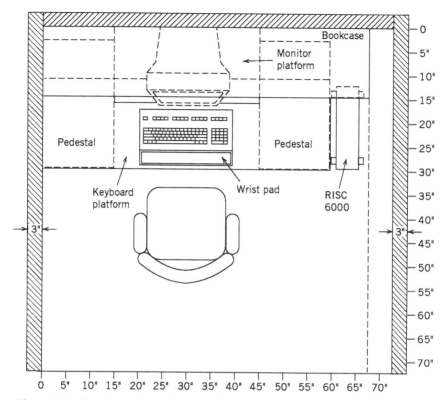

Figure 8.1 Top view of a work station for a large system. (Courtesy of L. Adams)

dealing with the many technical problems involved in the design of hardware and software.

Let me clarify this point with an example. "To design an adequate workspace," Bailey says in his 1982 book, "a designer should first consider the physical dimensions of the people who will use the system. . . Design workplaces that will accommodate at least 95 percent of the potential users."* This is a commendable and correct recommendation. But notice that, to follow it, the designer would first have to decide who will be using the system, then look up the physical dimensions of 95 percent of those users, and figure out what those dimensions mean in terms of physical equipment.

Figure 8.1 shows one of several drawings of a workstation for a certain system prepared by a human-factors professional. The dimensions of the workplace that satisfy the human requirements stated by Bailey are given in Table 8.1. The complete set of drawings (top, side, and front) and the dimensional data in Table 8.1

---

*In quoting Bailey, I do not intend to single him out for criticism. Comparable statements appear in any number of human-factors publications.

TABLE 8.1 Dimensions for the Work Station Shown in Figure 8.1*

Desk chair seat:	Width = 18″ minimum
	Depth = 17″ minimum
	Height (floor to seat) = 15″ to 20.5″ (user-adjustable with fingertip control)
	Pan angle = Within 0° to 10°
Desk Chair back:	Width = 12″ minimum
	Lumbar support width = 12″ minimum
	Lumbar support height = 9″ to 11″ from seat pan to center of support (adjustable with maximum back to seat angle of 90° to 105°)
Chair arm rest:	Inside width between arms = 18″ minimum
	Height adjustment = 3″ minimum
Other features:	1) Independent adjustable seat and back (free-floating)
	2) Synchronized adjustable seat and back
Keyboard platform with side extensions	Width = 27″ minimum when used with side extensions of at least 10″ widths. (*Note*: Keyboard platform plus side extensions must have a total width of at least 36″.)
	Depth = Keyboard platform = 12″ minimum
	Side extensions = 12″ minimum
	Height = User-adjustable from 23″ to 29″
Monitor platform	Width = width of keyboard platform
	Depth = 16″ minimum
	Height = User adjustable from 23″ to 29″
	Other: Must support 60 lbs.
Mobile pedestal file unit for personal storage	Width = 15″ minimum
	Depth = 19″ minimum
	Height = maximum of 1″ less than lowest keyboard platform side extension height

*My abridgement of an original specification.

were turned over to designers as workplace requirements for this specific system. The specification also contained requirements for the acoustical panels, overhead storage compartments, foot rest, and wrist pad. The names of suppliers that could provide equipment meeting these specifications, item numbers of acceptable equipment, and telephone numbers of suppliers were included. Figure 8.2 is a portion of a specification for another work station.

Notice what has been accomplished. In neither case must a designer decide about potential users, interpret the dimensions of 95 percent of them, or figure out where to get equipment that meets these human requirements. The human-factors professional has done that job. The designer now has only to be sure that the equipment purchased or constructed conforms to the specifications provided. The designer, in other words, now deals only with equipment, not with human factors.

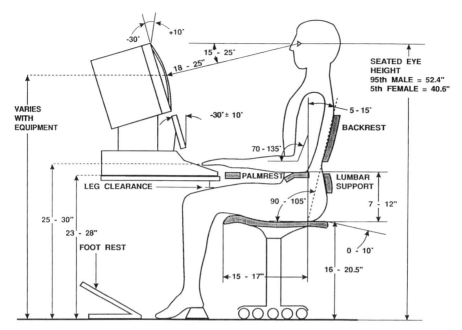

Figure 8.2 Human-factors design requirements for a computer workstation. (Reprinted with permission from Figure ID-25, IBM 1993)

System Requirements Must Be Precise

Although design standards, such as MIL-STD-1472D, NASA-STD-3000, and ANSI/HFS 100-1988, and design guidelines such as DOD-HDBK-761, and the recommendations and guidelines that can be found in many textbooks are admittedly useful and have been widely used, they have some important drawbacks. Foremost among them is that they are often too general for project-specific applications. [See, for example, Russell & Galer (1987), pp. 293–294; Wulff et al. (1994), and my discussion of this matter in Chapter 3, pp. 62–63). In the past, interpreting general recommendations has often been left to people who have little or no human-factors expertise. As a result, the same standard or guideline may be interpreted in different ways, some of which may not result in systems that are easily usable.

Moreover, short courses on design, military and industrial standards, design guidelines, and handbooks are useful, but cannot, in a few hours or days, be expected to provide designers with an adequate basis for designing systems with even moderately complex features. Guidelines do not design anything. Human-factors expertise is required to analyze the operational concept, trade-off alternatives, and translate them into project-specific requirements.

Consider, for example, the following recommendations:

- The temperature of the air at floor level and at head level should not differ by more than 5.5°C (10°F) (MIL-STD-1472D).

- The minimum diameter of a circular hatch shall be 760 mm (30 inches). The minimum dimensions of a rectangular hatch or passageway shall be 660 mm (26 inches) wide and 760 mm (30 inches) high (MIL-STD-1472D).
- The red light used for external lighting shall have a luminous intensity of 2.5 candelas, an x chromaticity coordinate not less than 0.650, and a y chromaticity coordinate not greater than 0.330 (NASA-STD-3000).
- The minimum design viewing distance [for VDTs] shall be equal to or greater than 30 cm (12 inches) (ANSI/HFS 100-1988).

A designer knows how to meet those requirements because they are precisely stated system requirements. Not only that, but an inspector can easily verify whether a machine or system has components that meet those requirements.

But now consider the following all from NASA-STD-3000:

- Control/display ratios for continuous-adjustment controls shall minimize the total time required to make the desired control movement (i.e., slewing time plus fine adjusting time) consistent with display size, tolerance requirements, viewing distance, and time delays.
- The relationships of a control to its associated display and the display to the control shall be immediately apparent and unambiguous to the operator.
- Neutral colors shall be used in work stations.
- The time lag between the response of a system to a control input and the display presentation of the response shall be minimized, consistent with safe and efficient system operation.

All these, and many more like them, are basically correct. Moreover, they are well intentioned. Notice, too, that they are all *shall* statements meaning, as I pointed out in Chapter 3, that they are mandatory requirements. The thing that is wrong with them is that they leave it to the designer, engineer, or programmer to decide exactly how to meet these requirements. What exactly must a designer do to find the control/display ratio that will minimize movement time under the conditions specified? Or, how is a designer to decide what movement relationships will be immediately apparent and unambiguous to an operator? Or, exactly what colors are neutral colors? Designers should not have to try to find exact answers to these and hundreds of other questions in books. (They almost certainly will not!) Nor should designers have to conduct their own studies to find the answers. After all, we, not engineers, are the experts on control/display ratios, control/display movement relationships, on subjective responses to colors, and on acceptable time lags. If we are not experts on all these questions, we know at least where to look for answers and, if we can't find answers in the literature, we know how to conduct the proper kinds of studies to find answers. When we are engaged in system development or are members of a development team, it is our responsibility to translate general guidelines into project-specific system requirements.

System Requirements Must Be Project-Specific

I have used the words *project-specific* several times in this book. To be sure you understand what I mean, look at the following requirement from NASA-STD-3000:

> Controls shall be designed and located so as to minimize susceptibility to being moved accidentally.

This is a general guideline, and you certainly can't argue with its intent. Notice once again that it is a *shall* statement, meaning that it is a mandatory requirement. However, there are several ways in which controls may be designed and located to avoid accidental activation. [See, for example, Chapanis & Kinkade (1972), or any of a number of human-factors textbooks.] The technique used in one situation might not be best to use in another. Here is an example of a project-specific requirement:

> On control panel X, the ON–OFF toggle switch shall be recessed such that the tip of the switch does not extend above the surrounding surface.

For another system a comparable requirement might read:

> On control panel Y the EMERGENCY MODE toggle switch shall be provided with a cover, spring loaded and hinged at the top, which must be raised before the switch can be used.

In other words, a general requirement may often be met in several or even many different ways, depending on the kind of tool, machine, system, or task that is involved. A project-specific requirement says exactly how a general guideline or general requirement will be met for a particular project, machine, or piece of equipment. Large systems may have multiple computer terminals, and the functions performed by these terminals may be different. In such cases, requirements have to be specified separately for each class of terminals in the system.

Requirements Must Be Stated in Verifiable Form

An important feature of system requirements is that they must be stated in verifiable form, that is, they should be phrased in ways that make it possible to test them for compliance. At various stages of system development, inspectors determine whether a design complies with specifications and requirements. These inspectors are typically not trained in human factors, and when they are faced with general guidelines, they cannot decide whether a design does or does not comply with the guidelines. Is this design easy to use? Is it usable? How does one decide whether it is or is not? With operationally defined system requirements this ambiguity is removed.

This involves a subtle, but very important orientation, that is often not appreciated by human-factors professionals. (Refer again to the discussion of this point in

278   8: SYSTEM REQUIREMENTS

Chapter 2, pp. 62–63). Let me illustrate with a term—*usable*—commonly associated with computer systems. Here is an operational definition.

> A system is usable if 19 of 20 randomly selected persons having the characteristics specified can complete a defined set of tasks within X units of time, with no more than Y errors, after Z units of instruction (or after T trials).

Designer and client may debate the characteristics specified, the set of tasks to be tested, and the precise values of X, Y, Z, and T, but once agreement has been reached on them, designers, clients, and inspectors all have an unequivocal way of determining whether the requirement has been met.

As an exercise, translate the following general guideline from MIL-STD-1472D—"Labels shall be as concise as possible without distorting the intended meaning or information and shall be unambiguous"—into a requirement for labels on an automated bank teller terminal.

Requirements Are Stated with Different Amounts of Specificity

From the discussions in earlier chapters, you should appreciate that system requirements are stated with different amounts of specificity, depending on the stage of development. For a system specification in early stages of development, requirements may be only general ones:

- Inputs to the system shall be through a keyboard.
- Transaction options shall be displayed on a CRT.

At lower segment or element levels, the input requirement becomes more detailed, specifying the size of the keyboard, the number of keys, their configuration, and labels. The output requirement also becomes more specific, for example, "Transaction options shall be displayed on a small CRT (X by Y inches) in the form of a menu."

What Should Requirements Cover?

Because they are project-specific, it is impossible to say exactly what requirements should cover for any particular system. Good sources of general guidance are the recommendation and guidelines in such documents as MIL-STD-1472D and NASA-STD-3000. For guidance about requirements for computer systems, consult ANSI/HFS 100-1988, MIL-HDBK-761A, or Smith and Mosier (1986). These documents are much more thorough and practical than the more discursive treatments you will find in most textbooks. Don't be misled by the military origins of some of these documents. Most of the guidelines and recommendations in them are general ones that apply to any product, whether military or commercial. As I have already said, some requirements in these several documents are stated with such precision that they may be quoted verbatim, provided they are relevant to the system under consideration. Most, however, will have to be translated into precise, project-specific requirements.

Table 8.2 is only a brief list of major areas that generally should be covered. Keep in mind that these are only major topic headings. Most could be expanded into a score or more specific requirements, and some—environmental requirements, interface requirements—into a hundred or more. The last item in Table 8.2 is common to most specifications, but is seldom seen in other human-factors documents. It requires the specification of methods, techniques, tools, facilities, personnel, and acceptance tolerance limits necessary to verify that system requirements are satisfied. Every requirement, that is, every *shall* statement, should be paired with a qualification requirement and a qualification method, for example, inspection, measurement, test. This ensures that requirements are verifiable. These qualification requirements and evaluations of their testability should be prepared at the same time as system requirements are first completed and approved, that is, well in advance of qualification itself. It is a way of guarding against the situation so often encountered in human factors in which there is disagreement about whether designs are, or are not, "usable." Establishing requirements in operationally defined ways and specifying ways of verifying that those requirements have been met, provides a common basis of understanding among developers, human-factors specialists, users, and customers.

Human-factors professionals do not necessarily, and most generally do not, write all the operator and maintainer requirements that apply to a system. Noise requirements may be prepared by acoustic engineers, training requirements by training experts, and nutritional requirements by experts in that area. Who writes requirements is determined, in part, by the kinds of expertise available on a design team.

Which System Requirements Are Relevant?

It should go without saying that not all the requirements apply to all systems. For example, nutritional and sanitation requirements do not apply to automated bank teller machines. Nor do designers of automobiles have to be concerned with work–rest cycles, duty hours, and personal equipment. So the list in Table 8.2 has to be tailored to the system.

All the methods described in Chapter 4 are sources of information about requirements that are relevant to a particular system. Operational-need documents, scenarios, functional analyses, and task analyses help establish the functions that a system must perform and so identify the system requirements that will enable those functions to occur. Activity analyses and workload studies help to define personnel requirements. Action information and operational-sequence analyses help to identify the user–system interfaces in a system and thereby the requirements for those interfaces. Critical incident studies and failure modes and effects analyses identify hazards in the system and so lead to safety requirements. Link analyses help to establish satisfactory interface configurations. Don't overlook useful information that may be obtained from the analyses of similar systems. Functions, equipments,

TABLE 8.2 Some Design Areas for Which System Requirements Should Be Prepared

System access
    method of activating the system or initiating actions
    security provisions to prevent unauthorized access
Personnel requirements, e.g.,
    operators
        numbers
        classifications
        skill levels
    maintainers
        numbers
        skill levels
    others, e.g.,
        supervisors
        managers
Training requirements, e.g.,
    training facilities
    training devices
    curricula
    trainers
Work environment, e.g.,
    lighting
    temperature
    noise
    air flow
    vibration
Workspace requirements, e.g.,
    workplace dimensions
    accesses, doors, windows
    corridors, ramps, ladders
Interface requirements, e.g.,
    input devices, controls
    output devices, displays
Job aids, e.g.,
    specialized tools
    checklists
Task requirements
    work–rest cycles
    duty hours
Interface configuration, e.g.,
    diagrams showing the placements, locations, and dimensions of input devices, output devices, workspaces, and auxiliary equipment
Safety requirements, e.g.,
    guards, interlocks
    warnings, labels, instructions

(continued)

TABLE 8.2 (*Continued*)

Personal equipment, e.g.,
   fire-resistant clothing
   radiation protection suits
   g-suits
Nutrional requirements
Sanitation requirements
Psychosocial requirements
Documentation, e.g.,
   operator manuals
   task handbooks
   position handbooks
   maintenance manuals
Qualification requirements, e.g.,
   demonstrations
   tests
   analyses
   inspections

and procedures that have been successful in predecessor systems may well be worth maintaining in the new one.

To all of the above are the standards and documents I have discussed in the previous section. Even the general guidelines and recommendations you find in most textbooks serve at least as checklists and reminders about topics that should be considered and, perhaps, rewritten as project-specific requirements.

The human-factors professional should be constantly alert to system requirements. In carrying out the various human-factors methods, and while consulting human-factors documents, keep in mind: "What implication does this have for the system requirements for my particular system?" This is where the experience, skill, and knowledge of the human-factors specialist is crucial, and there is, unfortunately, no way that can be imparted in a textbook.

TRADEOFFS

Throughout life we repeatedly have to make decisions about alternatives that involve tradeoffs. Some decisions are minor (hamburger versus hot dog); others are major (study engineering or medicine). Buying decisions often involve tradeoffs; cost versus utility, performance versus maintainability, appearance versus comfort. In system development, designers have to make numerous decisions about which of several alternative materials, components, or methods to use. Most of these tradeoff decisions (stainless steel versus zinc-plated carbon steel, glass versus plastic, diesel versus gasoline) do not concern human factors, but many do.

Some tradeoff decisions are quickly made on the basis of past experience or commonly available technical information. Others require more formal tradeoff or trade studies. A trade study is a systematic process for selecting a preferred concept from among a set of viable alternative concepts or for resolving conflicting requirements (Wulff et al., 1994). Trade studies may be performed at any phase in the system life cycle (concept exploration, system requirement definition, system design) when a decision is necessary and no candidate clearly dominates. As MIL-STD-499A on "Engineering Management" states it: "Desirable and practical tradeoffs. . . shall be continually identified and assessed."

One early activity that usually requires a documented trade study is the process that allocates system functions to users, for example, should the function be automated or performed by human operators? Once tasks are allocated to users, more detailed trade studies are continually required to define human–system interfaces:

- What level of automation should be provided?
- Should input devices be trackballs, joysticks, or direct pointing devices such as light pens or touch panels?
- If a keyboard is to be specified, should it be detachable or attached to the display as an integrated unit?
- Should two displays be used instead of windows in a single display?
- Should displays contain color or be monochrome?
- What is the optimum balance between skill levels required by selection versus training programs, to bring skill levels up to required levels?

In general, human factors is involved in trade studies whenever a decision or solution will affect:

- the numbers of personnel who will be required to operate and maintain the system;
- the skill levels operators and maintainers must have;
- the amounts of training that will be required to bring skill levels to acceptable levels of performance;
- how effectively the system can be operated and maintained;
- safety; or
- comfort and satisfaction operators and maintainers may derive from their work with the system.

Trade Study Methodology

Procedures for conducting trade studies have been described by several writers [for example, Meister (1971), and Blanchard & Fabrycky (1990)] and, although they do not agree in detail they follow similar patterns. The following draws on several sources:

1. Define the problem.
   a. Identify constraints (for example, the system should allow a user to be transported from Los Angeles to Washington, D.C., in six hours or less).
   b. Generate the problem statement.
2. Review requirements.
3. Select candidate concepts.
4. Choose and set-up methodology.
   a. Develop and quantify criteria.
   b. Determine weights to be applied to criteria.
5. Measure performance with each candidate.
6. Analyze the results.
   a. Evaluate the candidates.
   b. Perform a sensitivity analysis.
   c. Select preferred candidate.
7. Document the process and results.

Although these procedures are conceptually straightforward, they involve some of the most difficult questions for human factors. How do you measure and compare performance in comparable units? For example, suppose Function A can be performed faster on Computer System X, but Function B can be performed faster on Computer System Y. How do you weight these two functions to arrive at a measure of the *total* time saved with one or the other system, and how much is that saving in time worth in dollars and cents? Now extend the problem to 10, 20, or 30 functions on each system and the difficulties multiply. But, suppose the savings in time are accompanied by increases in error. How do you strike a balance between the savings in time versus the costs of errors? Does increased operator comfort increase productivity and, if it does, how can that be translated into dollar savings? Since more highly selected personnel require less training, is it better to spend more money on selection or on training?

These are only a few of the intractable questions the human-factors professional may be asked to answer in trade studies. Unfortunately, no study is like any other, because they are all for different systems, for unique questions, and usually for systems that have not yet been built. Textbooks are of little help because they seldom discuss or even mention these problems. The example that follows illustrates several of these difficulties and some ways they can be handled practically.

An Example

Figure 8.3 is a hypothetical trade study made to arrive at a decision about the number of crew members to use on a particular space vehicle. The options considered are one, two, or four crew members. The evaluation characteristics, or criteria, with their weighting factors, are listed across the top of the chart. The criteria would be arrived at jointly by members of the trade study team that in this example would certainly include astronauts as well as engineers and human-factors professionals

Trade options	Evaluation characteristics	Weighting Factor →									Score
		Cost (10)	Autonomy (5)	Response time (5)	Multimission/flexibility and growth (4)	Reliability (4)	Safety (3)	Rescue capability and EVA (3)	Replanning (3)	Operating weight (2)	Other (1)
One crew member	Cell-index factor	5	2	5	3	2	1	1	1	5	2	126
	Weight E.D. cell index factor	50	10	25	12	8	3	3	3	10	2	
Two crew members	Cell-index factor	4	4	4	4	3	4	4	4	4	5	(157)
	Weight E.D. cell-index factor	40	20	20	16	12	12	12	12	8	5	
Four crew members	Cell-index factor	1	4	3	5	3	4	4	4	4	1	122
	Weight E.D. cell-index factor	10	20	15	20	12	12	12	12	8	1	

Cell-index factor key
1 = unfavorable
2 = slightly unfavorable
3 = neutral
4 = favorable
5 = very favorable

- Evaluation characteristics are analyzed with respect to crew considerations only

Outcome: two crew member option

Weighting factor key
1 = low weight
10 = high weight

Figure 8.3 An example of a trade-off study. (Adapted from DOD-HDBK-763)

and would be based on a careful study of the operational need, operational concept, operational analysis and analysis of similar systems (see pp. 29–35), if there are any. Note that some of the criteria, cost and operating weight, are engineering ones, whereas others, autonomy, response time, and reliability, involve human considerations.

In the absence of any more mathematically precise methods for determining weighting factors, they are typically arrived at by consensus among the trade team members. This is the step that is likely to produce the greatest among of disagreement. Although ranking and psychophysical scaling methods help to quantify weighting factors, there is always some residual uncertainty about their validity. Is cost, for example, two-and-one-half times more important than multimission flexibility and growth, and three-and-one-third times more important than rescue capability? Despite the intangible nature of these questions, some agreement must be reached in order to complete the trade exercise.

The cell-index factors, or figures of merit, are a rough way of equating or comparing incommensurable quantities, such as dollars of cost, minutes of response time, and probabilities of mission success. These factors are used to compare candidates against system requirements. But they, too, involve a considerable amount of judgment. Some criteria, such as cost, response time, are precisely quantifiable; others, such as autonomy, multimission flexibility, and growth, are only roughly so. After deciding how each of the criteria can be measured or estimated, there is still the problem of how to classify each of the measures or estimates. How many dollars of cost or minutes of response time are unfavorable, slightly unfavorable, neutral, favorable, or very favorable?

Once these values have been determined, the rest of the trade study proceeds simply. Cell-index factors are multiplied by their weighting factors and summed. The option with the largest sum is the favored one.

As you may infer from this example, there is a considerable amount of subjectivity in carrying out most trade studies, particularly those involving human-factors issues. They also have some arbitrariness. For example, there is nothing sacred about having weighting factors go from 1 to 10 and cell-index factors go from 1 to 5. Some authors use values going from 0 to 1 for both quantities. For all that, trade studies serve a useful function in system development. They force everyone concerned to identify and agree on all the important ramifications of particular design options, to quantify, as well as possible, the consequences of alternative design options, and to provide customers with a rationale for choosing particular designs.

Sensitivity Analyses

Once the basic structure of the study has been formulated, if customers do not agree with elements of it (for example, the weighting factors or figures of merit), the study can be easily repeated with new values. Even if customers raise no objections or questions, it is usually a good idea to do a sensitivity analysis of the final outcome—to see how robust it is. Suppose, for example, that cost is not really three-and-one-third times more important than safety. Suppose that safety is really just as important as cost. How much would that affect the outcome? Or, suppose that reliability with four crew members deserves a very favorable rating rather than a neutral one. How much would that affect the outcome? By systematically varying the various terms in the analysis, one can find out how robust, how stable, the final solution is, or, conversely, how sensitive it is to changes in the underlying assumptions.

As an exercise, outline a trade study to make a decision about the use of a VDT or LED display to provide feedback to customers using an automated bank teller machine.

SUMMARY

This chapter showed that the proper role of the human-factors professional is writing design requirements in such a way that human-factors goals will be met. These requirements are for the externals of the interfaces between users and the system, with *what* interfaces should be and *how well* they should function. *How* those requirements are met is left to the creativity of the designer. System requirements tell designers exactly what must be done to make the system usable. They must be precise, project-specific, and stated in verifiable form. Requirements come from all the human-factors methods discussed in Chapter 4, from specific guidelines in standards, and from the translation of general standards into project-specific ones. Trade-off decisions have to be made throughout system development whenever there are conflicting requirements or viable alternatives. I describe the methodology for conducting a trade study, and give an example of how it could be done.

REFERENCES

ANSI/HFS 100-1988 (Feb. 4, 1988). *American National Standard for Human Factors Engineering of Visual Display Terminal Workstations.* Santa Monica, CA: Human Factors Society.

Bailey, R. W. (1982). *Human Performance Engineering: A Guide for System Designers.* Englewood Cliffs, NJ: Prentice-Hall.

Blanchard, B. S., & Fabrycky, W. J. (1990). *Systems Engineering and Analysis.* (2nd ed.). Englewood Cliffs, NJ: Prentice-Hall.

Burns, C. M., & Vincente, K. J. (1994). Designer evaluations of human factors reference information. In *Proceedings of the 12th Triennial Congress of the International Ergonomics Association* (pp. 28–31). Mississauga, Ontario, Canada: Human Factors Association of Canada.

Chapanis, A., & Budurka, W. J. (1990). Specifying human–computer interface requirements. *Behaviour and Information Technology 9,* 479–492.

Chapanis, A., & Kinkade, R. G. (1972). Design of controls. In H. P. Van Cott & R. G. Kinkade (Eds.), *Human Engineering Guide to Equipment Design* (pp. 345–379). (Rev. ed.). Washington, DC: Government Printing Office.

DOD-HDBK-761A (Sept. 30, 1989). *Human Engineering Guidelines for Management Information Systems.* Redstone Arsenal, AL: U.S. Army Missile Command.

DOD-HDBK-763 (Feb. 27, 1987). *Military Handbook: Human Engineering Procedures Guide.* Washington, DC: Department of Defense.

Meister, D. (1971). *Human Factors: Theory and Practice.* New York: John Wiley & Sons.

Meister, D. (1989). *Conceptual Aspects of Human Factors.* Baltimore: The Johns Hopkins University Press.

MIL-STD-499A(USAF) (May 1, 1974). *Military Standard: Engineering Management.* Washington, DC: Department of Defense.

MIL-STD-1472D (Mar. 14, 1989). *Military Standard: Human Engineering Design Criteria for Military Systems, Equipment and Facilities.* Washington, DC: Department of Defense.

NASA-STD-3000 (Oct. 1989). *Man–System Integration Standards (Rev. A).* Houston: National Aeronautics and Space Administration, Lyndon B. Johnson Space Center.

Russell, A. J., & Galer, M. D. (1987). Designing human factors design aids for designers. In G. Salvendy (Ed.), *Cognitive Engineering in the Design of Human–Computer Interaction and Expert Systems.* Amsterdam: Elsevier Science Publishers.

Smith, S. L. (1986). Standards versus guidelines for designing user interface software. *Behaviour and Information Technology, 5,* 47–61.

Smith, S. L., & Mosier, J. N. (Aug. 1986). *Guidelines for Designing User Interface Software* (Report Number ESD-TR-86-278 [MITRE 10090]). Bedford, MA: MITRE Corporation.

Wulff, L. A., Westgaard, R. H., & Rasmussen, B. (1994). "What is 'care', and how much is 'care'?" Ergonomic criteria in engineering design. In *Proceedings of the 12th Triennial Congress of the International Ergonomics Association* (pp. 25–27). Mississauga, Ontario, Canada: Human Factors Association of Canada.

Chapter **9**

Postscript

This chapter is about a few additional things you should know if you are involved, are or likely to be involved, in system development.

DOCUMENTATION

By now you should be impressed by the amount of paperwork involved in the development of systems. For example, the Human Engineering Design Document (HEDD) for the LAMPS Mark III System in 1978, was 477 pages long. Software requirements for the computer–human interface of FAA's new Tower Control Computer Complex in 1994, filled 350 pages. (Compare these numbers with the size of this book.) Keep in mind, too, that these are just two of the several documents that were or would be produced in each of these projects. Even more dramatic is that, in the 1950s, the technical manuals for the F-86 aircraft contained a modest 10,000 pages. In 1986 the corresponding figure for the B-1B had swelled to a staggering 1,000,000 pages! (Figure 9.1) According to Dooley (1986), the paperwork, proposals, charts, and reports used in developing and manufacturing a modern airplane weigh more than the finished plane itself.

To reinforce the message: Writing, using, and maintaining adequate documentation are critical for the development and operation of a system. Some engineers and designers, and human-factors professionals as well, tend to belittle documentation with flip remarks such as, "We're not running a paper factory." As a result, many problems in the design and development of systems are due to too little, rather than too much documentation. In the opinion of one authority, "The first rule of managing software development is ruthless enforcement of documentation requirements." (Royce, 1970, p. 5).

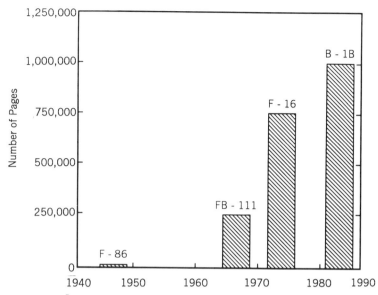

Figure 9.1 Numbers of pages in the technical manuals for selected U.S. Air Force aircraft. (From Chapanis, 1988. Reprinted with permission of the publisher.)

Users Who Need Documentation

Documentation serves four major classes of users:

1. *Designers and developers.* I include in this category human-factors professionals on the design team. I pointed out in Chapter 2 several documents that guide the development of systems. Operational-need and operational-concept statements provide a common basis of understanding about exactly why a system is needed in the first place, and the kind of system that everyone thinks will meet that goal. Both help to prevent later misunderstandings about the goals of the development project. This is especially the case with large projects that may take years. (Remember how long the USAF B-1 bomber was under development?) It frequently happens that some professionals leave a project part way through its development, and others may join long after the project was begun. Without proper documentation newcomers may not know the thinking that had gone into development up to that point and why certain decisions were made.

Detailed lists of system requirements (Chapter 8) define the specific goals that designs must meet and serve as criteria that can be used in final tests and evaluations. It is especially important to document the results of trade-off studies and the reasons why one or another design alternative was selected. For one thing, this kind of documentation helps human-factors professionals defend their decisions against challenges that may be raised in reviews and audits.

Finally, system documentation often serves to highlight ambiguities or inconsistencies that have to be resolved.

2. *Managers.* Managers must be kept informed so that they may supervise and administer a project intelligently. They need to make decisions about and verify allocations of resources and personnel to the project, determine that the work of various technical specialists (refer to Figure 2.4) is accomplished on time, and authorize subcontracting parts of the project if and when that seems necessary. Good written descriptions provide managers with tangible criteria that can be used in judging progress on a project, and prevent designers from hiding behind vague statements such as "I'm 90-percent finished" month after month.

3. *Operators.* In order to do their jobs, operators need job relevant, concise, well-organized documentation. This kind of documentation tells operators how to operate and care for the system. Task analyses, flow diagrams, and operational-sequence diagrams prepared earlier by human-factors personnel are useful in preparing this kind of documentation.

Unfortunately, procedures documents often leave much to be desired. Morgenstern and colleagues (1987), for example, studied operating procedures in 46 nuclear power plants, and concluded that, as a group, the documentation they inspected was minimally acceptable from the standpoint of its usability. Operating procedures were often written in vague terms, lacked specificity, failed to describe actions to be taken in a step-by-step manner, did not provide clear indicators of when a particular objective had been reached, and when a procedure had been completed. In addition, checklists, placekeeping aids, and other tools to enhance ease of use, were commonly missing. Appendix C of another report (U.S. Nuclear Regulatory Commission, 1989) contains a detailed list of commonly found glaring deficiencies in certain procedures manuals. Operators cannot be expected to perform safely and well if they are given vague or poorly written instructions.

4. *Maintainers.* Maintainers need to know how to care for the system, diagnose difficulties when they arise, and correct or change the system, if necessary. This kind of documentation is usually very detailed with exploded diagrams, parts lists, program listings, files, and the like.

No single kind of documentation can serve all classes of users equally well. Managers do not need a maintenance manual, nor do maintainers need the kind of documentation that is most useful to designers. Above all, documentation must be updated and kept current with modifications in design. There is almost nothing more frustrating than reading an operational manual, or a set of assembly instructions, only to find that the diagrams and instructions are for a slightly earlier version of the equipment which has been changed in the meantime.

Human-Factors Data File

Human-factors professionals in a development project have the responsibility for establishing and maintaining, in a Human-Factors Data File, all human-factors data generated in a program. These data are the Human Engineering Program Plan (HEPP, see pages 64–69); the results of analyses, trade-off studies, and design reviews; drawings; checklists; and other supporting background documents reflecting all human-factors decisions and the rationales for those decisions. This file should be available for inspection by the customer at the contractor's facility.

These data are typically reviewed at various contractor meetings such as design reviews, audits, and demonstrations; and during tests and evaluations. The data file should be organized to provide what is referred to as traceability—to show how the human-factors requirements identified initially during analysis were followed through design and development, and how final tests and evaluations of the approved design verified that the initial requirements had indeed been met.

Communication Skills—A Vital Asset

All this may perhaps help explain why the ability to communicate effectively in writing is such an important asset for the human-factors professional. A special kind of communication is required: the ability to communicate human-factors data, recommendations, and requirements in ways that can be understood and used by designers. I hope this book has taught you some ways to do that.

WHAT MAKES A SUCCESSFUL HUMAN-FACTORS PROGRAM?

Human-factors professionals are often frustrated because they cannot put into practice all the good things I've talked about in this book. Unfortunately, successful human-factors programs depend on several requirements that are not under human-factors control [see, for example, Meister (1993)].

Customer Support

First, and perhaps most important, the customer has to want human factors. The customer has to be genuinely convinced that usability, ease of use, safety, comfort, and efficiency are desirable design goals and that, in the case of commercial projects, they have positive market value. In some cases, that realization has come about only after a disaster, a market failure, or a law suit. The Three Mile Island nuclear power plant explosion was responsible for intensive human-factors efforts that had previously been entirely missing in the nuclear industry. Market failures costing millions of dollars have forced other manufacturers to confront the market realities of poor human factors [see, for example, Chapanis (1991)]. The recent flood of law suits involving stresses and strains attributed to poor workplace designs has made some kinds of ergonomics almost a growth industry. All that notwithstanding, if the customer doesn't want human factors, most likely little or none will be done.

Management Support

The second requirement is that management has to want human factors. Managers also have to be convinced that human factors is essential to the development of a successful system [see, for example, Meister (1993); U.S. Nuclear Regulatory Commission, (1989)]. In fact, enlightened managers have sometimes been able to convince customers about the importance and necessity for incorporating human factors in a development program even when it had not been specified by the customer. One thing is certain, however: a human-factors program will not succeed if management does not see the need for human factors, or if management merely pays lip service regarding human factors, when it was specifically required by the customer.

Adequate Funding

All the activities I've written about in this book take time and cost money. For a large project, several person-years of time may be required to carry out the full set of human-factors methods described in Chapter 4, to write all the documents described in Chapter 3, and to prepare the design requirements discussed in Chapter 8. Resources and facilities will certainly be required for simulations and for the various studies, analyses, and experiments that may be conducted. If adequate funding has not been allocated for all this work, the human-factors program is certain to suffer.

Dedicated Human-Factors Personnel

A successful human-factors program needs expertise specifically assigned to a development project. Borrowing human-factors personnel from some other department or organization for occasional inputs is not effective. A successful program requires continuity and continual interaction between human-factors personnel and other members of a design team that can be best provided by dedicated insiders. Although consultants can usually not provide that kind of service, they may be the only resource available for development projects by small companies that cannot afford full-time human-factors personnel.

A Systems-Disciplined Team

A system design team normally includes highly trained specialists with diverse technical backgrounds, vocabularies, professional orientations, values and aspirations. In working together it is easy for such individuals to become polarized, to think of "me" or "us" versus "them." Egos may be bruised when a basic design layout, or even minor details of design, constructed with conscientious and tedious effort, are rejected or criticized by other members of the team. A designer, or human factors professional, may also become ego involved and defensive after having assumed responsibility and having made a commitment for a design unsupported or only weakly supported by valid data and rationales. Misguided self-interest and ego-involvement conflicts occurring in the system design process are often responsible for discouragement among designers and high turnover rates among professional personnel.

To be successful it is necessary for the self-interest of individuals to be subordinated to the identity of the team and to the acceptance of group decisions. This requires a level of maturity and diplomacy on the part of every individual in the team that is difficult to achieve. It helps to have frequent discussion sessions in which the system mission is reviewed and every member is encouraged to present his or her proposed design solutions, each supported by a rationale. Documented rationales, reviewed calmly and objectively in terms of their contribution to the system mission, can create a spirit of cooperation that will overcome barriers of communication, specialization and differences in interests (Burgess, 1970).

Early Inclusion of Human Factors

As I said in Chapter 3, the major part of human-factors work has to be done early in the development project, long before any designs are committed to paper. All too often in the past, human factors has been brought in too late. Once major design decisions have been made, and basic architecture has been laid out, it becomes difficult, and sometimes prohibitively expensive, to make anything more than minor cosmetic changes. As a general rule, the earlier human factors becomes involved in a development project, the better.

User Participation

Something not stressed earlier is that users must participate in the development project. By *users* I mean people who are typical of those who will actually operate the system. Sometimes in the past, users have been interpreted to mean only the *buyers* of the system. It is certainly important to involve customers early and formally in the development process to obtain commitment on the operational need, operational concept, and system requirements, and to have customers sign-off successful reviews and audits. But it is also important to recognize that the people who contract for, or buy, a system are seldom the real *users* of it.

User participation in the design process has a number of advantages. In some cases, particularly in the case of projects involving the redesign of existing systems, such as those in offices and factories, it may help to introduce a more employee-centered style of management, with greater emphasis on open communication, opportunity for debate, and emphasis on personal responsibility. Users are also generally expert in the operation of antecedent systems and aware of all their faults and problems. This is invaluable experience, to be capitalized on in the development of a new system. Finally, participating in development gives users a sense of ownership in the new system, and makes it easier for them to transfer to it [see, for example, Mumford (1991)].

In projects for which there are no obvious antecedents (the very first space vehicle) and those designed for a mass market (automated banking terminals, new types of communication services), it is still valuable to involve users in the development process. A critical issue, however, is the selection of users: They should be a random sample of potential users. All too often, users chosen for participation in product development are highly selected or biased. One company had spent a considerable amount of time and effort and had involved users in the development of a computerized handprint recognition system. Tests run by the company showed that the system could recognize correctly over 90 percent of the samples of handprint submitted to it. After reviewing those tests I recommended another series of tests with people from the local population and in those tests recognition accuracy dropped to about 20 percent. The difficulty was that the users who had participated in the original development were all company employees and accountants as well!

Naïve users often contribute to product development in unexpected ways, because they see things differently than engineers and designers. Even human-factors professionals are not necessarily good judges of how most people will respond to new devices. (Remember my discussion of this point in Chapter 5?). Of several example, I have picked a simple one from a project involving the wording of instructions. The device in question had a tab which users had to depress before the device could be rotated to release it. The designers, and I myself, had agreed on the word *PUSH* to be placed on the tab, and the word *OPEN* with an arrow indicating the direction in which the device should be rotated. When I ran tests with people solicited from the local area, several suggested *PRESS* instead of *PUSH*, and *TURN* instead of *OPEN*. A second series of tests confirmed that these newer choices were indeed better. This is a simple example, to be sure, but is illustrative of an important principle to keep in mind.

Users Alone Cannot Design Usable Systems Having said all that, users, by themselves, cannot design usable systems. They often do not see design possibilities that any good designer or human-factors professional should be aware of. For example, naïve users may not be aware of all the possible types of controls (knobs, switches, levers, cranks, pedals, trackballs, mice) and their advantages and disadvantages. Nor are they likely to think of all the possible kinds of display that might be used to convey information from systems to operators and users. A strategy that Eason (1994) has found to be most successful is a mixed one, in which human-factors professionals design for users by designing with them.

THE FINAL WORD

Even the best efforts of human-factors professionals and system designers cannot guarantee that a system will be a success because of the many factors over which they have no control. For example, you may:

- Specify work configurations, work environments, and procedures for successful operation, only to have them thwarted or overruled by arbitrary management decisions.
- Prepare clear and understandable manuals and instructions, but have to depend on the willingness of operators and maintainers to read and comply with them.
- Specify the numbers of people and the kinds of skills needed to operate the

system, but find that the actual selection will be done by people who may not understand and meet your selection requirements.
- Develop a complete training program for the system, but discover that the actual job of training will be conducted by people who do not do a good job, because they do not understand the operational need and scenarios for which the system was designed.

If you find that set of difficulties discouraging, take heart! In spite of all the hurdles and uncertainties I've described in this book, usable systems *do* get built and there are enough success stories described in the literature. [See, for example, Chapanis (1991), Harris (1984), and Klemmer (1986)] to allow us to assert confidently: Human factors can make a difference!

Usability—the Ultimate Criterion

A curious thing has been observed by many human-factors professionals: When a tool, machine, or system is well designed, people don't seem to notice it. It fits naturally into their way of doing things, and they use it comfortably and almost instinctively. Do a poor job of human engineering and you are sure to hear about it, because, in the final analyses, we design things for only one purpose, and that purpose is to serve us. If your product does that safely, comfortably, and efficiently, nobody may ever compliment you. Nonetheless, you will be able to rest content with the inner satisfaction of knowing that you have done your job well. I hope this book will help you to earn and feel the inner glow that will be your reward for good design.

REFERENCES

Burgess, J. H. (1970). Ego involvement in the systems design process. *Human Factors, 12,* 7–12.

Chapanis, A. (1991). The business case for human factors in informatics. In B. Shackel & S. J. Richardson (Eds.), *Human Factors for Informatics Usability* (pp. 39–71). Cambridge, England: Cambridge University Press.

Dooley, B. (1986). Paperless office? Don't wait! *Management Information Systems Week, 7(26),* 40.

Eason, K. D. (1994). User centred design: For users or by users? In *Proceedings of the 12th Triennial Congress of the International Ergonomics Association. Volume 1: International Perspectives on Ergonomics* (pp. 78–80). Mississauga, Ontario, Canada: Human Factors Association in Canada.

Harris, D. H. (1984). Human factors success stories. In *Proceedings of the Human Factors Society 28th Annual Meeting* (pp. 1–5). Santa Monica, CA: Human Factors Society.

Klemmer, E. T. (1986). Some successful applications of human factors. In *Proceedings of the Human Factors Society 30th Annual Meeting* (pp. 456–46). Santa Monica, CA: Human Factors Society.

Meister, D. (1993). Non-technical influences on human factors. *Cseriac Gateway, 4(1)*, 1–3.

Morgenstern, M. H., Barnes, V. E., McGuire, M. V., Radford, L. R., & Wheeler, W. A. (Feb. 1987). *Study of Operating Procedures in Nuclear Power Plants: Practices and Problems* (Report Number NUREG/CR-3968. PNL-5648, BHARC-400/85/017). Washington, DC: U.S. Nuclear Regulatory Commission.

Mumford, E. (1991). Participation in systems design—what can it offer? In B. Shackel & S. J. Richardson (Eds.), *Human Factors for Informatics Usability* (pp. 267–290). Cambridge, England: Cambridge University Press.

Royce, W. W. (Aug. 1970). Managing the development of large software systems: concepts and techniques. *1970 WESCON Technical Papers, Vol. 14* (pp. 1–9).

U.S. Nuclear Regulatory Commission. (Apr. 1989). *Lessons Learned from the Special Inspection Program for Emergency Operating Procedures*. Washington, DC: Author.

Appendix A

Acronyms and Abbreviations

This appendix defines the acronyms and abbreviations I have used in this book, and some others you are likely to meet if you work with systems engineers or designers.

ACWP	Actual Cost of Work Performed
ADP	Automatic Data Processing
AF	Air Force
AFIPS	American Federation of Information Processing Societies
AFSC	Air Force Systems Command
AI	Action Item
AIAAE	American Institute of Aeronautical and Astronautical Engineers
AMSDL	Acquisition Management System and Data requirements control List
ANSI	American National Standards Institute
ARI	Air Conditioning and Refrigeration Institute
ARP	Availability, Reliability, Performance
ASCE	American Society of Civil Engineers
ASCII	American National Standard Code for Information Interchange
ASHRAE	American Society of Heating, Refrigerating and Air Conditioning Engineers
ASME	American Society of Mechanical Engineers
ASQC	American Society of Quality Control
ATC	Air Traffic Control
ATE	Automatic Test Equipment

ACRONYMS AND ABBREVIATIONS

B&P	Bid and Proposal
BCPE	Board of Certification in Professional Ergonomics
BIT	BInary digiT; Built in Test
BITE	Built in Test Equipment
B/L	Baseline
BOM	Bill of Materials
BSI	British Standards Institute
C&C	Command and Control
CADET	Computer Aided Design and Evaluation Technique
CAPE	Computerized Accommodated Percentage Evaluation
CAR	Crewstation Assessment of Reach; Corrective Action Report
CASE	Computer Aided Software/System Engineering
CASR	Cost and Schedule Reporting
CCB	Configuration/Change Control Board
CD&V	Concept Demonstration and Validation
CDR	Critical Design Review
CDRL	Contract Data Requirements List
CE	Concept Exploration
CFE	Contractor/Customer Furnished Equipment
CFI	Contractor/Customer Furnished Information
CHFP	Certified Human Factors Professional
CI	Configuration Item
CIC	Combat Information Center
CIE	Commission Internationale d'Éclairage
CIL	Configuration Identification List
CL	Control List
CM	Configuration Management
COMBIMAN	COMputerized BIomechnical MAN
COTR	Contracting Officer's Technical Representative
COTS	Commercial Off The Shelf
CPAF	Cost Plus Award Fee
CPCI	Computer Program Configuration Item
CPDP	Computer Program Development Plan
CPDS	Computer Program Design Specification; Computer Program Development Specification
CPE	Certified Professional Ergonomist
CPFF	Cost Plus Fixed Fee
CPIF	Cost Plus Incentive Fee

CPM	Critical Path Method
CPPS	Computer Program Performance Specification; Computer Program Product Specification
CPR	Cost Performance Report
CPU	Central Processing Unit
CR	Change Request
CRMA	Cost, Reliability, Maintainability, Availability
CRMM	Cost, Reliability, Maintainability, Manufacturability
CRT	Cathode Ray Tube
CSC	Computer Software Component
CSCI	Computer Software Configuration Item
CSERIAC	Crew System ERgonomics Information Analysis Center
CTAR	Critical Task Analysis Report
CUBITS	Criticality/Utilization/BITS of information
CWBS	Contract Work Breakdown Structure
DCA	Document Control and Approval
DCL	Digital Control Language
DCP	Decision Coordinating Paper
DCS	Data Collection System
DDD	Design Description Document
DEMP	Development Engineering Development Plan
DEP	Design Eye Point
DGD	Design Guidance Document
DH	Design Handbook
DI	Design Inspection
DID	Data Item Description
DIN	Deutsche Industrie Norm
DOD (DoD)	Department of Defense
DODD (DoDD)	Department of Defense Document
DODISS	Department Of Defense Index of Specifications and Standards
DOU	Document of Understanding
DP	Data Processing
DR	Design Rationale; Discrepancy Report
DRR	Design Rationale Report
DT&E	Development Test and Evaluation
EAC	Estimate At Completion
ECP	Engineering Change Proposal

EEPROM	Electronically Erasable Programmable Read Only Memory
EMC	ElectroMagnetic Compatibility
EMI	ElectroMagnetic Interference
EPROM	Erasable Programmable Read Only Memory
ERP	Eye Reference Point
ERS	Ergonomics Research Society
ETC	Estimate To Completion
F (f)	Function
FAA	Federal Aviation Administration
FAR	Federal Acquisition Regulation
FCA	Functional Configuration Audit
FE	Field Engineering
FI	Fault Isolation
FM	Figure of Merit
FMEA	Failure Modes and Effects Analysis
FM/FL	Fault Monitor/Fault Location
FOM	Figure of Merit
FOT&E	Follow on Test and Evaluation
FPC	Flow Process Chart
FQR	Formal Qualification Review
FQT	Formal Qualification Test
FRU	Field Replaceable Unit
FSD	Full Scale Development; Functional Sequence Diagram
FSED	Full Scale Engineering Development
FW	FirmWare
G (g)	Gravity
GFE	Government Furnished Equipment
GFI	Government Furnished Information
GFP	Government Furnished Property
GUI	Graphical User Interface
HCCB	Hardware Configuration Control Board
HCI	Human–Computer Interface
HDBK	HanDBook
HDE	Hardware Development Engineering
HE	Human Engineering
HEDAD-M	Human Engineering Design Approach Document-Maintainer

HEDAD-O	Human Engineering Design Approach Document-Operator
HEDGE	Human Engineering Data Guide for Evaluation
HEDSP	Human Engineering Dynamic Simulation Plan
HEPP	Human Engineering Program Plan
HEPR	Human Engineering Progress Report
HESAR	Human Engineering System Analysis Report
HETP	Human Engineering Test Plan
HETR	Human Engineering Test Report
HF	Human Factors
HFE	Human Factors Engineering
HFES	Human Factors and Ergonomics Society
HFT&E	Human Factors Test and Evaluation
HFTEMN	Human Factors Test and Evaluation MaNual
HODAC	Human Operator Data Analyzer/Collector
HOS	Human Operator Simulator
HP	Human Performance
HQ	Headquarters
HW	HardWare
HWCI	HardWare Configuration Item
HWDP	HardWare Development Plan
I	Input; Inspection
I&T	Integration and Test
ICD	Interface Control Document; Interface Control Drawing
ICI	International Commission on Illumination
ICWG	Interface Control Working Group
IDEF	Integrated computer-aided manufacturing DEFinition
IDD	Interface Design Document
IDS	Interface Design Specification
IEA	International Ergonomics Association
IEEE	Institute of Electrical and Electronic Engineers
IES	Illuminating Engineering Society
ILS	Integrated Logistics Support
I/O	Input/Output
IOC	Initial Operational Capability
IOT&E	Initial Operational Test and Evaluation
IPL	Initial Program Load; Indented Parts List
IPS	Integrated Product Support
IR&D (IRAD)	Independent Research And Development

302 ACRONYMS AND ABBREVIATIONS

IRS	Interface Requirements Specification
ISD	Instructional System Development
ISEA	Industrial Safety Equipment Association
ISO	International Standards Organization
ISPF	Interactive System Productivity Facility
IV&V	Independent Verification and Validation
JMSNS	Justification for Major System New Start
JPM	Joint Program Management
KISS	Keep It Simple, Stupid
KSLOC	[Kilo] thousand Source Lines Of Code
LCC	Life Cycle Cost
LCOM	Logistics COmposite Model
LED	Light Emitting Diode
LM	Labor Month
LOA	Letter of Agreement
LOC	Lines of Code
LORA	Level of Repair Analysis
LOS	Line of Sight
LRU	Line Replaceable Unit
LSA	Logistic Support Analysis
LSAR	Logistic Support Analysis Record
M&LA	Materiel and Logistic Affairs
MANPRINT	MANpower and PeRsonnel INTegration
M&S	Maintenance and Supply
MENS	Mission Element Needs Statement
MI	Management Instruction
MIL-SPEC	MILitary SPECification
MIL-STD	MILitary STandarD
MIPS	Million Instructions Per Second
MMH	Maintenance Man Hours
MOA	Memorandum of Agreement
MOE	Measure of Effectiveness
MPD	Multi-Purpose Display
MPDS	Modular Programmable Display System
MTBF	Mean Time Between Failures

MTBM	Mean Time Between Maintenance
MTBR	Mean Time Between Removals
MTM	Methods Time Measurement
MTP	Master Test Plan; Maintenance Test Program
MTTR	Mean Time to Repair
NAVSEA	NAVal SEA systems command
NBE	New Business Expense
NBS	National Bureau of Standards
NDI	Non-Development Item
NM	Nautical Mile
O	Output
O&M	Operation and Maintenance
O&S	Operation and Support
OCD	Operational Concept Document
OJT	On the Job Training
OMB	Office of Management and Budget
OND	Operational Need Document
OP	OPeration; OPerational; Operating Procedure
OPEVAL	OPerational EVALuation
OPR	Office of Primary Responsibility
OPS	OPerationS
OPREDS	Operational Performance Recording and Evaluation System
ORR	Operational Requirements Review
OSD	Operational Sequence Diagram
OT&E	Operational Test and Evaluation
OWLES	Operator WorkLoad Evaluation System
P&D	Production and Deployment
PCA	Physical Configuration Audit
PCR	Program Control Review; Program Correction Report
PDP	Program Development Plan
PDR	Preliminary Design Review
PE	Physical Element
PERT	Program Evaluation Review Technique
PM	Program Manager
PMD	Program Management Directive
PMP	Program Management Plan

304 ACRONYMS AND ABBREVIATIONS

PMR	Program Management Review
POC	Proof of Compliance
PPI	Personnel Planning Inventory; Plan Position Indicator
P³I (PPPI)	Pre-Planned Product Improvement
PQT	Preliminary Qualification Test
PROM	Programmable Read Only Memory
PRR	Production Readiness Review
PSL/PSA	Problem Statement Language/Problem Statement Analyzer
PTR	Program Trouble Report
PTS	Predetermined Time Standard
QA	Quality Assurance
QIT	Quality Improvement Team
QPMP	Quality and Productivity Measurement Plan
QQPRI	Qualitative and Quantitative Personnel Requirements Inventory
R&D	Research and Development
RAM	Random Access Memory; Requirements Allocation Matrix
RAS	Reliability, Availability, Serviceability
RD&E	Research, Development and Engineering
RFI	Request For Information
RFP	Request For Proposal
RMA	Reliability, Maintainability, Availability
ROM	Read Only Memory; Rough Order of Magnitude
RRR (R³)	Requirements Rationale Report
RTM	Requirements Traceability Matrix
SADT	Structured Analysis and Design Technique
SAE	Society of Automotive Engineers
SAINT	Systems Analysis of Integrated Networks of Tasks
SAR	Systems Assurance Review
SCL	Standards Compliance List
SCN	Specification Change Notice; System Change Notice
SDD	System Design Description; Software Design Document
SDMP	Software Development and Management Plan
SDR	System Design Review
SDRU	System Design Review Update
SE	Systems Engineering
SEB	Systems Engineering Board

SEM	Standard Electronic Module
SEMP	System Engineering Management Plan
SI	System Integration
SID	Society for Information Display
SIMWAM	Simulated Interactive Microcomputer Workload
SLOC	Source Lines of Code
S/N	Signal to Noise
SON	State of Need
SOSD	Spatial Operational Sequence Diagram
SOW	Statement of Work
SPO	System Program Office
SPS	Software Product Specification
SQA	Software Quality Assurance
SQL	Structured Query Language
SRP	Seat Reference Point
SRR	System Requirements Review
SSR	Software Specification Review
STAIRS	STorage And Information Retrieval System
S/W (SW)	SoftWare
SWAT	Subjective Workload Assessment Technique
SWIT	SoftWare Integration Test
SWORD	Subjective WORkload Dominance technique
T	Test
T&C	Terms and Conditions
T&E	Test and Evaluation
T&I	Test and Integration
TAT	Turn Around Time
TBD	To Be Determined
TECHEVAL	TECHnical EVALuation
TEMP	Test and Evaluation Management Plan
TM	Technical Manual
TMP	Technical Management Plan
TMU	Time Measurement Unit
TP	Technical Performance
TPM	Technical Performance Measurement
TRPP	Technical Risk and Performance Plan
TRR	Test Readiness Review
TRS	Test Requirements Specification

UI	User Interface
UL	Underwriters Laboratory
VCR	Video Cassette Recorder
VDU	Visual Display Unit
VLSI	Very Large Scale Integration
WBS	Work Breakdown Structure
WOSTAS	WORrkSTation ASessor

Appendix B

Some ANSI and International Standards*

ANSI STANDARDS

This is a compilation, but not an exhaustive one, of some additional ANSI standards that are relevant for human factors. As you see, they relate to a wide range of human-factors problems and, as you may infer, they contain a wealth of information. Although they are seldom mentioned in human-factors textbooks, these and other standards should be your primary sources of information for guidance and design requirements for projects you are working on.

ANSI A117.1-1986. Buildings and Facilities—Providing Accessibility and Usability for Physically Handicapped People.
ANSI A1264.1-1989. Safety Requirements for Floor and Wall Openings and (Nonresidential) Stair and Railing Systems.
ANSI B11.TRI-1993. Ergonomic Guidelines for the Design, Installation, and Use of Machine Tools.
ANSI B71.5-1984. Operator–Ear Sound Pressure Level Measurement and Rating Procedure for Powered Lawn and Garden and Snow Removal Equipment.
ANSI C95.1-1982. Electromagnetic Fields, Safety Levels with Respect to Human Exposure to Radio Frequency.
ANSI C95.2-1982(R1988). Radio Frequency Radiation Hazard Warning Symbol.
ANSI D16.1-1989. Manual on Classification of Motor Vehicle Traffic Accidents.
ANSI L1.1-1981. Safety and Health Requirements for the Textile Industry.
ANSI S3-W-39. Effects of Shock and Vibration on Man.

---

*The reference numbers, dates, and titles of these standards were up-to-date and correct at the time I prepared this manuscript. Since standards occasionally undergo revision, however, you should always be sure you consult the most recent versions.

308 SOME ANSI AND INTERNATIONAL STANDARDS

ANSI S3.1-1977(R1986). Permissible Ambient Noise During Audiometric Testing, Criteria for.

ANSI S3.2-1989. Method for Measuring the Intelligibility of Speech over Communication Systems.

ANSI S3.3-1960(R1990). Methods for Measurement of Electroacoustical Characteristics of Hearing Aids.

ANSI S3.4-1980(R1986). Procedure for the Computation of the Loudness of Noise.

ANSI S3.5-1969(R1986). Methods for the Calculation of the Articulation Index.

ANSI S3.14-1977(R1986). Rating Noise with Respect to Speech Interference.

ANSI S3.17-1975(R1980). Rating the Sound Power Spectra of Small Stationary Noise Sources, Method for.

ANSI S3.18-1979(R1986). Whole-Body Vibration, Guide for the Evaluation of Human Exposure to.

ANSI S3.19-1974(R1990). Method for Measurement of Real-Ear Protection of Hearing Protectors and Physical Attenuation of Earmuffs.

ANSI S3.20-1973(R1978). Psychoacoustical Terminology.

ANSI S3.29-1983(R1990). Guide to the Evaluation of Human Exposure to Vibration in Buildings.

ANSI S3.32-1982(R1990). Mechanical Vibration and Shock Affecting Man—Vocabulary.

ANSI S3.34-1986. Human Exposure to Vibration Transmitted to the Hand, Guide for Measurement and Evaluation of.

ANSI S3.40-1989. Guide for the Measurement and Evaluation of Gloves Which Are Used to Reduce Exposure to Vibration Transmitted to the Hand.

ANSI S3.41-1990. Audible Emergency Evacuation Signal.

ANSI S12.1-1983. Preparation of Standard Procedures to Determine the Noise Emission from Sources.

ANSI S12.4-1986. Method for Assessment of High-Energy Impulsive Sounds with Respect to Residential Communities.

ANSI S12.6-1984(R1990). Method for the Measurement of the Real-Ear Attenuation of Hearing Protectors.

ANSI S12.9-1988. Quantities and Procedures for Description and Measurement of Environmental Sound.

ANSI S12.10-1985(R1990). Methods for the Measurement and Designation of Noise Emitted by Computer and Business Equipment.

ANSI S12.23-1989. Method for the Designation of Sound Power Emitted by Machinery and Equipment.

ANSI X3.45-1982(R1989). Character Set for Handprinting, Specifications for.

ANSI X3.114-1984(R1991). Alphanumeric Machines, Coded Character Sets for Keyboard Arrangements.

ANSI X3.154-1988. Office Machines and Supplies—Alphanumeric Machines—Keyboard Arrangement.

ANSI X3.207-1991. Office Machines and Supplies—Alphanumeric Machines—Alternative Keyboard Arrangement.

ANSI Z535.1-1991. Safety Color Code.

ANSI Z535.2-1991. Environmental and Facility Safety Signs.
ANSI Z535.3-1991. Criteria for Safety Symbols.
ANSI Z535.4-1991. Product Safety Signs and Labels.
ANSI/ANS 10.5-1986. User Needs in Computer Program Development, Guidelines for Considering.
ANSI/ASHRAE 55-1981. Thermal Environmental Conditions for Human Occupancy.
ANSI/ASHRAE 62a-1991. Ventilation for Acceptable Indoor Air Quality.
ANSI/EIA 359-A-1984. Colors for Identification and Coding.
ANSI/EIA 359-A-1-1988. Special Colors—Standard Colors for Color Identification and Coding (addendum No. 1 to ANSI/EIA 359-A-1984).
ANSI/IEEE 1023-1988. Guide for the Application of Human Factors Engineering to Systems, Equipment, and Facilities of Nuclear Power Generating Stations.
ANSI/IES RP1-1982. Office Lighting, Practice for.
ANSI/IES RP3-1988. Guide for Educational Facilities Lighting.
ANSI/IES RP7-1990. Industrial Lighting, Practice for.
ANSI/IES RP16-1986. Illuminating Engineering, Nomenclature and Definitions for.
ANSI/NFPA 101-1991. Life Safety Code.
ANSI/NFPA 101M-1988. Systems Approaches to Life Safety.
ANSI/RIA R15.02/1-1990. Human Engineering Design Criteria for Hand-Held Robot Control Pendants.
ANSI/RIA R15.06-1986. Industrial Robots and Industrial Robot Systems, Safety Standard for.
ANSI/SAE ARP 268G. Location and Actuation of Flight Deck Controls for Transport Aircraft.
ANSI/SAE ARP 571C. Flight Deck Controls and Displays for Communication and Navigation Equipment for Transport Aircraft.
ANSI/SAE ARP 4101. Flight Deck Layout and Facilities.
ANSI/SAE ARP 4102. Flight Deck Panels, Controls, and Displays.
ANSI/SAE ARP 4102/7. Electronic Displays.
ANSI/SAE ARP 4102/8. Flight Deck Head-Up Displays.
ANSI/SAE ARP 4107. Aerospace Glossary for Human Factors Engineers.
ANSI/SAE ARP 4153. Human Interface Criteria for Collision Avoidance Systems in Transport Aircraft.
ANSI/SAE ARP 4155-OCT90. Human Interface Design Methodology for Integrated Display Symbology.
ANSI/SAE J826-MAY87. Devices for Use in Defining and Measuring Vehicle Seating Accommodations.
ANSI/SAE J833-MAY89. USA Human Physical Dimensions.

INTERNATIONAL STANDARDS

Following is a small set of the many international standards you should be sure to consult if you are working on products that will be used internationally.

IEC 964. Design for Control Rooms of Nuclear Power Plants. Geneva, Switzerland: International Electrotechnical Commission.

ISO 226:1987 (E). Acoustics—Normal Equal-Loudness Level Contours. Geneva, Switzerland: International Organization for Standardization.

ISO 532-1975 (E). Acoustics—Method for Calculating Loudness Level. Geneva, Switzerland: International Organization for Standardization.

ISO 1503-1977 (E). Geometrical Orientation and Directions of Movement. Geneva, Switzerland: International Organization for Standardization.

ISO 7243:1989 (E). Hot Environments—Estimation of the Heat Stress on Working Man, Based on the *WBGT*-Index (Wet Bulb Globe Temperature). Geneva, Switzerland: International Organization for Standardization.

ISO 7730:1994 (E). Moderate Thermal Environments—Determination of the PMV and PPD Indices and Specification of the Conditions for Thermal Comfort. Geneva, Switzerland: International Organization for Standardization.

ISO 7731-1986 (E). Danger Signals for Work Places—Auditory Danger Signals. Geneva, Switzerland: International Organization for Standardization.

ISO 7933:1989 (E). Hot Environments—Analytical Determination and Interpretation of Thermal Stress Using Calculation of Required Sweat Rate. Geneva, Switzerland: International Organization for Standardization.

ISO 8995:1989 (E). Principles of Visual Ergonomics—The Lighting of Indoor Work Systems. Geneva, Switzerland: International Organization for Standardization.

Author Index

Adams, J. A., 109, 139
Adams, L., 273
Adams, O. S., 199, 202
Ainsworth, L. K., 29, 56, 81, 140
Angiolillo-Bent, J. S., 155, 202
Aoki, K., 141
Aoki, T., 141

Bailey, R. W., 173, 175, 202, 273, 286
Baker, C. A., 218, 257
Barnes, R. M., 252, 256
Barnes, V. E., 296
Beare, A. N., 254, 256
Bell, G. E., 249, 258
Berger, E. H., 189, 204
Blanchard, B. S., 21, 25, 26, 28, 29, 55, 101, 139, 262, 268, 282, 286
Boehm, B. W., 27, 39, 56
Boff, K. R., 140, 211, 220, 222, 250, 251, 256
Booher, H. R., 140
Booth, J. M., 219, 257
Bradtmiller, B., 203
Brennan, L., 37, 56
Brigham, F., 139
Buckley, D. S., 203
Burdurka, W. J., 22, 26, 27, 56, 272, 286
Burgess, J. H., 293, 295

Burns, C. M., 53, 56, 272, 286

Cabrera, E. F., 140
Casey, S., 8, 20
Caylor, J. S., 245, 257
Chao, B. P., 54, 56
Chapanis, A., 9, 16, 18, 20, 52, 56, 83, 87, 88, 91, 92, 118–121, 124–126, 139, 141, 145, 147, 199, 200–201, 202, 203, 205, 229, 231, 237, 238, 249, 250, 256, 257, 258, 261, 268, 272, 277, 286, 289, 291, 295
Charness, N., 152, 203, 241, 257
Chiles, W. D., 199, 202
Christensen, J. M., 87, 88, 139
Churchill, E., 149, 150, 151, 204
Churchill, T., 203
Ciriello, V. M., 173, 203
Clark, D. W., 29, 56
Clauser, C. E., 203
Cleveland, D. E., 204
Coren, S., 234, 257
Cramer, M. L., 29, 56
Crossman, E. R. F. W., 241, 257
Cushman, W. H., 29, 56, 262, 268

Dain, S. J., 221, 258
Damon, A., 167, 204

Damos, D., 139
Daniels, G. S., 149, 151, 203
de Jong, J. R., 242, 257
Deutsch, S., 124, 140
Devoe, D. B., 255, 257
de Vries, G., 126, 139
Donchin, E., 138, 140
Dooley, B., 288, 295
Dorris, R. E., 256

Eason, K. D., 294, 295
Egan, J. P., 214, 257
Eggemeier, F. T., 136, 138, 139, 140, 141
Eisner, H., 56
Endsley, M. R., 100, 139
Ericsson, K. A., 152, 203, 241, 257

Fabrycky, W. J., 21, 25, 26, 28, 29, 55, 101, 139, 262, 268, 282, 286
Fancher, P. S., 204
Farrell, R. J., 219, 257
Fitts, P. M., 90-92, 139
Fleishman, E. A., 242, 251, 257
Fletcher, H., 229, 257
Fox, J. G., 202
Fox, W. L., 245, 257
Freedman, M., 151, 203
Fuld, R. B., 101, 139
Furukawa, T., 141

Galer, M. D., 275, 286
Garner, W. R., 203, 229, 256
Gates, D., 123, 139
Gebhard, J. W., 212, 213, 215, 217, 258
Geldard, F. A., 216, 257
Gilbert, D. K., 140
Glass, J. T., 123, 139
Gopher, D., 138, 140
Gordon, C. C., 163, 203
Gordon, S. E., 260, 268
Grether, W. F., 218, 257
Guttman, H. E., 113, 141

Haig, K. A., 125, 140
Happ, A. J., 226-227, 258
Harris, C. M., 203, 204
Harris, D. H., 187, 189, 295
Hart, S. G., 138, 140
Haynes, H., 197-198, 203, 223, 257

Hennessy, R. T., 124, 140
Hertzberg, H. T. E., 169, 170, 171, 174, 203
Hinchcliffe, R., 153, 203
Hoecker, D. G., 140
Hoffman, M. S., 56
Hopkins, C. O., 1
Howell, E., 17, 20
Huey, B. M., 135, 140
Huey, R. W., 152, 203
Hughes, G. J., 203

Israelski, E. W., 227, 257

Jeffress, L. A., 204
Johansson, G., 254, 257
Jones, E. R., 124, 140
Jones, R. E., 90-92, 139

Kaucharik, D., 57
Kaufman, J. E., 197-198, 203, 223, 257
Kaufman, L., 140, 211, 256
Kelly, M. J., 250, 257
Kennedy, K. W., 166, 203
Kibler, A. W., 254, 259
Kinkade, R. G., 170, 178, 203, 213, 257, 258, 259, 265, 268, 277, 286
Kinoshita, S., 141
Kirk, F. G., 29, 56, 262, 268
Kirwan, B., 29, 56, 81, 114-115, 140
Klein, G., 239, 257
Klemmer, E. T., 57, 125, 140, 241, 257, 295
Kline, D. W., 152, 204
Klinger, D., 239, 257
Kostyniuk, L. P., 204
Kozinsky, E. J., 256
Kroemer-Elbert, K. E., 164-165, 180, 185, 203
Kroemer, H. J., 164-165, 180, 185, 203
Kroemer, K. H. E., 164-165, 172, 180, 185, 203
Kryter, K. D., 187, 203, 210, 231, 232, 250, 251, 257
Kumar, S., 176, 204
Kurke, M. I., 131-134, 140
Kutche, G. B., 116, 141

Lalomia, M. J., 226-227, 258

Laubach, L. L., 146, 203
Ledgard, H., 238, 258
Lerner, N. D., 203, 253, 258
Lincoln, J. E., 211, 220, 222, 250, 251, 256
Lindahl, L. G., 242–243, 258
Lockhead, G. R., 241, 257

McCauley, M. E., 235, 258
McClelland, I. L., 253, 258
McConville, J. T., 203
McCormick, E. J., 213, 234, 258
McCulloch, W., 143
McFarland, R. A., 5, 167, 204
McGuire, M. V., 296
McLeod, R. W., 136, 140
McQuistion, L., 7, 20
Mark, R., 4, 20
Martin, D. K., 221, 258
Martin, L., 155
Meister, D., 81, 101, 102, 106, 140, 272, 282, 287, 291, 292, 296
Melville, B. E., 98–99, 140
Michaels, S. E., 246, 258
Miles, W. R., 154, 204
Miller, D. C., 140
Miller, G. A., 236, 258
Mital, A., 176, 204
Morgan, C. T., 203, 229, 256
Morgenstern, M. H., 290, 296
Moser, H. M., 249, 258
Mosier, J. N., 64, 77, 278, 287
Moulden, J. V., 147, 203, 237, 256
Mowbray, H. M., 212, 213, 215, 217, 258
Mumford, E., 293, 296
Munger, S. J., 107, 113, 140
Munson, W. A., 229, 257

Naisbitt, J., 7, 20
Neff, R., 9, 20
Nixon, C. W., 189, 204
Norman, D. A., 9, 20
Noro, K., 141
Nussbaum, B., 9, 20

O'Donnell, R. D., 138, 140
Olson, P. L., 195, 204, 253, 258

Paquette, G., 57
Parker, J. F., 258

Parsons, H. M., 52, 56
Payne, D., 140
Pelsma, K. H., 59, 77
Pew, R. W., 88–89, 140
Phillips, M. D., 98–99, 140
Price, H. E., 101, 140

Quaintance, M. K., 242, 251, 257

Radford, L. R., 296
Raggio, L., 166, 168, 204
Rasmussen, B., 272, 287
Rhodes, W., 121–122, 140
Richardson, S. J., 20, 56, 296
Rodgers, M. D., 100, 139
Rodgers, S. H., 172, 204
Roebuck, J., 162, 166, 168, 204
Rogers, W. A., 87, 140
Rosenberg, D. J., 29, 56, 262, 268
Royce, W. W., 38–39, 56, 288, 296
Ruffell Smith, H. P., 248, 258
Rumar, K., 254, 257
Russell, A. J., 275, 287

Safren, M. A., 91, 141
Salvendy, G., 287
Sanders, M. S., 13, 20, 213, 234, 258
Schaie, K. W., 152, 204
Scherzinger, P., 38, 57
Schieber, F., 152, 204
Schneider, L. W., 204
Shackel, B., 20, 56, 256, 296
Shafer, J. B., 83, 116, 136, 137, 139, 141, 261, 268
Shaffer, M. T., 116, 141
Sheppard, W. T., 18, 20
Sherwood-Jones, B. M., 136, 140
Simmons, C., 17, 20
Simpson, C. T., 253, 258
Sinaiko, H. W., 139, 154, 204
Singer, A., 238, 258
Singleton, W. T., 202
Sivak, M., 253, 258
Sloan, L. L., 145, 204
Smith, K., 166, 168, 204
Smith, L., 13, 20
Smith, R. W., 140
Smith, S. L., 64, 77, 272, 278, 287
Snook, S. H., 203

Staplin, L., 151, 203
Starbuck, A., 253, 258
Stoudt, H. W., 167, 204
Swain, A. D., 113, 141
Swink, J. R., 253, 258

Tanaka, H., 141
Tarrants, W. E., 20, 139
Taylor, J. E., 245, 257
Taylor, J. H., 221, 258
Tebbetts, I., 203
Thomas, J. P., 140, 211, 256
Topmiller, D. A., 254, 259
Trager, J., 4, 20

Van Cott, H. P., 170, 178, 203, 212, 213, 215, 217, 257, 258, 259, 286
van Gelderen, T., 139
Vincente, K. J., 53, 56, 272, 286
von Gierke, H. E., 193, 204
von Neumann, J., 143, 144, 204

Walker, J. H., 140
Walker, R. A., 203
Ward, W. D., 193, 204
Warrick, M. J., 212, 213, 215, 217, 254, 258, 259

Wasserman, A. S., 40–42, 57
Wechsler, D., 144, 148, 154, 204
Wells, H. G., 155
West, V. R., 258
Westgaard, R. H., 272, 287
Wheaton, G. R., 265, 268
Wheeler, W. A., 296
White, R. M., 150, 204
Whiteside, J., 238, 258
Whitfield, D., 202, 238
Wickens, C. D., 135, 138, 140
Wierwille, W. W., 136, 141
Willett, B., 57
Wilson, G. F., 138, 139
Winters, J. A., 124–126, 141
Wisner, A., 154, 205
Wolfle, D., 242–243, 259
Wulff, L. A., 272, 275, 282, 287
Wymore, A. W., 21, 57, 262, 269
Wyndham, C. H., 154, 205

Yajima, K., 141
Yamanoi, N., 123, 141
Young, L., 8, 20

Zador, P., 151, 203
Zaloom, V., 123, 139

Subject Index

Abbreviation(s), 92
Absolute judgments, 217–218
Absolute thresholds, 212–213
Abuse of equipment, 10, 143
Acceleration(s), see also g-forces
  in buildings, 193
  effects of on the body, 194–197
  sense of, 213
Accesses, 60, 62, 71, 158
Accessibility, 121
Accident(s), 90, 240
  air and space transport, 9, 18
  automobile, 9
  computer-caused, 8
  home, 9
  industrial, 9
  investigation, 85, 109
  medical, 8
  nuclear, 1, 8–9
  provocative situations, 123
  reports, 86
  ship, 8
  studies, 30
Accidental activation, 92, 277
Acclimatization, 176, 182
Achromat(s), 227
Action-information analysis, 83, 98–100, 101, 109, 113, 279

Action requirements, 107
Activity:
  analysis, 83, 85, 87–89, 93, 97, 98, 116, 279
  relationship chart, 121–123
Acuity, see Visual acuity
Adjustment errors, 92
Advanced Rotocraft Technology Integration (ARTI), 137
Advertising, 55
Aesthetics, 15, 16
AFQT, see Armed Forces Qualification Test
Age, 7, 149
  and anthropometric dimensions, 162, 166, 167
  effects of on vision, 151, 220
  and hearing losses, 151, 153, 229, 234
  and learning, 244
  and mental abilities, 152
  and mobility, 172
  and reaction time, 252–253
  and strength, 174, 176
Age Discrimination in Employment Act of 1967, 7, 154
Agriculture, 8
AI, see Articulation Index

SUBJECT INDEX

Air:
  flow, 183
  movement, 181–185
  pressure, 207
  temperature, 183–184
Aircraft, 5, 17, 23, 29, 85, 163, 221
  accidents, 9, 18
  B-1B bomber, 17, 288, 289
  B-25C, 91
  B-2, 38
  cockpits, 14
  controls, 90–92
  C-47, 91
  C-17 airlifter, 17
  F-86, 288
  Lockheed Electra, 130
  simulators, 235, 266, 267
  technical manuals, 288–289
Airline pilots, 261
Airsickness, 235
Air traffic:
  controllers, 261
  control system(s), 11, 23, 25, 29, 64, 75, 98, 100, 210, 230
Alarms, 210–211. *See also* Warning(s)
Allocation:
  of functions, 36, 42, 45, 46, 73, 74, 82, 83, 84, 86, 87, 98, 100–105, 137, 138
  models, 15
Alphabetic keyboard, 246
Alphabet, word-spelling, 249
American Institutes for Research, 107
American National Standards Institute, 59
Americans with Disabilities Act of 1990, 7
Amplified speech, 233
Analysis of similar systems, 83, 85–87, 93, 97, 98, 137, 279, 285
Anatomical:
  landmarks, 159–162
  planes, 160
Anatomy, 142
AND gate, 94, 111–113
AND/OR gate, 94
Ankle movement, 171
ANSI, *see* American National Standards Institute
ANSI/HFS 100-1988, 61, 77, 275, 276, 278, 286

ANSI Z53.1-1979, 223–224, 256
Anthropology, 12, 142
Anthropology Research Project, 163
Anthropometric(s), 15, 144
  data:
    and the "average" man, 148–151
    ethnic differences in, 162, 166
    interpreting, 161–162
    representative, 163–165
Anthropometry, 14, 75, 158–170
  anatomical planes, 160
  functional, 168–170
  landmarks, 159–162
  postures in, 159
  static, 168
  structural, 168
Appliances, home, 9, 16, 31, 38, 262
Application engineering, 28
Application-specific requirements, 3, 277–281
Applied:
  experimental psychology, 12
  physiology, 14, 15
  versus basic research, 80–81
Architect(s), 18
Architectural view of a system, 40
Armed Forces Qualification Test, 245
ARTI, *see* Advanced Rotorcraft Technology Integration
Articulation Index (AI), 230–232, 250
Association chart, 122
Association Française de Normalization, 59
Astigmatism, 220
Astronaut(s), 6
AT&T, 11
A-320 simulator, 267
Attending, 209–211
Attention, lapses of, 157
Attitudes, 144, 207
Audit(s), 37, 38, 289, 291
Audition, 187–189, 209, 212–213, 215–217, 228–234. *See also* Hearing
Auditory:
  alarms, 210–211
  defects, 234
  sense, *see* Hearing
  signal(s), 229–230
Automation, 50, 282

Automobile(s), 4, 11, 14, 17, 23, 29, 31, 48, 54, 60, 85, 91, 97, 99–100, 163, 262, 279
  accidents, 9
  simulator(s), 235
Autonomic nervous system, 179
Autonomy, 284–285
Availability, system, 16, 19, 28, 31
Availability heuristic, 240
Average person, myth of, 148–151
A-weighting function, 187
AZERTY keyboard, 246

Bandwidth, 229
Bank:
  proof machine operation, 241
  teller machine, 23, 29, 32–35, 43–49, 69, 71, 87, 98, 100, 102, 126, 131, 278, 286
Banking systems, 23–25
Basic versus applied research, 80–81
Binaural versus monaural listening, 229
Bioastronautics, 12
Biological rhythms, 198–200
Biomechanics, 12, 15
BOCA National Building Code, 63
BOCA National Mechanical Code, 63
Body dimensions, *see* Anthropometry
Boeing, 11
Bolt, Beranek and Newman, Inc., 88–89
B-1B aircraft, 17, 288, 289
Boredom, 16, 157, 207
Breadth, in anthropometry, 159
Brightness, 221–225
British Standards Institution, 59
B747-400 cockpit methods trainer, 266
B-2 aircraft, 38
B-25C aircraft, 91
Building(s), acceleration in, 191–193
Building Officials and Code Administrators International, 63, 77
Build-to specification, 71, 75
Built-in-test (BIT) equipment, 264
Buzzer, reaction time to, 253

Can opener(s), 38
Carbon dioxide, 177
Carbon monoxide, 186–187, 228
Cardiac function, 14, 135, 177, 180

Carpal tunnel syndrome, 178–179
Carrying, 178
Caution displays, 210
CD players, 126
Central nervous system, 179
CFF, *see* Critical flicker fusion
C-47 aircraft, 91
CFR, *see* Code of Federal Regulations
Chair(s), design of, 14, 158
Checklists, 63, 75, 79, 238
Chernobyl nuclear power station, 8
Chesapeake Bay piloting, 200–202
Chroma, 224–226
Chunking, in memory, 236–237
CIC, *see* Combat information center
CIE system, 223–234
Cincinnati Milacron Marketing Company, 24
Circadian rhythms, 198–199
Circulatory system, 179
Circumference, in anthropometry, 159
Classification of signals, 235
Clothing:
  design, 158–159
  effect of on comfort, 182–183
  effect of on work, 176
  effects of on body dimensions, 167–168
Cockpits, 14
Code(s), 58, 62, 63, 270–271
  color, 7
  position, 7
Code of Federal Regulations, 60, 77–78
Coding, 209, 236–237
  system, 92
Cold, tolerance to, 182
Collision avoidance system, 133–135
Color(s), 15, 218, 220, 276, 282
  blindness, 6
  coding, 7
  on computers, 50, 226–227
  for marking physical hazards, 223–234
  vision, 215, 221–228
    deficiency, 61, 144, 149, 227–228
  in workstations, 276
Combat Information Center (CIC), 117–121
Comfort, 16, 282–283
  thermal, 182–185

SUBJECT INDEX

Commissioning, 29
Commission Internationale de l'Éclairage, *see* CIE system
Communication, face-to-face, 230–233
  effects of noise on, 188, 232–233
  skills, importance of, 291
Component, in systems, 25
Computer(s), 9, 13, 23, 29, 48, 91, 143, 208, 247, 283
  assisted interviews, 88–89
  caused accidents, 8
  colors for, 50, 226–227
  commands, 10
  consoles, 14
  date entry to, 254, 278
  design, 12, 61
  home, 262
  impact of on society, 7–8
  installation of, 52
  personal, 5, 16–17
  programmers, 8
  programming, 240
  in ship collision avoidance, 133–135
  software configuration item, 25
  systems, 7, 11
  and visual defects, 221
  workstation(s), 61, 158, 163, 273–275
Concept(s), 29, 127
  demonstration and validation, 30, 36, 39, 70
  exploration, 30, 36, 39, 70, 73
Conceptual design, 29
Concurrent engineering, 52–54
Conduction, in heat exchanges, 182, 185
Configuration management, 28
Conservation, 37
Consumer products, 16, 262
Contrast, in vision, 217, 220
Control(s), 5, 10, 60, 74, 75, 168, 207, 277, 282, 294
  design, 92
  display:
    ratio(s), 62, 276
    relationship(s), 276
  errors, 90–92, 112
  panels, 116, 123
  points, 37
  resistance, 178
  room(s), 210
    design, 1, 53

Controlled experimentation, 82, 83, 129–131, 137, 138
Convection, in heat exchanges, 182, 185
Coopers and Lybrand, 155
Copier(s), 40–42
Correlation matrix, 122
Cost(s), 31, 102, 284–285
  engineering, 28
Crawling posture, 169–170
Crewmembers, for space vehicles, 283–285
Critical design review, 30, 70, 75
Critical flicker frequency (CFF), 144–145
Critical incident technique, 30, 83, 85, 89–92, 98, 279
Critical item, 25
Crotch height, 149–150
CRT, 97
Crowding, index of, 121
C-17 airlifter, 17
Cubital tunnel syndrome, 178–179
Cumulative trauma disorder(s) (CTDs), 178–179
Curvature, in anthropometry, 159
Customer support, 291
Cutaneous senses, 234

Danger signals, *see* Safety signs; Warning signals
Dark adaptation, 145
Data:
  entry, 254–255
  devices, 74
  file, 290–291
Data Item Description(s) (DIDs), 64, 67, 68, 72, 77, 262–263, 268
Data Systems Modernization (DSM) system, 106–111
Decibel(s), 187–188, 228–229
Decision-action analysis, 83, 97–99, 101, 109, 113
Decision-logic diagram(s), 97
Decision making, 13, 209, 214–215, 238–240
  availability heuristic, 240
  bias in, 240
  gambler's fallacy, 240
Decoding, 209
Definition, system, 29

Delivery of products/system, 29
Demo II, 128
Deployment, 30, 37, 39
Depth, in anthropometry, 159
Depth perception, 218–219
De Quervain's disease, 178–179
Description versus prediction, 81
Design, 29, 75
  approach, 68
  description(s), 70, 72, 74, 76
  document(s), 270
  job, 5, 7
  machine, 5, 158–159
  rationale(s), 70, 72, 74
  requirements, 33
  rule(s), 272
  -to-cost, 28
  -to specification(s), 71, 74
Designer(s), 16, 18
Detail design, 29, 67
Detection of signals, 214–215, 235
Deuteranopia, 227
Deutsches Institut für Normung, 59
Dialogues, computer, 74
DI-CMAN-80008A, 262–263, 268
DID, *see* Data Item Description
Differences, among people, 144–156
DI-HFAC-:
  80740A, 64, 77
  80743A, 68, 77
  80744A, 68, 77
  80745A, 67, 77
  80746A, 72, 77
  80747A, 72, 77
DIN, *see* Deutsches Institute für Normung
Disassembly, human factors in, 37
Disaster control center(s), 210
Display(s), 1, 5, 13, 62, 74, 75, 90, 207, 210, 220, 271, 282, 294
  formats, 74
  panels, 7, 123
Disposal, of systems, 29, 37, 54–55
Distance, in anthropometry, 159
Diurnal cycle(s), 198–199
Documentation, 2, 3, 49, 51, 74, 76, 127, 288–291
  handbooks, 51, 275
  instructions, 51
  manuals, 13, 51, 74
  requirements, 281
DOD-HDBK-:
  761A, 275, 286
  763, 28, 56, 81, 101, 102, 122, 133, 136, 139
Dominant wavelength, 223
Door(s), 60, 62, 71, 158
Drawing(s), 75, 144
Driving, 135, 207–208, 210
  reaction times in, 253–254
DSM, *see* Data Systems Modernization system
Dvorak keyboard, 246
Dynamic force, 173
Dynamic range, of microphone/input devices, 62

Ear muffs, 189, 232
Earplugs, 189, 232
Ease, of learning, 16, 126–127
  of production, 54
  of use, 16, 126–127
Eastman Kodak Company, 200–201
Education, 44, 149, 155–156
Effective temperature, 182–184
Egress, means of, 60, 62, 71, 158, 276
EIA Standard RS-359, 225
Electrocardiograph, 123
Electromagnetic capability, 28
Electromyographic activity, 14
Element, in systems, 25
Energy, efficiency, 54
  expenditure, 180
Engineering, 1, 15, 53
  concurrent, 52–54
  development, 30, 36, 39, 70
  industrial, 13
  mechanical, 13
  psychology, 12
  safety, 67
  system(s), 21–55
    academic departments of, 13
    defined, 21–23
    as a discipline, 21
    objectives, 15–16
    time-and-motion, 4, 251–252
Environment(s), 5, 6–7, 14, 32, 45, 71, 73, 85, 86
  requirements, 280

Environment(s) (*Continued*)
  simulated, 126, 128
  thermal, 180–185
  vibration in, 191, 193
  working, 16, 49, 60, 75, 107, 207
Environmental medicine, 14
Epicondylitis, 178–179
Equal-loudness level(s), 229
Equipment design, 67, 158–159
Ergonomics, 12–13
Error(s), 16, 19, 71, 74, 90, 98, 109–115, 283
  control, 90–92
  human, 8–9
  maintenance, 18
  medication, 91–92
  messages, 128
  near, 90
  provocative situations, 9, 123
  rates, 13
  susceptibility to, 31
Ethnic differences, 154–155, 162, 163, 166
Evaluation, 29
Evaporation, 182, 185
Excitation purity, 223
Exercise, 176
  and mobility, 172
Experimentation, *see* Controlled experimentation
Expert(s), everyone as, 157–158
  performance, 152
  systems, 239
  use of, 138
Expertise, human factors, 292
Extra vehicular activity, in space, 5, 60
Eyeglasses, 220–221

FAA, *see* Federal Aviation Administration
Face-to-face, communication, 230–233
  interviews, 88–89
  speech, 230–231
Factory (factories), 193
Failure(s), 85
  foreseeable, 113
  hard, 264
  modes and effects analysis, 83, 109, 113–115, 279
  soft, 264

Falling, sense of, 212
Familiarization training, 265, 267
Farm machinery, 11
Farsightedness, *see* Hyperopia; Presbyopia
Fatigue, 16, 157, 176–178, 207
Fault tree analysis, 30, 83, 109–113
Fax machine(s), 97
Fechner's law, 216
Federal Age Discrimination in Employment Act of 1967, 7
Federal Aviation Administration, 10, 12
Federal Rehabilitation Act of 1973, 7
Federal Standard Number 595a, 223, 225–226, 257
Federal System Company (FSC), 25
Feedback, 62
  kinesthetic, 178
  in learning, 243–244, 246
Feeling, sense of, 234
  of tiredness or fatigue, 177
F-86 aircraft, 288
Fidelity of simulation, 247
Figure of merit, in trade studies, 285
Flexibility, 170–172
Flicker, 215, 218
Flight, navigator activities in, 87–88
Flow sheeting, 29
Fluorescence, 226
Flying, 240
  helicopters, 31, 33, 94–97, 115–116, 130, 137, 210
Font design, 61
Force(s), exertable, 173–178
  grip, 174
  in maintenance, 124
  opening and closing, 62
  push, 124–125, 145–146
Forgetting, 236
  errors, 92
Fovea, 145, 209, 220
Function (functional), allocation, 36, 42, 45, 46, 73, 74, 82, 83, 84, 86, 87, 98, 100–105, 137, 138
  analysis, 45–46, 67, 68, 82, 83, 84, 86, 87, 93–98, 101, 103, 109, 113, 138, 279, 290
  anthropometry, 168–170
  block diagrams, 94

flow, analysis, 83, 101, 113, 127
    diagrams, 45, 94, 290
    requirements, 84
    view of system, 40-43
Functionality, 15, 127

Gambler's fallacy, 240
Gate(s), logic, 94, 111-113
Generalized Application Simulator (GAS) 2.0, 128
g-forces, 5, 181, 195-196. *See also* Acceleration
Glare, 10, 151
Gloss, 61, 226
Grip forces, 174
Guidelines, 2, 58, 63-64, 79, 275, 281

Habit interference, 246
Hand, vibration to, 191
Handbook(s), 51, 275. *See also* Documentation
Handicapped, designing for, 7
Handprinting, 254-255
Handsets, telephone, 125-126
Hard failures, 264
Hardware, 3, 38-39, 73
    configuration item, 25
    design, 39, 49-51, 72
    development, 39, 51
    engineering, 28, 54
    specifications, 75
Hatch(es), 276
Hazardous materials, 60
Hazards, 107, 211
    colors for marking, 223-224
    hidden, 8, 181
Health, 14, 60
    requirements, 71
Hearing, 187-189, 209, 212-213, 215-217, 228-234
    aids, 250
    as a function of sex and age, 151, 153, 229, 234
    impairment, 234
    loss(es), 187
    protectors, 189
Heart, function, 14
    rate, 135, 177, 180, 198-199
Heat, 182-185
HEDAD-M, *see* Human Engineering Design Approach Document—Maintainer
HEDAD-O, *see* Human Engineering Design Approach Document—Operator
Height, in anthropometry, 159, 166
Helicopter(s), 31, 33, 80, 130, 137
    flight, 94-97, 115-116, 210
Helmets, 221
    noise attenuating, 189
Help, facilities in computer systems, 74, 128
    messages, 238
HEPP, *see* Human Engineering Program Plan
Heuristics, 239-240
Hidden hazards, 8, 181
Highway signs, 151-152, 224-225
Hip movement, 171
Home appliances, *see* Appliances, home
Hospitals, vibration in, 193
Household appliances, *see* Appliances, home
Hue, 221-226
Human:
    capacities, 144
    characteristics, 2, 10, 142-259
    engineering, defined, 12
    Engineering Design Approach Document—Maintainer, 72
    Engineering Design Approach Document—Operator, 72, 288
    Engineering Program Plan (HEPP), 19, 64-69, 270
    error(s), 8-9
    factors, 28
        data file, 290-291
        data stores, 136
        defined, 11-16
        engineering, 11
        historical development of, 3-11
        methods, 2, 79-141
        objectives, 15-16
        professionals, 18
        requirements, 27
        technical disciplines in, 13-15
        terms related to, 12-13
    Factors and Ergonomics Society, 13, 15
    reliability, 109, 113

Humidity, 14, 181–185, 207–208
Hyperopia, 220
Hyperoxia, 186
Hypoxia, 186, 228

IBM, 11, 56, 77, 108, 109, 110, 111, 131, 140, 275
  5153 color display, 226–227
  PC, 247
  Wheelwriter, 247
ICI system, *see* CIE system
Icons, 127, 128
Identification of signals, 235
Illumination, 14, 197–198, 208
  and readability of highway signs, 151–152
  and visual acuity, 220
ILS, *see* Integrated logistic support
Impact acceleration, 195–197
Impulsive sounds, 193
Index, of crowding, 121
  of walking, 121
Indicator(s), 90. *See also* Displays
Individual differences, 144–156
Indoctrination training, 265–266
Industrial:
  accidents, 9
  design, 15
  engineering, 13, 18, 81
Information:
  flow charts, 97
  processing, 210
  requirements, 99, 105, 107
  transmission, 209
Input device(s), 50, 282. *See also* Control(s)
Installation, of equipment, 18, 39, 52, 54
Instant Replay, 128
Instructions, 90. *See also* Documentation
  translation of, 155
  wording of, 294
Integrated Logistic Support (ILS), 16, 23, 28, 67
Integration, 29
  and test, 39, 51
Intelligence, 146–148
  in learning, 244–245
Intelligibility, measurement of, 230–231
  of speech, 230–234, 250
Interface(s), 73

human-computer, 127
human-system, 48–49, 72, 75, 99
  requirements, 36, 71, 74, 106, 271–275, 280
  studies, 73
user—system, 46–51
  configuring, 51
  evaluating, 50, 127
Internal Revenue Service, 11
International Civil Aviation Organization (ICAO), 233, 249
International Commission on Illumination, *see* CIE system
International Organization for Standardization, 59
International Standards Organization (ISO), 59
Interview(s), 30, 85, 88–89, 90
  computer-assisted, 88–89
  face-to-face, 88–89
ISO, *see* International Organization for Standardization
Isometric force, 173

Job, aids, 107, 280
  design, 5, 7
Joint movement, mobility, 170–172
Just Noticeable Difference(s) (JNDs), 216–217, 224

Key(s), design of, 61
Keyboard(s), 71, 74, 247, 282
  alphabetic, 246
  AZERTY, 246
  Dvorak, 246
  QWERTY, 246
Keying tasks, 254–255
Kinesthesis, sense of, 212, 216
Kinesthetic feedback, 178
Kitchen appliances, 9, 13, 16, 31, 38, 262
Kneeling posture, 169–170
Knee movement, 171
Knowledge, of performance, 246
  of results, 243–244, 246

Labels, 10, 92
Laborforce, composition of, 7–8
LAMPS Mark III System, 94–97, 131, 288

Language:
    ability, 249
    conventions in standards, 61
    skills, 156
Lawn mowers, 17, 60, 262
Layout:
    control (instrument) panels, 116
    workstations, 273–275
Learning, 13, 127, 240–247
    and mental ability, 244–245
    motor, 244–246
Life:
    cycle(s), 26
        control points, 37
        costs, 28
        system, 2, 28–38, 270
    Safety Code, 63
    sciences engineering, 12
    support, 12
Lifting, 178
    capacities of males and females, 175
Light, reaction time to, 252–253. *See also* Illumination
Lighting, 197–198, 276
Lightness, 221–226
Link analysis, 72, 83, 116–123, 279
Lockheed-Martin, 11, 64
Lockhead Electra aircraft, 130
Logic:
    flow charts, 98
    gates, 94, 111–113
Loom operation, 145
LORAL, 11
Loudness, 187–188, 228–229
Luminance, and visual acuity, 220–221

Machine(s), 1, 3, 22–23
    bank teller, 23, 29, 32–35, 43–49, 69, 71, 87, 98, 100, 102, 126, 131, 278, 286
    copy, 40–42
    design, 5, 158–159
    guarding, 60
    letter sorting, 103–105
    maintainability, 19
    model, 206–207
    oriented view of system, 40–41
    performance, 19
    in systems, 22–23
    tool(s), 23–24
Macroclimate, 182

Mail sorting, 103–105
Maintainability, 16, 19, 28, 86, 102
Maintainer(s), 16–17, 19, 23, 26, 31, 35, 43, 168
    needs, 17
    requirements, 72, 280
    training, 264
Maintenance, 17, 23, 29–30, 32, 36, 37, 39, 52, 80, 124
    errors, 18
    features, 75
    requirements, 71–74
Manability, 26
Management, science, 13
    support, 291–292
Manhole covers, 38
Man-machine system design, defined, 12
Manpower requirements, 86, 282
MANPRINT, 12, 262, 269
Manuals, 13, 51, 74
Manufacturer(s), 53
Manufacturing, 52
    engineering, 28
Marketing, 29, 53, 55
Masking, of sounds, 229, 230
Mathematics skills, 156
Matrix layout, 128
Meals, providing, 94
Meaningfulness of material, 237
Mechanical engineering, 13
Medical instrumentation, 13
Medication errors, 91–92
Medicine, environmental, 14
Memory, 209, 235–238
    long-term, 236–238
    for numbers, 146
    sensory, 236
    short-term, 236–237
    span, 144
Mental ability (abilities), 144, 146–148, 249
    and aging, 152
Mental model(s), 247
Metabolic system, 179–180
Method(s), 2, 14, 271
    action—information analysis, 83, 98–100, 101, 109, 113, 279
    activity analysis, 83, 85, 87–89, 93, 97, 98, 116, 279

324 SUBJECT INDEX

Method (*Continued*)
  analysis of similar systems, 83, 85-87, 93, 97, 98, 137, 279, 285
  controlled experimentation, 82, 83, 129-131, 137, 138
  critical incident study, 30, 83, 85, 89-92, 98, 279
  decision-action analysis, 83, 97-99, 101, 109, 113
  failure modes and effects analysis, 83, 109, 113-115, 279
  fault tree analysis, 83, 109-113
  functional allocation, 36, 42, 45, 46, 73, 74, 82, 83, 84, 86, 87, 98, 100-105, 137, 138
  function analysis, 45-46, 67, 68, 82, 83, 84, 86, 87, 93-98, 101, 103, 109, 113, 138, 279, 290
  link analysis, 72, 83, 116-123, 279
  mission analysis, 84
  operational analysis, 83, 84-85, 93, 97, 285
  operational sequence analysis, 83, 131-135, 138, 279, 290
  prototyping, 36, 45, 50, 67, 74, 127-128
  simulation, 28, 50-51, 83, 85, 123-128, 129, 136, 137, 138
  task analysis, 30, 67, 68, 72, 82, 83, 101, 103, 105-111, 113, 115, 116, 137, 138, 264, 279, 290
  time line analysis, 83, 115-116, 137
  workload assessment, 82, 83, 101, 103, 115, 123, 135-138, 279
Microclimate, 182
Microphone(s), 62
Microwave ovens, 262
Milestones, 37
MIL-HDBK-761A, 278
MIL-H-46855B, 64, 77, 106
Military systems, 10
MIL-STD-:
  490A, 69, 77
  499A, 282, 287
  721C, 22, 56
  882B, 22, 56
  1472D, 60, 64, 77, 183, 184, 192, 204, 213, 258, 275, 276, 278, 287
  1521B, 69, 73, 77
Mission analysis, 84
Mission Control Complex, 106-109

Mistakes, 90
Misuse of equipment, 10, 143
Mobility, human, 170-172
Mockups, 74, 75, 123, 265
Model(s), 123, 144
  allocation, 15
  machine, 206
  mental, 247
  operator-machine, 207-209
  queuing, 15
  waterfall, 38-40, 52-53
Modeling, 28
Monaural versus binaural listening, 229
Monotony, 16
Motion economy, 251
  principles of, 4, 252
Motion sickness, 194, 235
Motivation, 207
  in learning, 244
Motor:
  abilities, 144
  learning, 242-243, 244-246
  responses, 251-255
Movement, body, 194
  direction of control, 276
  human, 12
  sense of, 212, 213
Multi-channel listening, 210
Munsell system, 223-225
Muscle(s), 172-178
  power, 12
  strength, 12, 145, 173-178
  tension, 135
Myopia, 220

NASA, *see* National Aeronautics and Space Administration
NASA-STD-3000, 60, 62, 64, 77, 160-162, 168, 169, 172, 173, 186, 189, 191, 211, 213, 258, 276, 277, 278, 287
National Aeronautics and Space Administration, 12, 14
National Bureau of Standards, 224
National Center for Education Statistics, 155
National Electrical Safety Code, 63
National origin, 149, 154-155, 162
National Safety Council, 9, 20
Navigator(s), 87-88
NCR Corporation, 11

Near accidents (errors, mistakes), 90
Nearsightedness, 220
Need(s), human, 26
  identification of, 29
  maintainer, 17, 30
  operator, 30
  user, 17, 30
Nervous system, 179
Noise, 10, 14, 187-190, 208, 229, 230, 231
  attenuating helmets, 189
  criteria curves, 189-190
North Sea offshore oil platform, 114-115
Noxious gases and fumes, 14, 207
Nuclear:
  energy, 5
  power plant(s), 1, 4, 31, 53, 55, 60, 64, 75, 85, 91
    operators, 261
    simulators, 254, 268
  submarines, 80
Nuclear Regulatory Commission, 11, 290, 292, 296
Numbers, meaningfulness of, 237
  memory for, 146-147, 237
Nutrition, 5, 14, 60
  requirements, 281

Objectives, 15-16
  formation of, 29
  human factors (and systems), 15-16
Observations, structured, 85
Occupational Safety and Health Administration, 60, 189, 204
Odors, 234
Office(s), 16
  layout, 121-122
  vibration in, 193
Offshore oil platform, 114-115
Operability, 76, 86, 260-262
  evaluation of, 37
Operation(s), 29-30, 37, 39, 52, 73, 90
  decision diagrams, 97
  methods of, 72
  modes of, 74
  requirements, 74
Operational:
  analysis, 83, 84-85, 93, 97, 285
  concept, 30-35, 73, 82, 107, 264, 285, 289
  document, 2, 39, 43, 45, 58, 70
  decision analysis, 83, 97-98, 101, 109, 113
  need, 29-31, 36, 39, 73, 264, 285, 289
    determination, 70
    document, 2, 39, 43, 45, 58, 69, 70, 279
    statement, 29-30
  performance, 31
  profile, 33, 84
  readiness review, 70, 73
  requirements, 84
  review, 30
  scenario, 264, 279
  sequence:
    analysis, 83, 131-135, 138, 279, 290
    diagrams, 45-49, 138, 290
  view of a system, 40-43
Operations research, 13, 15, 28
Operator(s), 16-17, 19, 23, 26, 31
  machine model, 207-209
  model, 207
  requirements, 72, 280
  satisfaction, 37
  skill levels, 264
  as a system component, 206
  in systems, 19
OR gate, 94, 111-113
OSD, see Operational sequence diagram
OSHA, see Occupational Safety and Health Administration
Output devices, 50. See also Displays
Oxygen:
  in air, 186-187
  consumption, 14
  deficiency, see Hypoxia

Packaging of equipment, 15
Pain, sense of, 212, 216
Parasympathetic nervous system, 179
Pascals, 187-188
Passageway(s), 158, 276
Peak clipping of speech, 233, 250
Perceiving, 209
  versus sensing, 235
Perception, 194, 235
Performance, 19, 127
  and age, 152, 154
  case study, 80
  effects of noise on, 188-189

Performance (*Continued*)
  machine, 19
  motor, 13
  standards, 59, 61
  system, 16, 19
Peripheral nervous system, 179
Personal computers, 17
Personal equipment, design of, 159
  protective, 60
    requirements, 281
Personality, 144, 207
Personnel:
  accommodations, 75
  constraints, 85
  requirements, 2, 16, 36, 49, 68, 71-74, 87, 99, 102, 115, 263, 280
  selection, 5, 6, 13, 23, 50, 72, 86, 87, 102, 260-262, 271, 282-283
  subsystem, 260-269
    design, 12
Phaseout, 29
Phonetically-balanced (PB) words, 230
Physically handicapped persons, 7
Physical view of system, 40-43
Physiology, 12, 135, 142
  applied, 14
Piloting, aircraft, 240, 261
  ships, 200-202
Pitch, 228
Planning, product, 29
Politics, in allocation decisions, 102
Position:
  coding, 7
  sense of, 212, 213
Postal:
  service, 11
  sorting, 103-105
    automated, 104
    manual, 103-105
  system, 23, 25
Posture(s), in anthropometry, 159-160
Power, human output, 173-174
  muscle, 12
  plants, 4, 17, 53, 55, 60, 75, 85
  requirements, 102
Prediction versus description, 81
Preliminary design, 29
  review, 30, 70, 74
Presbyopia, 7, 220

Pressure, breathing, 186
  sense of, 212, 213, 216
Pressurization, of living and working areas, 186
Prime item, 25
Probabilities, combining in systems, 112-113
Procedural training, 265-267
Procedure(s):
  development, 67-68
  operating, 123, 290
  requirements, 101
Processing information, *see* Information processing
Producibility, 102
Product improvement review, 30
Production, 29, 30, 37, 39
  ease of, 54
  economy of, 16
  line, 30, 86
  readiness review, 30
  systems, 23
Productivity, 283
  records, 86
Profile, operational, 31
Programmers, computer, 8, 18
Project-specific requirements, 277
Prompts, 238
Prone position, 169-170
Proof-of-compliance, 75
Protective equipment, 60
Protanopia, 61, 227
Prototype(s), 85
Prototyping, 36, 45, 50, 67, 74, 127-128
  tools, 128
Psychology, 12-14, 142
Psychomotor performance, 13, 144
Psychosocial requirements, 281
P-3 weapon system trainer, 130
Punch card system, 112
Push forces, 124-125, 145-146

Qualification requirements, 279, 281
Questionnaires, 30, 85
Queuing model(s), 15
QWERTY keyboard, 246

Radar, 4
Radiation, in heat exchanges, 182, 185

SUBJECT INDEX

Railroad(s), 23, 253–254
Range Control Complex, 106–109
Rapid prototyping, *see* Prototyping
Rationale reports, 70, 72, 76, 270
R^3 reports, 70, 72
Reach, in anthropometry, 159
 envelopes, 92, 168–169
 inability to, 92
Reaction time, 144, 194–195, 251–254
Reasoning, 209
Recall, 209, 237–238
Receiver operating characteristic (ROC), 214–215
Recognition, 238
Recommendations, 2, 63, 79
Recycling, 37
Reflectance, specular, 61
Rehearsal, 237
Relative humidity, 181–185
Reliability, 16, 28, 31, 284
 data stores, 109, 113
 human, 109
Remembering, *see* Memory
Remote Tracking Station, 106–109
Request(s) for proposal (RFP), 2, 25–26, 29, 63, 66, 84
Requirements, 2, 3, 42, 43, 58, 66, 127, 270
 action, 107
 analysis and elaboration, 39–45
 application-specific, 3, 277–281
 design, 33
 dimensional, 36
 documentation, 281
 health, 71
 how-well, 271–272
 human factors, 27, 271
 human-system, 271–281
 information, 99, 105, 107
 interface, 36, 71, 74, 106, 127, 280
 maintainer, 72
 maintainability, 36, 71–74
 maintenance, 84
 manpower, 86, 115
 operational, 84
 operator, 71, 72
 personnel, 2, 16, 36, 49, 68, 71–74, 87, 99, 102, 115, 263, 280
 power, 102
 procedures, 101
 project-specific, 277–281
 psychosocial, 281
 qualification, 279, 281
 rationale reports, 70, 72, 76, 270
 safety, 36, 71, 280
 sanitation, 281
 skill, 106, 280
 space, 71
 in specifications, 26
 staffing, 71, 86, 87, 101, 106, 107
 support, 71, 102
 system, 37, 42–43, 139, 271–281, 289
 task, 280
 test and evaluation, 74
 training, 16, 33, 36, 68, 72–74, 101, 106, 107, 279, 280
 usability, 74
 user, 19, 26, 36, 272
 what, 271–272
 workplace, 45, 71, 75, 274, 280
Resistance, of controls, 178
Respiration, 14, 135
Respiratory system, 179
Responding, 209, 247–255
Response time, 251–254, 284–285
Rest periods, 198–202
Retirement, system, 37
Reversal errors, 92
Review(s), 2, 3, 37, 38, 73–76, 106, 139, 289, 291
 critical design, 30, 70, 75
 operational readiness, 70, 73
 operational requirements, 30
 preliminary design, 30, 70, 74
 product improvement, 30
 production readiness, 30
 software specification, 70, 74
 system design, 30, 70, 73–74
 system requirements, 30, 70, 73
 test readiness, 70, 75–76
RFP, *see* Request for proposal
RMA, *see* Reliability, maintainability, availability
Robot(s), 102
ROC, *see* Receiver-operator-characteristic
Rotation(al), acceleration, 196
 sense of, 212

Safety, 12, 16, 37, 62, 75, 102, 282, 284–285
  color(s), 223–225, 228
  engineering, 28, 67, 81
  requirements, 36, 71, 280
  standards, 59, 61
  studies, 73
Sale(s), 55
  personnel, 53
  product/system, 29
Sanitation requirements, 281
SAT, *see* Scholastic Assessment (formerly Aptitude) Test
Satellites, space, 80, 106–109
Satisfaction, 282
  operator, 37
Saturation, in color, 221–226
Scenarios, 31, 45, 46, 84, 114–115, 279
Schematics, 75
Scholastic Assessment (formerly Aptitude) Test (SAT), 155, 157
Seagirt Marine Terminal, 11
Seasickness, 235
Seat(s), 10, 14
  design of, 158
  reference point (SRP), 168–169
Segment, of systems, 25
Selection, 2, 5, 6, 13, 23, 50, 72, 86, 87, 102, 260–262, 271, 282–283
Selective attention, 209–210
Sensing, 209, 211–215, 235
Sensitivity analysis, in trade studies, 285–286
Sensory capacities, 13, 144
Serial position effect, 146
Service:
  personnel, 43
  providers, 53
  systems, 23
Servicing systems, 54
Sex:
  and color vision defects, 149, 227
  and hearing losses, 151, 153
  differences, 149
    in anthropometric dimensions, 162, 166, 167
    in mobility, 172
    in strength, 174–176
Shades of gray, 217

Shift work, 200–202
Ship(s), 17, 23
  accidents, 8–9
  collision avoidance system, 133–135
  piloting, 200–202
Sidetone, 232
Sign(s), 13
  highway, 151–152, 224–225
Signal:
  audibility, 230
  detection, 214–215
  meaning, 230
  recognition, 230
Signal-to-noise ratio (S/N), 231
Simulation(s), 28, 50–51, 83, 85, 123–128, 129, 136, 137, 138
  plan, 68
Simulator(s), 74, 123, 124, 235, 246–247, 254, 265–268
  design, 5
Situation awareness, 100
Skeletal:
  muscles, 172–179
  system, 170, 172
Skill, acquisition, 244–246
  levels, 32, 72, 86, 87, 155, 156, 260–264, 282, 294
  requirements, 106, 280
  training, 265–267
Skin resistance, 198–199
Smalltalk/V 286, 128
Smell, sense of, 212, 213, 216, 234
Social Security Administration, 11, 88–89
Sociology, 142
Soft failure(s), 264
Software, 3, 38
  architecture, 50
  configuration item, 25
  design, 49–51, 66–67, 72
  development, 39, 51
  engineering, 28, 54
  psychology, 12
  specification(s), 75
  review, 70, 74
Sonar, 4, 80, 130–131
Sonobuoys, 80, 130–131
Sorting mail, 103–105
Sound(s), 228–229

pressure level(s), 187–188, 229
    reaction time to, 252, 253
SOW, *see* Statement of Work
Space:
    activities in, 60
    adaptation syndrome, 235
    requirements, 71
    satellites, 80, 106–109
    systems, 10, 64
    vehicles, 85, 106–109, 283–285
Speaking, 248–251
Specification(s), 2, 3, 10–11, 27, 39, 58, 69–72, 76, 106, 107, 127, 139, 270–271
    build-to, 71, 75
    of color, 223–226
    defined, 69–70
    design-to, 71, 74
    requirements, 26, 271–281
    system, 36
    system/segment, 263
    tree, 70–71
    workstation, 274
Spectrum:
    colors, 223
    speech, 231–232
Specular:
    gloss, 226
    reflectance, 61
Speech, 228–234, 248–251
    amplified, 230, 233
    communication, 188, 230–234
    comprehension, 231–232
    compression, 250
    distortion, 233
    intelligibility, 232–233, 250
        measurement of, 230–231, 250
    interference level (SIL), 250
    to noise ratio (S/N), 231
    packaging, 232
    peak clipping, 233
    recognition, 102
    side tone, 232
    spectrum, 231–232
    synthetic, 230, 233–234
    vocoded, 230, 233–234
Spine (spinal), 172
Sports equipment, 13
SRP, *see* Seat reference point

Staffing requirements, 86, 87, 101, 106, 107
Standard observer, 223
Standardization in design, 246–247
Standard(s), 2, 10–11, 58–63, 69, 76, 79, 270–271, 275, 281
    defined, 59
    drafting organizations, 59
    language conventions, 61
    limitations of, 61–63
    tailoring, 63, 66
State transition diagrams, 87
Statement of work (SOW), 26, 29, 63
Static force, 173
Statistics, 15
Stature, in anthropometry, 159, 166
    NASA flight crews, 168
Steam irons, 126
Stereo acuity, 218–219
Stopping distance(s), 194–195
Stoves, 17
Strength, 173–178
Stress, 14, 16, 19, 86, 87, 207
Structured observation(s), 85
SUBACS, 108
Subcontractor(s), 67
Submarine(s), 80, 124
Substitution errors, 92
Subsystem(s), 24–26
    vendors, 28
Success stories, 3, 295
Supervisor(s), 35, 43
Supply function, 23
Support, 29
    integrated logistic, 16
    personnel, 263
    requirements, 102
Surfaces, walking, 60
Symbology, 74, 132, 133
Sympathetic nervous system, 179
Synthetic speech, 233–234
System(s), 1, 3, 21–57, 270
    access, 280
    air traffic control, 11, 23, 25, 29, 64, 75, 98, 100, 210, 230
    analysis, 67
    automated bank teller, 23, 29, 32–35, 43–49, 69, 71, 87, 98, 100, 102, 126, 131, 278, 286

System(s) (*Continued*)
  availability, 19
  banking, 24-26
  complexity, 263-264
  defined, 21-26
  design, 39, 45-49, 70
    review, 30, 70, 73-74
  development, 63
    audits, 37
    process, 38-52
    reviews, 37, 72-76
  engineering, 1, 10-11, 13, 21-57, 270
    defined, 26-28
    objectives, 15-16
    process, 38-52, 270
  engineers, 18
  fire, 23
  functions, 45-46
  hierarchy, 24-26
  information processing, 23
  integration and test, 39, 51
  life cycle(s), 2, 28-38, 270
  maintainability, 19
  military, 10
  performance, 16, 19
  police, 23
  postal, 23, 25, 103-105
  production, 23
  requirements, 37, 42-43, 139, 271-281, 289
    review, 30, 70, 73
  retirement, 30, 37
  scenarios, 45
  service, 23
  space, 10, 64
  specifications, 36
  three views of, 40-43
  transportation, 23
  users, 43-44
  weapons, 23
System Effectiveness Organization, 77

Tables, 14
Tactile sense, *see* Touch
Tailoring of documents, 63, 66
Task analysis (analyses), 30, 67, 68, 72, 82, 83, 101, 103, 105-111, 113, 115, 116, 137, 138, 264, 279, 290
  requirements, 280

Taste, sense of, 212, 213, 216, 234
Team performance, 14
Technical manuals, 288-289
Telecommunication systems, 11
Telephone(s), 4, 5, 9, 11
  handsets, 125-126
  sets, 126
  systems, 23, 25, 31, 230
Television, 4, 262
Temperature, 14, 181-185, 207-208
  effects of on strength, 176
  sense of, 212, 213
  of work spaces, 275
Tendinitis, 178-179
Tenosynovitis, 178-179
Test:
  equipment, 28
  and evaluation, 19, 28, 36, 39, 62, 68, 291
  interfaces, 73
  readiness review, 70, 75-76
  requirements, 74, 76
Testing and verification, 29
The **PROTOTYPING** System (PROTO), 128
Theory (theories), 80
Thermal, comfort, 182-185
  environment, 181-185
Three-Mile-Island nuclear power plant, 1, 8, 20, 55, 291
Threshold(s):
  absolute, 212-214
  difference, 216-217
  upper, 215
Time-and-motion engineering, 4, 251-252
Time:
  lag(s), 276
  line analysis, 83, 115-116, 137
  sharing, 210
  zone changes, 199-200
TMI, *see* Three-Mile-Island
Tool(s), 1, 3, 5, 17, 23, 38, 107
  design of, 158-159
  link analysis, 123
  machine, 23-24
  prototyping, 128
Touch, sense of, 212, 213, 215, 234
Tower Control Computer Complex, 288

SUBJECT INDEX **331**

Traceability, 291
Trade:
   offs, 3, 6-7, 27, 45, 46, 50, 75, 105, 264, 281-286
   studies, 36, 49-50, 72-74, 94, 138, 289
Traffic signals, 6-7
Training, 2, 5, 12-13, 19, 23, 32, 37, 39, 50, 54, 67, 68, 72, 73, 86, 87, 102, 123, 241, 260-263, 282-283. *See also* Learning
   aids, 265
   constraints, 85
   devices, 123, 246-247, 263, 265-267
   equipment, 75, 265
   familiarization, 265-267
   indoctrination, 265-267
   procedural, 265-267
   programs, 264, 294
   requirements, 16, 33, 68, 72-74, 101, 106, 107, 280
   skill, 265-267
   systems, 28
   transfer of, 246-247
      habit interference, 246-247
      negative, 246-247
      positive, 246
   transition, 265-267
Transfer of training, *see* Training, transfer of
Trans-illuminated displays, 220
Translation of instructions, 155
Trigger finger, 178-179
Typing, 255
Typewriters, 262

Unintentional activation, 92, 277
United Airlines, 266, 267
U.S. Department of Education, 155
U.S. Labor Department, 156
Usability, 37, 76, 127
   evaluation of, 126-127
   laboratories, 127
   requirements, 74
   testing, 126
Usable, defined, 278
Use, ease of, 16
User(s), 16-17, 19, 26, 35, 38-39
   acceptance, 16
   characteristics, 42-44

   needs, 17, 30
   oriented view of a machine, 40-42
   participation, 293-294
   requirements, 19, 26, 36, 272
   system interface(s), 45-51
   in systems, 23
USS Louisville, 117-121
Utilization, 29

Value, of color, 224-226
VCR(s), 5, 9, 91, 262
VDT(s), *see* Visual display terminal(s)
Vehicle stopping distances, 194-195
Verbal abilities, 152, 154
Vestibular senses, 235
Vibration, 5, 189-194, 208, 215, 217
   building, 193
   effects of, 191-193
   exposure limits, 192-193
   sense of, 212, 234
Vibratory alarms, 211
Vision, 209, 212-213, 216-228
   and aging, 151, 220
      presbyopia, 151
   color, 215, 221-228
   dark adaptation, 145
Visual:
   acuity, 44, 218-221
   alarms, 210-211
   angle, 218-221
   display terminal(s), 61, 276
Vocabulary (vocabularies), 233
   and speech intelligibility, 250
Vocoded speech, 233-234

WAIS, *see* Wechsler Adult Intelligence Scale
Waist circumference, 149-150
Walking, index of, 121
Warning:
   displays, 210
   signals, 253-254
Washing machines, 60, 262
Waste disposal, 5
Water vapor, 181-185
Waterfall model, 38-40, 52-53
Weapon systems, 23
Weber, fraction (law), 216

Wechsler Adult Intelligence Scale (WAIS), 146–148, 152, 154
Weight-lifting capacity, 173–175
Weightlessness, 5
Westinghouse, 11
Wheelwriter, 247
Wolverine VS Series Vertical CNC Copy and Profile Milling Machine, 24
Word processor, 247
Word-spelling alphabet, 233, 249
Work, 14
  efficiency, 180
  energy expenditure and heart rate in, 179–180
  force, changes in the composition of, 7–8
  products, 2, 64, 69–76, 80
Work-rest cycles, 5, 177, 198–202, 279, 280
Working:
  capacity, 173–178
  conditions, 107
  environment, 16, 49
Workload, 86, 87
  assessment, 82, 83, 101, 103, 115, 123, 135–138
Workshops, vibration in, 193
Workspace(s), 45, 71, 75
  requirements, 280
Workstations, design of, 14, 62, 116, 117, 158–159, 163, 273, 275

Xerox 1075 copier, 40–42

9-5-97